55 Structure and Bonding

Editors:
M. J. Clarke, Chestnut Hill · J. B. Goodenough, Oxford
J. A. Ibers, Evanston · C. K. Jørgensen, Genève
J. B. Neilands, Berkeley · D. Reinen, Marburg
R. Weiss, Strasbourg · R. J. P. Williams, Oxford

Transition Metal Complexes – Structures and Spectra

With Contributions by
M. Bacci   J. Fischer   M. H. Gubelmann
B. Koreň   F. Mathey   M. Melník   J. H. Nelson
P. Sivý   F. Valach   A. F. Williams

With 93 Figures and 21 Tables

Springer-Verlag Berlin Heidelberg GmbH 1983

ISBN 978-3-662-15750-3   ISBN 978-3-540-38751-0 (eBook)

DOI 10.1007/978-3-540-38751-0

Library of Congress Catalog Card Number 67-11280

© Springer-Verlag Berlin Heidelberg 1983

Originally published by Springer-Verlag Berlin Heidelberg New York Tokyo in 1983

2152/3140-543210

Table of Contents

The Structure and Reactivity of Dioxygen Complexes of the Transition Metals

Michel H. Gubelmann and Alan F. Williams

Département de Chimie Minérale, Analytique et Appliquée, Université de Genève, 30 quai Ernest Ansermet, CH-1211 Geneve 4, Switzerland

This article gives a review of complexes in which a dioxygen ligand is bonded to a transition metal. Three aspects of these complexes are discussed in detail: the structure, the electronic structure, and the reactivity. The structural section summarises the recent X-ray crystal structure determinations, and the structural data obtained by other methods. The electronic structure is first considered in qualitative terms which allow the rationalisation of the different structures observed, and this qualitative model is compared with the results of calculations and with spectroscopic data. The reactivity of the complexes is discussed separately for each structural class in terms of the electronic structure. An attempt is made to compare the results obtained in historically different areas of research. Our objective is to give a clear summary of current knowledge of these compounds for workers interested in their application to catalysis and in their rôle in biochemical systems.

A. Introduction

The study of dioxygen complexes of the transition metals is generally accepted to have begun with the report by Frémy in 1852 of the oxygenated ammoniacal salts of cobalt[1]. A satisfactory explanation of his results had however to await the development of a general theory of coordination compounds and the dioxygen bridged complexes of cobalt(III) figured among the many complexes studied by Werner at the turn of the century[2]. In the nineteen thirties the mechanism of the auto-oxidation of metal ions was studied and the first synthetic oxygen carriers discovered. In 1936, Pauling and Coryell proposed the first of many theoretical models to explain the iron dioxygen interaction in haemoglobin[3]. The increasing availability of physical methods allowing the ready characterisation of dioxygen complexes and the determination of their molecular structures, coupled with a better understanding of the electronic structures, has given considerable encouragement to the study of these compounds in recent years. The early work tended to concentrate on specific types of complex, and we may distinguish three basically different areas of research:

(i)   Complexes of cobalt with Schiff bases and nitrogen-containing ligands.
(ii)   Complexes of group VIII metals in low oxidation states.
(iii) Biological systems where transition metals (especially iron and copper) are known
       to be intimately involved in reactions with molecular oxygen. This field includes
       innumerable simpler "model" complexes, and covers systems which act as oxygen
       carriers as well as those acting as redox systems.

The distinctions between these topics have become somewhat blurred with the passing of time, and the increasing availability of good crystal structure data has brought to light many similarities between apparently different complexes. In 1976 Vaska[4] published an important paper classifying dioxygen complexes according to their molecular geometry, and showing that the complexes for which data were available fell into four closely related categories, and that the well known peroxo complexes of the early transition metals[5] were also structurally similar to many complexes of dioxygen.

From a practical point of view the study of the chemistry of dioxygen complexes has considerable interest. Complexation of molecular oxygen by a transition metal has been widely adopted by biological systems as a means of reducing the considerable kinetic barrier to the reduction of O_2. Quite apart from the inherent interest of the biological systems, the transition metal complexes offer the possibility of efficient catalysis of auto-oxidation reactions, and have recently attracted interest as possible catalysts for the reduction of O_2 in fuel cells[6].

In this review we shall take Vaska's structural classification as the basis for the examination of the electronic structure and reactivity of dioxygen complexes. We wish to follow Vaska's unifying approach to the chemistry of these systems and to give as general a coverage as possible, and it is therefore impossible to discuss all the published work in a review of this length. Many reviews on the chemistry of these complexes have been published, dealing with general properties[4, 7, 8], complexes of cobalt with Schiff bases and nitrogen ligands[9, 10], complexes of group VIII metals[11, 12], catalysis by dioxygen complexes[13–16] and biological subjects[17, 18] including oxygen carriers[19–21] and redox systems[22–26]. These reviews may be consulted for more detailed discussion of particular topics.

B. Properties of Molecular Oxygen

For the purposes of this review we shall reserve the term molecular oxygen for free gaseous O_2; in cases where the O_2 entity is bound to other atoms we shall use the term dioxygen. This is a *structural* definition which requires only the continued existence of O-O bonding in the complex molecule, and gives no information on the bonding of the O_2 species.

The electronic structure of molecular oxygen in its ground state $^3\Sigma_g^-$ is well known (Fig. 1). The two lowest excited states of molecular oxygen, the $^1\Delta_g$ and $^1\Sigma_g^+$ are obtained by pairing the spins of the two electrons in the π_g^* orbital, and lie respectively at 94.2 and 156.9 kJ/mol above the ground state $^3\Sigma_g^-$ [27].

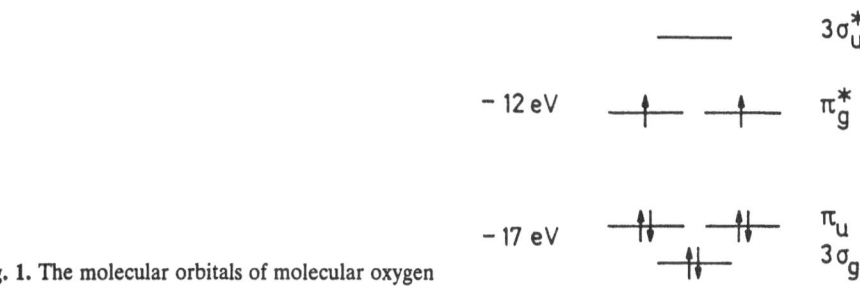

Fig. 1. The molecular orbitals of molecular oxygen

The triplet ground state of molecular oxygen provides a considerable kinetic barrier to the auto-oxidation of normally diamagnetic organic molecules where reactions involving change of spin are generally very slow, and where products formed in triplet states are unstable. This barrier may be circumvented in three ways:

(i) the formation of the lowest singlet state (by photochemical activation) where the spin conservation barrier is removed[28];

(ii) reaction with radical species or the free electron in two distinct steps to give diamagnetic products;

(iii) reaction with a heavy element such as a transition metal where greater spin-orbit coupling considerably reduces the kinetic barrier to change of spin, and where the formation of a metal dioxygen complex may itself provide sufficient energy to pair the spins.

The ions arising from simple one electron reductions or oxidations of molecular oxygen are well characterised and their properties summarised in Table 1.

The possible importance of partially reduced dioxygen species in biochemical reactions has led to a reinvestigation of their equilibria in aqueous solutions[45–47] and the values obtained are summarised in Table 2.

Several interesting points arise from these data. The potential for the first one electron reduction of molecular oxygen is unfavourable and gives the unstable superoxide anion. This and the other possible intermediates (H_2O_2 and HO) are notably more reactive than free molecular oxygen and most biological systems appear to have taken steps to eliminate them. O_2^- is eliminated by superoxide dismutase[47–49] which catalyses the highly favourable reaction:

Table 1. Properties of some dioxygen species

Species	Bond order	Compound	O-O distance		ν_{O-O}		Bond energy	
			Å	Ref.	cm^{-1}	Ref.	kJ/mol	Ref.
O_2^+	2.5	O_2AsF_6	1.123	30	1858	31	625	32
$O_2(^3\Sigma_g^-)$	2	O_2	1.207	30	1555	32	490	19
$O_2(^1\Delta_g)$	2	O_2	1.216	29	1484	33	396	32
O_2^-	1.5	NaO_2	1.33	34				
		KO_2	1.32–1.35	35	1146	37		
		$O_2^-(g)$	1.34	36				
O_2^{2-}	1	$H_2O_2(g)$	1.475	38				
			1.467	39				
		$H_2O_2(l)$			880	37		
		$H_2O_2(s)$	1.453	39				
		Na_2O_2	1.50	40	794	42	204	19
			1.49	41	738^a			
		$Na_2O_2 \cdot 8H_2O$			842	42		
		BaO_2	1.49	43				
		CdO_2	1.49	44				

^{a} There are two different peroxide ions in the unit cell

Table 2. Equilibria in aqueous solution

Reaction	Electrode potential (volts)	
	pH = 0	pH = 7
$O_2 + e^- \rightarrow O_2^-$	−0.33	−0.33
$O_2^- + 2H^+ + e^- \rightarrow H_2O_2$	+1.69	+0.87
$H_2O_2 + H^+ + e^- \rightarrow H_2O + HO$	+0.793	+0.38
$HO + H^+ + e^- \rightarrow H_2O$	+2.76	+2.33
$O_2 + 4H^+ + 4e^- \rightarrow 2H_2O$	+1.23	+0.82

Values calculated for aqueous solutions at 25 °C with
$P_{O_2} = 1$ atmosphere, $[O_2^-] = [H_2O_2] = [HO] = 1\,M$

$$2O_2^- + 2H^+ \rightarrow O_2 + H_2O_2$$

and ensures that the molecular oxygen liberated is exclusively in the less reactive $^3\Sigma_g^-$ state. H_2O_2 is eliminated by peroxidases. The OH radical, generally thought to be formed in acidic aqueous solutions of Fenton's reagent (hydrogen peroxide and a ferrous salt) is extremely reactive towards organic substrates[50, 51] and probably reacts in a non-specific way before it may be eliminated. The very reactivity of these intermediates justifies the use by biological systems of metal ion mediated pathways of molecular oxygen reduction.

Much of the experimental work on oxygen chemistry has been carried out in non-aqueous solvents and the thermodynamic data of Table 2 may not be applied to such systems. Molecular oxygen is appreciably more soluble in organic solvents than in water (the solubilities differ by a factor of ten between water and diethyl ether)[52]. Groves[51] has shown that Fenton's reagent in non-aqueous solutions does not produce OH radicals but rather a reactive Fe(IV) (ferryl) species. A change of solvent can thus have a considerable effect on the mechanism of oxygen reduction.

C. Structural Classification of Dioxygen Complexes

In his review[4], Vaska showed that every dioxygen complex whose structure was then known fell into one or other of 4 structural types. This structural classification has proved extremely useful in discussing the properties of dioxygen, and in this section we discuss the classification and review structural data published since 1976. Vaska's four structural types are the first four entries in Table 3; he grouped the four structures into the superoxo compounds (types I a and I b) where the O-O distance is roughly constant (~ 1.3 Å) and close to the value reported for the superoxide anion (Table 1), and the peroxo compounds (types II a and II b) where the O-O distance is close to the values reported for H_2O_2 and O_2^{2-} (~ 1.48 Å). The a or b classification distinguishes complexes where the dioxygen is bound to one metal atom (type a) or bridges two metal atoms (type b). In this review we shall use a "hapto" nomenclature in which the structures are classified by the number of atoms of dioxygen bound to the metal ion; although this does not distinguish explicitly between types I b and II b, it does avoid assigning a possibly misleading oxidation state to the dioxygen (see Sect. D) and may readily be applied to the structural types discovered since Vaska's review.

Table 3. Structural classification of dioxygen complexes

Structure type	Structural designation	Vaska classification	Example
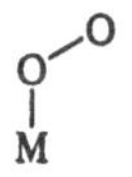 η^1dioxygen structure	η^1dioxygen	Type I a (superoxo)	$[Co(CN)_5O_2]^{3-}$
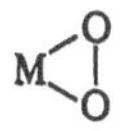 η^2dioxygen structure	η^2dioxygen	Type II a (peroxo)	$(Ph_3P)_2PtO_2$
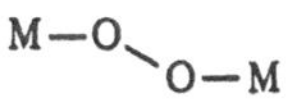 $\eta^1{:}\eta^1$dioxygen structure	$\eta^1{:}\eta^1$dioxygen	Type I b (superoxo)	$[(H_3N)_5CoO_2Co(NH_3)_5]^{5+}$
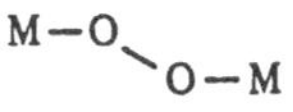 $\eta^1{:}\eta^1$dioxygen structure	$\eta^1{:}\eta^1$dioxygen	Type II b (peroxo)	$[(H_3N)_5CoO_2Co(NH_3)_5]^{4+}$
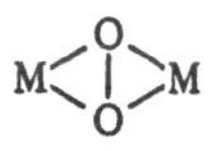 $\eta^2{:}\eta^2$dioxygen structure	$\eta^2{:}\eta^2$dioxygen	–	$[(UO_2Cl_3)_2O_2]^{4-}$
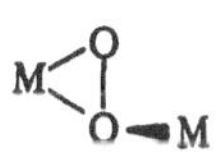 $\eta^1{:}\eta^2$dioxygen structure	$\eta^1{:}\eta^2$dioxygen	–	$[(Ph_3P)_2ClRh]_2O_2$

The $\eta^2 : \eta^2$ structure with a "sideways" bound dioxygen bridging two metal atoms has been suggested for the complex [{Rh(diene)}$_2$O$_2$][53] and a crystal structure showing this geometry has been reported for the uranium complex [(UO$_2$Cl$_3$)$_2$O$_2$]$^{4-}$ [54] and a complex of La^{3+} [55]. The $\eta^1 : \eta^2$ structure (Fig. 2) is known only for [RhCl(O$_2$)(PPh$_3$)$_2$]$_2$ [56].

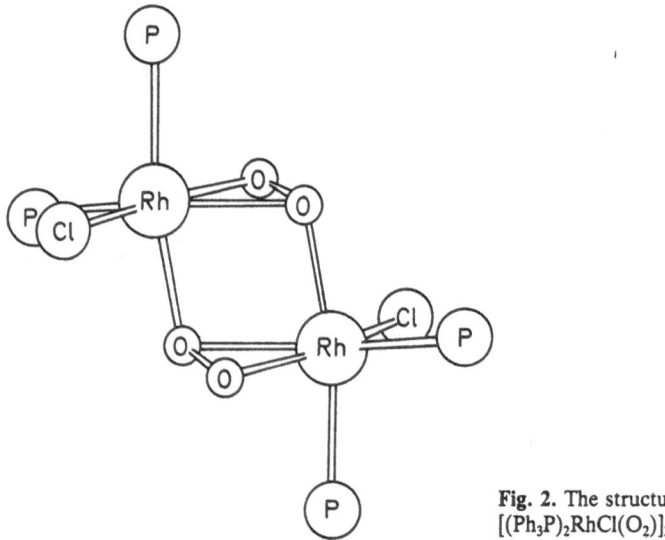

Fig. 2. The structure of [(Ph$_3$P)$_2$RhCl(O$_2$)]$_2$ (Ref. 56)

The only completely unambiguous method of structure determination has proved to be X-ray diffraction. Vaska noted however that the stretching frequencies attributed to the O-O vibration were closely related to the structural type[4]. Type I complexes show O-O stretching vibrations around 1125 cm^{-1} and type II around 860 cm^{-1}. This sharp difference enables the O-O stretching frequency as measured by infra-red or Raman spectroscopy to be used to assign the structure type, provided (as is usually the case) the formation of a dinuclear species can be confirmed or excluded by other means.

If X-ray diffraction gives an unambiguous description of the structure, it should nevertheless be noted that the accurate determination of bond lengths and angles for the coordinated dioxygen is not always easy. When the dioxygen is bonded closely to a very

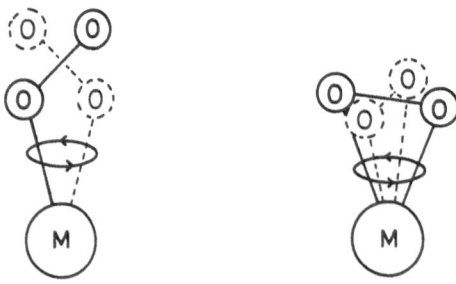

(a)   (b)   Fig. 3a, b. Disorder in dioxygen complexes

heavy transition metal (for example Ta or Ir), precise location of the oxygen atoms is difficult, and independent determinations of the same structure do not always agree[57]. A second problem is the possible presence of disorder in the dioxygen bonding, especially in the non-bridging systems. In $[Co(CN)_5O_2]^{3-}$ the dioxygen was found to show rotational disorder as shown in Fig. 3 a[58], and a recent n.m.r. study of complexes of the type $Ti(porphyrin)O_2$ has shown a low barrier to rotation as shown in Fig. 3 b[59]. Bond angles and distances which deviate sharply from the average values should therefore be interpreted with prudence.

Tables 4–6 report the structural parameters of complexes studied by X-ray diffraction since the publication of Vaska's review, together with some of the more significant results published beforehand. Structural information obtained by other methods is discussed separately.

Structural investigations using methods other than X-ray diffraction have also provided much useful information, and in the remainder of this section we review some of the results obtained from such investigations.

Titanium Complexes

An η^1 dioxygen complex has been prepared as an intermediate in the auto-oxidation of Ti^{3+} complexes, and E.P.R. evidence for such a species has come from studies of the auto-oxidation of Ti(TPP)F trapped in a matrix of $Ti(TPP)O^{124)}$, however, in contrast with the η^1 dioxygen complexes of cobalt, the spin density is only weakly localised on the dioxygen, and is quite high on the porphyrin ligand. E.P.R. spectroscopy has also been used to demonstrate an η^1 dioxygen species formed during the auto-oxidation of a Ti(III)

Table 4. Structural details of η^1 dioxygen (Vaska type I a) complexes

Complex	O-O (Å)	M-O (Å)	M-O-O angle (deg.)	Ref.
$[Co(bzacen)pyO_2]$	1.26	1.86	126	60
$[Co(acacen)pyO_2]$		1.95		61
$[Co(salen-C_2H_4-py)(O_2)]$	1.1		136	58
$[Co(CN)_5O_2]^{3-}$	1.240	1.904	153	58
$[Co(t-Bu-saltmen)(BzIm)O_2]$	1.27[a]	1.882	117.5	62
	1.26[b]	1.974	118.5	62
$[Co(3 F-saltmen)(MeIm)O_2]$	1.302[c]	1.881	117.4	63
$[Co(saltmen)(BzIm)O_2]$	1.277	1.889	120.0	64
$[Co(3-t-Busalen)(py)O_2]$	1.35	1.87	116.5	248
oxycobaltmyoglobin	1.26	1.89	131	65
$[Fe(TpivPP)(MeIm)(O_2)]$	1.16	1.75	131	66
oxymyoglobin	1.25[d]	1.83	115(5)	67
oxyerythrocruorin	1.25	1.8	170(30)	68
$[Fe(TpivPP)(2 MeIm)(O_2)]$	1.22	1.898	129	69, 70
oxyhaemoglobin		1.67, 1.83	156(10)	71

[a] at −152 °C;   [b] at room temperature;   [c] at −171 °C,   [d] constrained to this value.
For ligand abbreviations, see Table 7.

Table 5. Structural details of $\eta^1 : \eta^1$ dioxygen complexes

Complex	O-O (Å)	M-O (Å)[a]	M-O-O (deg.)[a]	Dihedral angle (deg.)[b]	Ref.
(i) Vaska type I b complexes					
[(H₃N)₅Co(O₂)Co(NH₃)₅]⁵⁺ᶜ	1.31	1.89	118	175	72, 73
[(H₃N)₅Co(O₂)Co(NH₃)₅]⁵⁺ᵈ	1.32	1.89	117.3	180	74
[(en)₂Co(O₂; NH₂)Co(en)₂]⁴⁺	1.35	1.88	119	23.4	75, 76
[(H₃N)₄Co(O₂; NH₂)Co(NH₃)₄]⁴⁺	1.320	1.867	120.9	0.0ᵉ	77
[(en)₂Co(O₂; OH)Co(en)₂]⁴⁺	1.339	1.875	119.9	22.0	78
[(NC)₅Co(O₂)Co(CN)₅]⁵⁻ᶠ	1.29	1.92	120.7	180	79
[(NC)₅Co(O₂)Co(CN)₅]⁵⁻	1.24	1.94	121.2	165.9	79
(ii) Vaska type II b complexes					
[(H₃N)₅Co(O₂)Co(NH₃)₅]⁴⁺ᵍ	1.473	1.883	112.8	145.8	80
[(H₃N)₅Co(O₂)Co(NH₃)₅]⁴⁺ʰ	1.469	1.879	110.8	180	81
[(H₃N)₅Co(O₂)Co(NH₃)₅]⁴⁺ᵈ	1.472	1.886	110.9	180	82
[(dien)(en)Co(O₂)Co(en)(dien)]⁴⁺	1.488	1.896	110.0	180	83
[(tren)(H₃N)Co(O₂)Co(NH₃)(tren)]⁴⁺	1.511	1.889	111.5	180	84
[(O₂N)(en)₂Co(O₂)Co(en)₂(NO₂)]²⁺	1.529	1.887	110.0	180	85
[(pydpt)Co(O₂)Co(pydpt)]⁴⁺	1.456	1.891	114.3	162	86
[(pydien)Co(O₂)Co(pydien)]⁴⁺	1.489	1.876	112.5	180	87
[(papd)Co(O₂)Co(papd)]⁴⁺	1.486	1.924	111.9	180	88
[(NC)₅Co(O₂)Co(CN)₅]⁶⁻	1.447	1.938	111.8	180	89
[(H₂O)(salen)Co(O₂)Co(salen)]₂	1.31	1.97	118	122	90
[(DMF)(salen)Co(O₂)Co(salen)(DMF)]	1.339	1.910	120.3	110.1	91
[(salptr)Co(O₂)Co(salptr)]	1.45	1.93	118.5	149.3	92
[(pip)(salen)Co(O₂)Co(salen)(pip)]	1.383	1.912	120.1	121.9	93
[(dmtad)Co(O₂; OH)Co(dmtad)]³⁺	1.43	1.89	108.6	68.0	435
[(en)₂Co(O₂; OH)Co(en)₂]³⁺ⁱ	1.465	1.863	110.3	60.7	78
[(en)₂Co(O₂; OH)Co(en)₂]³⁺ʲ	1.460	1.873	109.3	64.5	94
[(tren)Co(O₂; OH)Co(tren)]³⁺	1.462	1.863	110.8	60.7	95
[(tren)Co(O₂; tren)Co(tren)]⁴⁺	1.49	1.90	115.8	19.8	436
[(Ph₃P)₂Pt(O₂; OH)Pt(PPh₃)₂]⁺	1.55	2.01	102.8	79.0	96

[a] average of two values; [b] angle between two M-O-O planes; [c] acid sulphate; [d] nitrate; [e] the Co-N-O-O ring is nearly planar; [f] two independent ions in the unit cell; the first ion is centrosymmetric; [g] sulphate; [h] thiocyanate; [i] racemate; [j] *meso* form

tartrate complex[125], and also in the auto-oxidation of Ti(III) bound to a pyridine-based polymer[126]. An η^2 dioxygen complex with dioxygen bound to a TiN_4 square pyramidal fragment has been reported[127].

Vanadium Complexes
Complexes formed by the action of hydrogen peroxide on vanadium compounds have been known for many years and have recently been studied by [51]V n.m.r. spectroscopy[129]. A report of the reversible formation of a dioxygen complex by vanadium(IV) catecholates has been re-investigated recently and refuted[130].

Table 6. Structural details of η^2dioxygen (Vaska type II a) complexes

Complex	O-O (Å)	Ref.
Ti(OEP)(O$_2$)	1.458	97
	1.445	59
Ti(dipic)(OH$_2$)$_2$O$_2$	1.469[a]	98
	1.477[b]	98
	1.458[b]	99
[{Ti(NTA)(O$_2$)}$_2$O]$^{4-}$	1.469	100
	1.481	98
[Ti(pic)$_2$(HMPT)O$_2$]	1.419	101
[Nb(O$_2$)F$_5$]$^{2-}$	1.17	102
Nb(η^5-C$_5$H$_5$)(O$_2$)Cl	1.47	103
[Ta(O$_2$)F$_5$]$^{2-\,c}$	1.39	104
[{Ta(O$_2$)F$_4$}$_2$O]$^{4-\,c}$	1.64	104
[Ta(O$_2$)F$_4$(MepyO)]$^-$	1.55	105
[MoO(O$_2$)F(dipic)]$^-$	1.46	106
Mo(T(p-Me)PP)(O$_2$)$_2$	1.40	107
[MoO(O$_2$)$_2$(py2c)]$^-$	1.462	108
	1.467	108
MoO(O$_2$)(pycc)	1.447	108
[{MoO(O$_2$)(pycc)}$_2$F]	1.43	109
MoO(O$_2$)(PhCON(Ph)O)$_2$[d]	1.21	110
MoO(O$_2$)$_2$((s)-MeCH(OH)CONMe$_2$)	1.459	111
	1.451	111
[WO(O$_2$)F$_4$]$^{2-}$	1.20	102
[Co(2=phos)$_2$(O$_2$)]$^+$	1.42	112, 118
L$_3$(NC)$_2$Co(μCN)Co(CN)L$_2$(O$_2$), L=PPhMe$_2$	1.44	113
[Co(R, R-C$_{24}$H$_{38}$As$_4$)(O$_2$)]$^+$	1.424	114
[Rh(PPhMe$_2$)$_4$(O$_2$)]$^+$	1.43	115
[RhCl(O$_2$)(PPh$_3$)$_2$]$_2^{\,e}$	1.44	56
RhCl(O$_2$)(PPh$_3$)$_2$	1.413	116
[Rh(O$_2$)(dppa)$_2$]$^+$	1.419	117
[Rh(2=phos)$_2$(O$_2$)]$^+$	1.43	118
[Rh(AsPhMe$_2$)$_4$(O$_2$)]$^+$	1.46	119
[Ir(2=phos)$_2$(O$_2$)]$^+$	1.38	118
[Ir(O$_2$)(dppe)$_2$]$^+$	1.52	57
[Ir(PPhMe$_2$)$_4$(O$_2$)]$^+$	1.49	120
[Ir(O$_2$)(dppm)$_2$]$^+$	1.45–1.50	121
IrCl(PPh$_2$Et)$_2$(CO)(O$_2$)	1.47	122
Pd(PPh(t-Bu)$_2$)$_2$O$_2$	1.37	123
Pt(PPh(t-Bu)$_2$)$_2$O$_2$	1.43	123

[a] Triclinic form; [b] Orthorhombic form; [c] both ions found in K$_6$Ta$_3$(O$_2$)$_3$OF$_{13}$ · H$_2$O; [d] disorder in oxide and peroxide positions; [e] η^1 : η^2 complex

Table 7. Abbreviations of ligands

Ligand	Ligand
acacen	N,N'-bis(acetylacetone)ethylenediamine
[14]ane N_4	see cyclam
bipy	2,2'-bipyridine
bzacacen	N,N'-bis(benzoylacetone)ethylenediamine
bzIm	benzimidazole
cyclam	1,4,8,11-tetra-aza-tetradecane
dien	diethylenetriamine
dipic	dianion of dipicolinic acid
DMF	N,N-dimethylformamide
dmtad	4,7-dimethyl-1,4,7,10-tetraazadecane
dppa	bis(diphenylphosphino)amine
dppe	1,2 bis(diphenylphosphine)ethane
dppm	bis(diphenylphosphino)methane
EDTA	ethylenediaminetetraacetate
en	ethylenediamine
F-salen	N,N'-bis(3-fluorosalicyldene)ethylenediamine
3 F-saltmen	N,N'-bis(3-fluorosalicyldene)tetramethylethylenediamine
HMPT	hexamethylphosphortriamide
IDA	dianion of iminodiacetic acid
Im	imidazole
J-en	N,N'-ethylenebis(2,2 diacetylethylideneamate)
L	any unspecified ligand
MeIm or 1-MeIm	N-methylimidazole
MepyO	2-methylpyridine-N-oxide
NTA	nitrilotriacetate
OEP	octaethylporphinate
papd	1,5,8,11,15-pentaazapentadecane
Pc	phthalocyanine
2=phos	cis-1,2-bis(diphenylphosphino)ethene
pic	picolinate
pip	piperidine
Porph	porphinate
py	pyridine
pycc	(=dipic) dipicolinate
py2c	picolinate
pydien	1,9-bis(2-pyridyl)-2,5,8-triazanonane
pydpt	1,11-bis(2-pyridyl)-2,6,10-triazaundecane
salen	N,N'-bis(salicylidene)ethylenediamine
S-salen	N,N'-bis(thiosalicylidene)ethylenediamine
salen-C_2H_4-py	N,N'-bis(salicylidene)-1-(ethyl-2-pyridine)ethylenediamine

Table 7 (continued)

Ligand	Ligand
salptr	N,N''-bis(salicylidene)dipropanetriamine
saltmen	N,N'-bis(salicylidene)tetramethylethylenediamine
S-Me₂en	N,N'-dimethylethylenediamine
t-Bu-saltmen	N,N'(3-tert-butylsalicylidene)tetramethylethylenediamine
TpivPP	meso-tetra(α,α,α,α-o-pivalamidophenyl)porphinate (picket fence porphinate)
T(p-Me)PP	meso-tetra(p-methylphenyl)porphinate
TPP	meso-tetraphenylporphinate
tren	tris(2-aminoethyl)amine
trien	triethylenetetramine

Chromium, Molybdenum and Tungsten Complexes

The infra-red spectrum of Cr(TPP)py(O_2) obtained by the oxygenation of a chromium(II) porphyrin complex suggests the presence of an η^1 dioxygen species[131]. The photolysis of chromium complexes such as $[Cr_2(\eta^5\text{-}C_5H_5)_2(CO)_6]$ in the presence of O_2 gives paramagnetic complexes which may be studied by E.P.R.[132, 297]. The g-values lie close to 1.99 and isotopic labelling studies suggest that the two oxygen atoms are equivalent (implying an η^2 geometry) with the spin localised essentially on the metal atom. Photolysis of Cr(CO)₆ in an O_2-doped Ar matrix leads to the dioxo-compound $CrO_2(CO)_2$[133]. Cr^{2+} in a zeolite matrix has been reported to show reversible uptake of O_2[134].

An η^1 complex is reported to be formed by the reaction of MoO(TPP)(NCS) with superoxide anion to give MoO(TPP)(O_2)[135]; the complex has been studied by E.P.R. and U.V. – visible spectroscopies. The peroxide complexes $MO(O_2)_2 \cdot LL'$ (M = Mo, W; L = amine oxide, tertiary phosphine or arsine oxide; L' = L, OH₂) have been studied by infra-red spectroscopy. The O-O stretching frequency lies between 811 and 930 cm⁻¹, the arsine oxides generally give the lowest frequencies[136]. A similar study has recently been reported for $[MoO(O_2)_2(C_2O_4)]^{2-}$[137]. The complex $[Mo_2Al(O\text{-}iPr)_4]_2(CH_3COO)_2$ which contains a quadruple Mo-Mo bond is reported to add dioxygen irreversibly, giving a product with $\nu_{\text{O-O}} = 840$ cm⁻¹[138].

Manganese Complexes

Although manganese-dioxygen complexes have been intensively studied there is very little definite structural information and no crystal structure is available. A review of the subject has appeared fairly recently[139], and we give below some of the more recent work in this field.

The phthalocyanine complexes of manganese have been studied for many years[140]. The most recent work has established the formation of an η^1 dioxygen complex on oxygenation of [Mn(Pc)] in N,N-dimethylacetamide solution[141], and a similar result has been obtained for a tetra-sulphonated phthalocyanine derivative[142]. The porphyrin complex Mn(TPP)O_2 was assigned an η^2 structure on the basis of E.P.R. data[143] and this assignment has recently been supported by infra-red data using ¹⁸O_2[144]. Schiff base complexes of Mn(II) were reported to form dinuclear $\eta^1 : \eta^1$ complexes on oxygenation in

1970[145], and similar complexes have been reported recently, either by oxygenation of Mn(II)[146] or by addition of superoxide ion to Mn(II) complexes[147, 148]. Addition of two equivalents of superoxide to Mn(TPP)Cl gives a product identified as Mn(II)(TPP)(O_2^-)[309].

There have been several other reports of manganese dioxygen complexes, but unambiguous spectroscopic evidence is lacking. The photolysis of $Mn_2(CO)_{10}$ and its derivatives in the presence of dioxygen gives complexes identified as $Mn(CO)_5O_2$. E.P.R. studies, including isotopic labelling, suggest this complex to have an η^1 structure[149]. There have been several studies on the interaction of Mn(II) catechol complexes with dioxygen[150, 151] but it is not clear whether oxygen attack occurs on the metal or the ligand as has been proposed for similar iron complexes[152]. The complexes $MnLX_2$ (L = tertiary phosphines) are reported to show reversible oxygenation[153] but little structural information is currently available, and the oxygenation reaction is the subject of some controversy[154].

Iron and Ruthenium Complexes

Most work published on iron-dioxygen complexes has been concerned with analogues of haemoglobin and myoglobin. Attention has been focussed on methods of preventing dimerisation of the dioxygen complexes formed by methods such as bonding the deoxygenated complex to a polymer[155, 156] or the construction of suitably hindered porphyrins or similar macrocycles[66, 69, 70, 157-159]. The dioxygen complexes are assumed to have η^1 structure, and, since many reviews on the subject are available[17, 19-21, 160] we will concentrate here on the other dioxygen complexes of iron. There is, however, one possible exception to this uniformity of structure since it has been reported that a capped porphyrin with *two* coordinated 1-methyl imidazole groups is also capable of binding dioxygen[161], thereby giving an apparently 7 coordinate species. A recent development in this field is the synthesis of η^1 complexes in which the axial ligand is a sulphur donor as is thought to be the case in cytochrome P 450[162].

Less information is available for other types of dioxygen complex. ^1H n.m.r. was used to establish the formation of the $\eta^1 : \eta^1$ dioxygen complex P-Fe-O_2-Fe-P (P = porphyrin) in toluene at low temperatures[163, 495]. This species, which appears to contain two high spin Fe(III) ions antiferromagnetically coupled through a peroxide bridge, deoxygenates reversibly at low temperatures, but decomposes on warming to an μ-oxo species. The $\eta^1 : \eta^1$ dioxygen complex appears to be very unstable, and careful work has shown that a previous claim to have isolated an analogous complex with phthalocyanine replacing the porphyrin is unjustified[164]. More recently relatively stable $\eta^1 : \eta^1$ species have been reported to be formed on oxygenation of ferrous complexes of a pentacoordinate nitrogen macrocycle[165] and of the N_2S_2 ligand S-salen[166]. In the biological oxygen carrier oxyhaemerythrin[167] a similar Fe(III)-(O_2^{2-})-Fe(III) structure is generally accepted. The two iron atoms are in slightly different environments and are thought to be bridged by a second ligand, possibly an oxo group[168]. A careful study of the electronic spectra supports this assignment[169].

A recent study of the interaction of superoxide anion with Fe(II) porphyrins in dimethyl sulphoxide or acetonitrile has suggested the formation of an η^2 complex (porphyrin)FeO_2, which is formulated on the basis of infra-red, U.V.-visible, n.m.r., and E.P.R. spectroscopic measurements as (porphyrin)-Fe^{3+}(high spin)-O_2^{2-} [170, 309]. The E.P.R. spectrum differs from that of other high spin ferric porphyrin complexes but is similar to that of rhombic high spin ferric complexes such as [Fe(III)EDTA]$^-$. The complex Mn(porphyrin)O_2 discussed earlier also shows a rhombic E.P.R. spectrum and is thought to have an η^2 structure. The formation of an unstable complex between [Fe(III)EDTA]$^-$ and H_2O_2 has previously been reported and is known to contain high spin Fe^{3+} [171, 172]. The resonance Raman spectrum of this complex shows an O-O stretch at 824 cm^{-1} in agreement with this assignment[173].

In more unusual conditions, the η^1 complex Fe(TPP)O_2 has been prepared by condensation of Fe(TPP) in an O_2 doped argon matrix[174] and the condensation of iron atoms in a similar matrix has also been studied[175]. The photolysis of [Fe(η^5-C_5H_5)(CO)$_2$]$_2$ in the presence of oxygen is reported to give [Fe(η^5-C_5H_5)(CO)$_2O_2$][149, 297].

Ruthenium dioxygen complexes are less popular and most published work has been concerned with dioxygen addition to Ru(O) complexes, although the formation of a dioxygen complex analogous to the dioxygen-ferrous porphyrins has been reported [176]. Ru(O) complexes such as Ru(NO)(PPh$_3$)$_2$Cl[177] and Ru(CO)(CNR)(PPh$_3$)$_2$[178] add dioxygen to give η^2 dioxygen complexes similar to the adducts of Ir(I) complexes (see below). The crystal structure of Ru(NO)(PPh$_3$)$_2$Cl(O_2) has been reported as similar to that of the dioxygen adduct of Vaska's compound[179]. A report of dioxygen complex formation by RuCl$_2$(AsPh$_3$)$_3$ has recently been shown to involve oxidation of the arsine rather than complex formation[180].

Recently two reports of interaction of dioxygen with Ru^{3+} have appeared. The complex of Ru^{3+} with EDTA appears to form an $\eta^1 : \eta^1$ dioxygen complex with a bridging OH ligand similar to many Co(III) complexes[181] and Ru^{3+} in a zeolite has been shown to react with dioxygen to give Ru^{4+} and superoxide[182].

Cobalt Complexes

The many dioxygen complexes of cobalt represent by far the best structurally characterised family of dioxygen complexes and many references to crystal structure determinations are given in Tables 4–6. These complexes typically contain an octahedral cobalt atom with the dioxygen ligand bonded in an η^1 fashion and frequently acting as a bridging ligand to another cobalt atom. The other ligands around the cobalt are typically polydentate amines or Schiff bases, porphyrins or macrocycles, and recently polydentate ligands containing sulphur and oxygen donors as well as nitrogen donors have been used[183–185]. These complexes have been reviewed extensively[9, 10, 19, 186] and we will discuss here only those complexes of particular structural interest. The numerous spectroscopic studies on these complexes will be discussed at the end of Sect. D.

Dimerisation of monomeric η^1 complexes to give $\eta^1 : \eta^1$ complexes (types I b or II b) may be restricted by working in non-polar solvents, at low temperatures or by using ligands which sterically hinder the dimerisation such as the picketfence porphyrins mentioned for iron. This appears to be relatively easy for cobalt complexes, and simple methylation of the nitrogen atoms of ethylenediamine allows the monomeric η^1 complex [Co(S-Me$_2$en)$_2$XO$_2$]$^{n+}$ to be studied in solution[187]. Resonance Raman spectroscopy has been used to follow the dimerisation reaction[188]. A recent paper reports the formation of η^1 and $\eta^1 : \eta^1$ complexes in aqueous DMF solution on oxygenation of Co(II) complexes of a water soluble Schiff-base[189]. Cobalt(II) phosphine complexes add dioxygen readily, but phenyl groups on the phosphorus ligands prevent dimerisation: Co(CN)$_2$(PPhMe$_2$)$_3$ forms an η^1 dioxygen adduct detectable by E.P.R.[190] which decomposes to the cyanide bridged η^2 complex [Co(PPhMe$_2$)$_3$(CN)$_2$-μ(CN)-Co(CN)(PPhMe$_2$)$_2$(O_2)][113], while

$[Co(2=phos)_2]^{2+}$ adds dioxygen to give $[Co(2=phos)_2(O_2)]^+$ [191], an η^2 complex where the phenyl groups of the phosphine prevent dimerisation[192]. This compound may also be formed by addition of O_2 to a Co(I) complex.

Some workers have deliberately synthesised ligands likely to favour the formation of an $\eta^1:\eta^1$ dimer by placing two multidentate ligands in proximity: two terdentate amino ligands joined by a xylene bridge form a template for the formation of an $\eta^1:\eta^1$ (type II b) complex[193] and two porphyrin groups held "face-to-face" by bridging side chains show the same effect[194]. In the second case the formation of the $\eta^1:\eta^1$ complex could be supressed by separating the two porphyrin rings sufficiently. Oxygenation of the dicobalt(II) complex of the dinuclear ligand 2,6bis[bis(2-pyridylmethyl)aminomethyl]-4-methylphenol is reported to give an $\eta^1:\eta^1$ complex[195].

Other studies have been concerned with trapping the Co-O_2 monomeric complex in an environment where dimerisation is impossible. Nakamoto and co-workers have published a series of papers on Co(II) complexes trapped in O_2-doped argon matrices[196-199], and a report has appeared of Co^{2+} complexed by a nitrogen macrocycle in a micellar phase[200]. Co^{2+} complexed by ethylenediamine in zeolite cages forms an η^1 mononuclear complex at low concentration but $\eta^1:\eta^1$ dinuclear complexes are observed at higher concentrations. There is some evidence for formation of free O_2^- ion[201]. Similar results were found for Co(II) ethylenediamine complexes absorbed on a cation exchange resin[202].

An early report of the formation of an η^1 dioxygen complex $Co(CO)_4O_2$ on sublimation of $Co_2(CO)_8$ in the presence of dioxygen[203] has been followed by studies with metal atom vapours[204] which support the existence of $Co(CO)_4O_2$ and also suggest the formation of $Co(CO)_nO_2$ (n = 1 to 3) in which the dioxygen appears from infra-red spectroscopy to form an η^2 complex.

Rhodium and Iridium Complexes

Most of the work in this field has been concerned with the η^2 complexes formed on addition of dioxygen to square planar M(I) complexes such as Vaska's compound, and a number of references to more recent work are given in Table 6. ^{31}P n.m.r. spectroscopy has been used recently to characterise the dioxygen adducts of a series of Rh(I)-chelating diphosphine complexes[205]. The NQR spectrum of the dioxygen adduct of Vaska's compound $[Ir(CO)Cl(PPh_3)_2O_2]$ shows the two oxygen atoms to be non-equivalent[206].

There has been relatively little effort made to synthesise rhodium and iridium analogues to the bridged cobalt dioxygen complexes although it has been known for some time that a μ-peroxo $\eta^1:\eta^1$ complex is formed on oxygenation of $[RhH(CN)_4(OH_2)]^{2-}$ [207] and μ-peroxo and μ-superoxo $\eta^1:\eta^1$ complexes are formed on ozonolysis of solutions of $[Rhpy_4Cl_2]^+$ [208]. More recently η^1 (mononuclear) and $\eta^1:\eta^1$ (binuclear) dioxygen complexes have been reported as having been formed after the photolysis in presence of oxygen of $[Rh(NH_3)_4H(OH_2)]^{2+}$ [209] and $[Rh(en)_2(NO_2)_2]^+$ and similar compounds[210]. The dimeric Rh(II) porphyrin complex $[Rh(OEP)]_2$ reacts with dioxygen to give $Rh(OEP)O_2$ which later dimerises to $[Rh(OEP)]_2O_2$ [211]. E.P.R. spectroscopy has been used to study the paramagnetic η^1 complexes. The only equivalent report for iridium concerns the unstable complex formed by dioxygen with the sterically hindered complex of Ir(II) with 3-methoxy-2-di-t-butylphosphinophenolate[212]. Condensation of rhodium atoms with Rh gives the complexes $Rh(O_2)$ and $Rh(O_2)_2$. Bimetallic species have also been found[218].

Finally we may recall that rhodium is at present the only transition metal known to form an $\eta^1 : \eta^2$ complex[56] and that it also forms an $\eta^2 : \eta^2$ complex[53].

Nickel, Palladium and Platinum Complexes

Most of the work involving the dioxygen complexes of these metals has been concerned with the planar η^2 $L_2M(O_2)$ complexes obtained by addition of dioxygen to M(O) species. A recent crystal structure determination for L_2MO_2 (L = PPh(t-Bu)$_2$, M = Pd, Pt) shows that an opening of the L-M-L angle arising from the presence of the bulky phosphine destabilises the dioxygen adduct for Pd (L-M-L = 115.4°) in comparison with the Pt complex (L-M-L = 113.1°)[123]. The complex Pt(PPh$_3$)$_2$O$_2$ may be protonated to give a μ-hydroxo-μ($\eta^1 : \eta^1$dioxygen) species[213] and the crystal structure of this complex has been determined[96]. Reaction of the di-μ-chloro complexes [LL'PdCl]$_2$ with superoxide is reported to form the $\eta^2 : \eta^2$ dinuclear complex [LL'Pd]$_2$O$_2$[214]. An EPR study of Ni(I) complexes fixed on substituted polymers and reacted with dioxygen reports the formation of Ni^{2+}-O$_2^-$-complexes[215].

Ozin and co-workers have reported infra-red spectra for Ni, Pd and Pt atoms condensed in an O$_2$/Ar[216] and an O$_2$/N$_2$/Ar matrix[217]. With dioxygen alone the complexes MO$_2$ and M(O$_2$)$_2$ were identified, whereas with nitrogen present M(O$_2$)(N$_2$) and M(O$_2$)(N$_2$)$_2$ were found. In all cases the dioxygen was bound in an η^2 mode.

Copper, Silver and Gold Complexes

Copper resembles manganese in that it is known to play an important role in biological processes involving dioxygen, but there is very little reliable structural data available. In recent reviews some of the properties of the bimetallic species haemocyanin and laccase were discussed[167, 219]. In the oxygen carrying protein oxyhaemocyanin EXAFS[220] and resonance Raman spectroscopy[221] suggest the formation of a doubly bridged species with an $\eta^1 : \eta^1$ bridging dioxygen (Fig. 4) but there is some disagreement as to the nature and disposition of the other ligands bound to copper. The oxidase and mono-oxygenase tyrosinase is thought to show a similar coordination of dioxygen[222]. The enzyme superoxide dismutase is thought to use only one copper for binding O$_2^-$[223]. An Fe-O-O-Cu bridging system has been suggested for cytochrome c oxidase by several authors, but no firm evidence is available.

Most of the recent preparative work has concentrated on trying to mimic the supposed coordination site[224–227] by careful design of ligands. Although reversible dioxygen binding has been observed in two cases[224, 225] and in one case oxygenation of an aromatic residue of the ligand was observed[226], structural data is still lacking. The stoichiometry of dioxygen uptake indicates formation of a dinuclear complex. There is also abundant

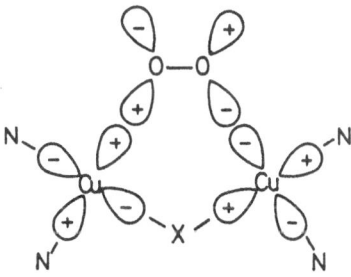

Fig. 4. The structure of oxyhaemocyanin (Ref. 220)

evidence of formation of a 2 : 1 (Cu : O_2) complex from the kinetics of auto-oxidation and Cu catalysed oxidations[228]. A monomeric EPR silent complex has been observed to be formed on addition of O_2^- to a DMSO solution of a Cu^{2+} macrocycle complex[229].

The condensation of silver, gold and copper atoms with O_2 has been studied[331]. Copper gives the η^2 complexes CuO_2 and $Cu(O_2)_2$ and gold gives only $\eta^2 AuO_2$. Silver gives an apparently ionic compound $Ag^+ O_2^-$ and a poorly characterised AgO_4 species.

Group II B Metals, Lanthanides and Actinides

Zinc tetraphenylporphyrinate forms a weak complex with O_2^- in non-aqueous solutions. The bonding in this complex appears to be essentially ionic[230]. We have already mentioned crystal structure determinations of lanthanide[55] and actinide[54] compounds. There is every reason to suppose that these elements have a rich dioxygen complex chemistry[5], and this is confirmed by two recent papers[231, 232]. For reasons of space, however, we shall not discuss the dioxygen complex chemistry of these elements.

D. The Electronic Structure of Dioxygen Complexes

In the discussion of the electronic structure of dioxygen complexes we seek to understand how the dioxygen ligand binds to a transition metal and the factors determining the mode of coordination which is adopted. Any picture of the bonding in these complexes should enable us to rationalise the properties of the complexes, and we will be particularly interested in the spectroscopic properties which allow us to test the electronic model of the bonding, and in the chemical reactivity of the dioxygen ligand resulting from its binding to a transition metal. In this section we shall consider the various qualitative descriptions of binding in the complexes, the results of calculations and their agreement with the qualitative models, and the spectroscopic data relevant to the discussion of the bonding.

I. Qualitative Models of Dioxygen Binding

The tendency of early workers to concentrate on one type of dioxygen complex is also reflected in the earliest discussions of the bonding. In their original paper [3] Pauling and Coryell proposed a linear Fe-O-O structure for oxyhaemoglobin, but Pauling later changed his prediction[233] to the bent η^1 structure which has since been found experimentally. J. S. Griffith[234] concluded that a linear Fe-O-O group would be unstable in oxyhaemoglobin, but η^1 or η^2 geometries were possible. He favoured an η^2 structure in which the filled π_u bonding orbital acts as a donor to the metal rather than a non-bonding sp^2 hybrid. Following these two papers, the η^1 and η^2 structures have frequently been referred to as Pauling and Griffith structures respectively. In 1964 Weiss proposed that oxyhaemoglobin should be regarded as a $Fe^{3+} O_2^-$ complex[235].

The bonding in peroxychromates was initially discussed in completely different terms. Tuck and Walters considered the primary interaction to be that of the π_u bonding orbitals of the peroxide ligand acting as a donor to the metal ion in a way similar to the Dewar-Chatt-Duncanson model of olefin binding[236]. W. P. Griffith favoured a "bent" bonding arrangement arising from overlap between the metal acceptor orbital and a sp^2 hybrid pair on each oxygen atom[237]. In a review of the early crystal structure determinations of the peroxychromates[238], Stomberg pointed out that an interaction between the occupied π_g^* orbitals of the peroxide ion and the d orbitals of the chromium atom would explain the unusual pentagonal pyramidal coordination found around the chromium atom, and a π_g^*-3d interaction was also proposed by Swalen and Ibers[239].

Werner correctly identified the $\eta^1:\eta^1$ dinuclear complexes that Vaska classifies as type II b as μ-peroxo complexes of two Co(III) ions, but it was only with the advent of modern physical techniques that it was possible to show by E.P.R. that the unpaired electron in Vaska type I b complexes such as $[(H_3N)_5CoO_2Co(NH_3)_5]^{5+}$ is localised on the dioxygen ligand[240], leading to their classification as μ-superoxo complexes. X-ray structural data (Table 5) show that the O-O bond lengths in type I b complexes are significantly shorter (and closer to the value for free O_2^-) than those in type II b complexes which lie close to the values obtained for O_2^{2-}.

The assignment of oxidation states for the chromium and cobalt complexes discussed above is relatively straightforward, but as more η^1 cobalt dioxygen complexes and dioxygen adducts of Vaska type complexes were prepared, the assignment of oxidation states to dioxygen became more complicated. The coordinated dioxygen was variously considered as neutral dioxygen (possibly in its first excited state, $^1\Delta_g$[241]), superoxide or peroxide. Thus an η^1 cobalt complex such as $[Co(CN)_5O_2]^{3-}$ could be regarded as Co(II)-O_2 or Co(III)-O_2^-. In his review[4], Vaska noted that η^1 complexes generally had O-O bond lengths close to 1.3 Å and O-O stretching frequencies around 1125 cm^{-1} whereas the equivalent values for η^2 complexes were 1.45 Å and 860 cm^{-1}. By analogy with the free superoxide and peroxide ions (Table 1) he proposed that η^1 complexes should be considered as superoxo and η^2 complexes as peroxo compounds, but this classification was by no means universally accepted. With the passage of time however, it has become accepted that the metal-O_2 bond, although polarised (with the dioxygen carrying at least a partial negative charge), generally has an appreciable covalent character and the assignment of formal oxidation states is to some extent a semantic problem. The subject of the oxidation states of η^1 complexes has recently been discussed in some detail[242].

Apart from the unambiguous and useful classification of the dinuclear cobalt complexes as μ-peroxo or μ-superoxo species, we shall not use formal oxidation states frequently in this review, and we limit the discussion here to the following points:

(i)   the assignment of formal oxidation states to the metal and to the dioxygen may be of use in establishing an electron count for the molecule;

(ii)   although physical measurements such as stretching frequencies or hyperfine coupling constants may suggest an assignment of oxidation states, the values so obtained should be considered as spectroscopic oxidation states as defined by Jørgensen[243] and not as an accurate description of the electron distribution in these complexes;

(iii)   the oxidation state assigned to dioxygen, however it is obtained, is not, in general, a useful means of classifying the chemical reactivity of these complexes, and a correlation with the properties of neutral dioxygen, O_2^-, or O_2^{2-} is not justified.

Most recent authors have discussed the bonding in dioxygen complexes in terms of molecular orbital theory. Thus Drago in his "spin-pairing" model for η^1 complexes of cobalt proposes a M.O. diagram arising from the interaction of one π_g^* orbital of dioxygen with a d_z^2 orbital of cobalt and a weaker π interaction of the second π_g^* orbital with a cobalt d_{xz} orbital[244]. Mingos[245] has used second order Jahn-Teller arguments to predict whether an η^1 or η^2 complex will be stable to distortion to the other conformation. The most detailed treatment published so far has been given by Hoffmann and his co-workers[246] in a general paper on the interaction of a diatomic ligand with the metal-ligand fragment ML_4 or ML_5. The intention of the authors was to establish a Walsh diagram which would enable one to determine which of three possible coordination geometries (linear, or bent η^1, η^2) would be adopted by the ligand as a function of the total number of valence electrons on the ligand and the metal ion. The qualitative predictions were supported by a number of Extended Hückel (EHMO) calculations, and were remarkably successful in predicting whether dioxygen would be bound in an η^1 or an η^2 conformation. Recently we have discussed the electronic structure of dioxygen complexes using a qualitative M.O. approach similar to that of Hoffmann et al. and Drago. Since we feel that this model provides a useful basis for discussion of spectra and reactivity we summarise here the principal features. For further details the reader is referred to the original papers[247].

We assume, as do Hoffmann et al.[246], that the bonding between the transition metal and dioxygen may be discussed in terms of the interaction between the molecular orbitals of dioxygen (Fig. 1) and the frontier orbitals of the fragment ML_n (where L_n represents the other ligands bound to the metal). The most readily accessible orbitals of O_2 are the π_g^* orbitals which photoelectron spectroscopy shows to have an ionisation energy near 12 eV, much lower than the $3\sigma_g$ and π_u which have ionisation energies close to 17 eV[249]. Any interaction of the metal with the filled π_u or $3\sigma_g$ orbitals is assumed to be of secondary importance and will be neglected. As the dioxygen ligand is a better acceptor than donor, any interaction with the empty metal s and p orbitals may similarly be neglected in a first order approximation. The problem is thus reduced to the consideration of the interaction of a π_g^* orbital of dioxygen with a metal d-orbital. The three favourable overlaps possible are shown in Fig. 5.

Overlap (i) is that proposed by Drago[244] in his spin-pairing model and overlap (iii) that proposed by Stomberg[238] for the peroxychromates. On overlap grounds alone the η^2 complex (overlap (iii)) will clearly be more stable, and, indeed, η^2 complexes are the

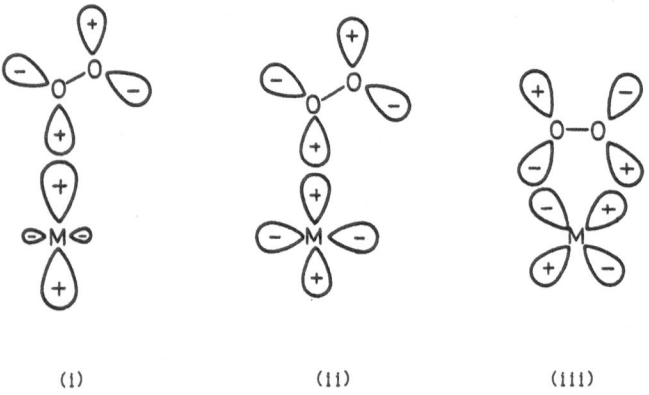

(i)   (ii)   (iii)

Fig. 5. Possible overlaps between metal d-orbitals and a π_g^* orbital of dioxygen

most commonly encountered, but an η^1 complex may be formed if steric constraints due to the other ligands present hinder overlap (iii) or if the d_z^2 orbital is favoured energetically.

The second π_g^* orbital of dioxygen, shown in Fig. 6, lies perpendicular to the MO_2 plane and will henceforth be denoted as $\pi_g^*(\perp)$. For an η^2 complex $\pi_g^*(\perp)$ has δ symmetry with respect to the M-O_2 bond, and any interaction with the metal d-orbitals will be weak and may be ignored (although it has been suggested that such an interaction is important in limiting rotation about the M-O_2 axis (Fig. 3b))[250, 251]. For η^1 geometry, however, the $\pi_g^*(\perp)$ orbital has π symmetry with respect to the M-O_2 bond, and a secondary π interaction is possible.

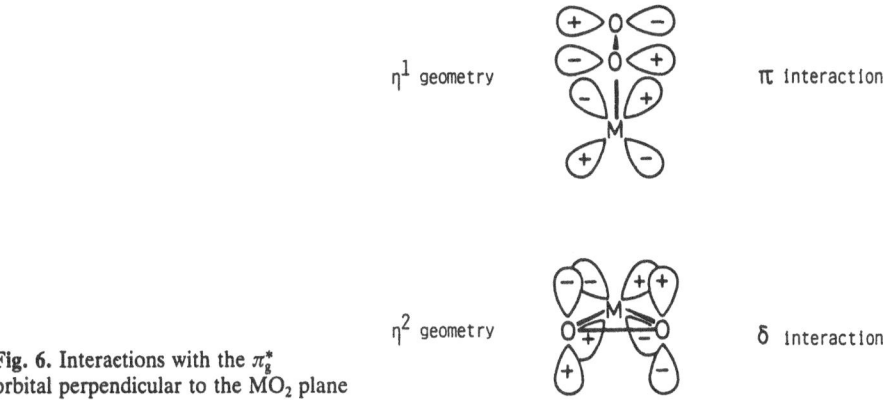

η^1 geometry π interaction

η^2 geometry δ interaction

Fig. 6. Interactions with the π_g^* orbital perpendicular to the MO_2 plane

We may now draw the simple M.O. diagrams for η^1 and η^2 geometry shown in Fig. 7. The orbital interacting with π_g^* via one of the overlaps in Fig. 5 is labelled d_σ and that able to interact with $\pi_g^*(\perp)$ in an η^1 complex is labelled d_π.

If we leave aside for the moment the question of the relative energies of d and π_g^* orbitals, then we may see that for η^1 geometry the system will be most stable with 4 electrons, 2 from dioxygen and 2 in the fully occupied d_π. Addition of one electron weakens the bonding as this electron must enter d_π-$\pi_g^*(\perp)$ which is M-O_2 antibonding. Addition of a second electron to d_π-$\pi_g^*(\perp)$ will result in the secondary π bonding interaction becoming antibonding, and we might expect a distortion. For an η^2 geometry, however, any interaction of $\pi_g^*(\perp)$ with the metal d-orbitals will be very weak, and the most favourable state will be attained if d_σ initially contained two electrons.

In practice, for all known η^2 complexes the d_σ orbital may be considered as donating two electrons to the M-O_2 bond. For the well characterised η^1 complexes of iron and cobalt the d_σ orbital contributes respectively 0 and 1 electrons to the bond. It has been suggested that the heats of oxygenation of cobalt complexes are slightly less negative than those of the equivalent iron complexes as a result of the partial occupation of the d_π-$\pi_g^*(\perp)$ antibonding orbital[252] but it is not certain that the data available may be compared directly[244]. There is some evidence that addition of one electron to an η^1 cobalt complex (e.g. by outer-sphere reduction) leads to formation of an η^2 complex[113, 253, 254].

We may now return to the question of the relative energies of the metal d-orbitals and the dioxygen π_g^* orbitals. Dioxygen is unusual amongst ligands in that its high electronegativity causes it to act more as an electron acceptor than as an electron donor. The

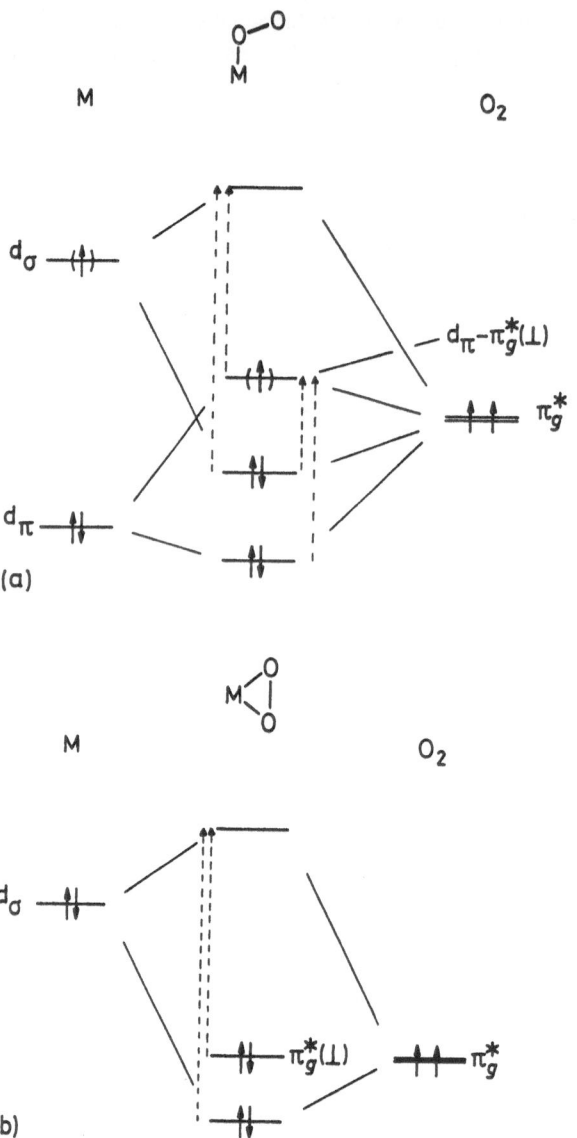

Fig. 7a, b. Molecular orbital diagrams for (a) η^1 complexes, and (b) η^2 complexes. The electron in parentheses is absent in η^1 FeO$_2$ complexes, but is present in those of cobalt

extent to which it will act as an electron donor will depend on the energy of the d-orbitals relative to the π_g^* orbitals which the photoelectron spectrum suggests to lie at -12 eV. It is well known that the d-orbitals become progressively more stable on crossing the transition series, and Hartree-Fock-Slater calculations on neutral atoms show the d-orbital binding energies to rise from 7 eV for scandium to 17 eV for zinc[255]. Although close comparison of these energies with the O$_2$ π_g^* is not justified, one may reasonably surmise that electron transfer will be easy for the early transition metals, but that the later transition metals will not readily donate electrons to dioxygen. The early transition metals form numerous dioxygen complexes (usually formulated as peroxy-complexes of the d^0 or d^1 ion), whereas the later transition metals form dioxygen complexes only in their lower oxidation states. It is reasonable to suppose that dioxygen complexes of the early transition metals in their lower oxidation states will be unstable with respect to electron transfer from the metal d-orbitals to the $3\sigma_u^*$ orbital of dioxygen, leading to breakage of the O-O bond and formation of an oxo-complex. Norman[256] first pointed out the rôle of the other ligands bound to a group VIII metal, and a survey of the crystal structure data confirms that dioxygen will only bind to the later transition metals when the d-orbital interacting with dioxygen is destabilised by one or more of the other ligands present. Furthermore, the correlation of the stability of the dioxygen adduct with the donor powers of the other ligands present is explained by their rôle in destabilising the d-orbitals.

A few examples will illustrate the points made above:

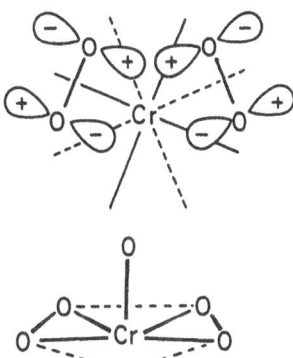

Fig. 8. Above: the overlap of d_{xy} (——) and $d_{x^2-y^2}$ (----) orbitals with dioxygen in CrO_5; below: the structure of the CrO_5 unit

(i) CrO_5 (Fig. 8) may be considered as two dioxygens bonded to a CrO fragment. The two dioxygens interact with the d_{xy} and $d_{x^2-y^2}$ orbitals of chromium (the other d-orbitals are Cr-O antibonding) via type (iii) overlaps to give a planar CrO_4 unit perpendicular to the CrO axis. As Stomberg pointed out[238], the 45° angle between the lobes of the d_{xy} and $d_{x^2-y^2}$ orbitals accounts for the relative orientation of the dioxygen ligands.

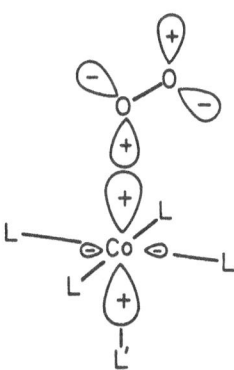

Fig. 9. The structure of $CoL_4L'\eta^1(O_2)$

(ii) $CoL_4L'(O_2)$ (Fig. 9). The CoL_4L' fragment is assumed to have a square-based pyramidal structure (as in e.g. $[Co(CN)_5]^{3-}$). The most accessible d-orbital is the d_z^2 (containing one electron) which may interact with dioxygen via a type (i) overlap. The M.O. diagram (Fig. 7 a) is essentially that used by Drago in his spin-pairing model[244] of the interaction between the electron in the d_z^2 orbital and one of the unpaired electrons of ground state molecular dioxygen. As cobalt lies to the right of the transition metal series the d-orbitals are quite strongly bound, and the *trans* ligand L' plays an important role in destabilising the d_z^2 orbital. The correlation between the donor power of L' and the stability of the complex is well documented[19, 186]. The generally accepted picture of the bonding in oxyhaemoglobin is similar to this but the system contains one less valence electron and the orbital d_{xz}-$\pi_g^*(\perp)$ is empty.

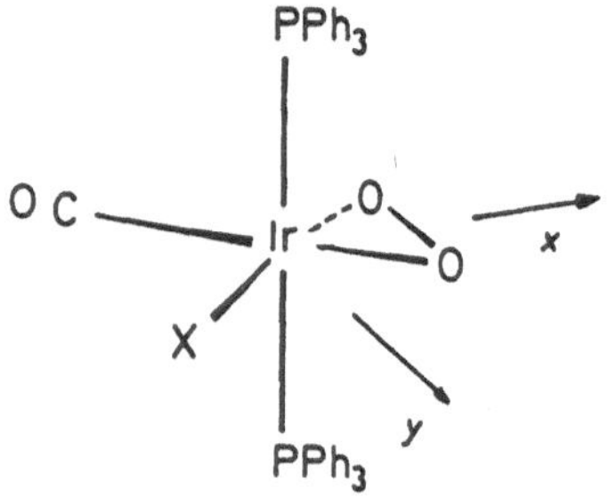

Fig. 10. The structure of $Ir(PPh_3)_2X(CO)\eta^2(O_2)$

(iii) $Ir(CO)X(PPh_3)_2O_2$ (Fig. 10). The square planar complex $Ir(CO)X(PPh_3)_2$ distorts in such a way as to destabilise the d_{xy} orbital sufficiently to allow a type (iii) overlap with dioxygen. An increase in the donor power of X will increase the destabilisation of d_{xy} and stabilise the dioxygen complex. Such an effect has been observed experimentally[257]. The alternative structure for this complex, in which the dioxygen adds directly in an η^1 fashion to the square planar complex via a type (i) overlap with the d_z^2 orbital is unlikely for two reasons: the d_z^2 orbital, in the absence of an axial ligand, is not sufficiently destabilised to act as an electron donor, and secondly, the interaction of $\pi_g^*(\perp)$ with the filled metal d_π orbitals would have a net antibonding effect and, as mentioned before, this would favour a distortion to η^2 geometry.

The model considering only the interactions between the π_g^* orbitals of dioxygen and the metal d-orbitals gives a good general account of the bonding and the effect of other ligands in the mononuclear complexes. The definition of oxidation states in these complexes is essentially a matter of localising the occupied M.O. on the metal ion or the dioxygen. The approach may be extended to dinuclear complexes by considering the interaction of a second ML_n fragment with the π^* orbitals of dioxygen bound in a mononuclear complex L_nMO_2. The possible structures are shown in Fig. 11.

Let us first consider the overlap of the second metal ion with the π^* orbital in the MO_2 plane of the first fragment. In the (1, 2) $\eta^1:\eta^1$ geometry the second metal overlaps with one of the lobes of π^* on the distal oxygen of the first complex leading to a *cis* or *trans* planar structure. In (1, 1) $\eta^1:\eta^1$ complexes the overlap is with the lobe of π^* on the proximal oxygen atom. This geometry has only been reported for a hydroperoxide formed on protonation of the (1, 2) $\eta^1:\eta^1$ complex $[(en)_2Co(\mu\text{-}NH_2)(\mu\text{-}O_2)Co(en)_2]^{3+}$ [258] although a similar dinuclear aluminium complex has recently been reported[259]. In view

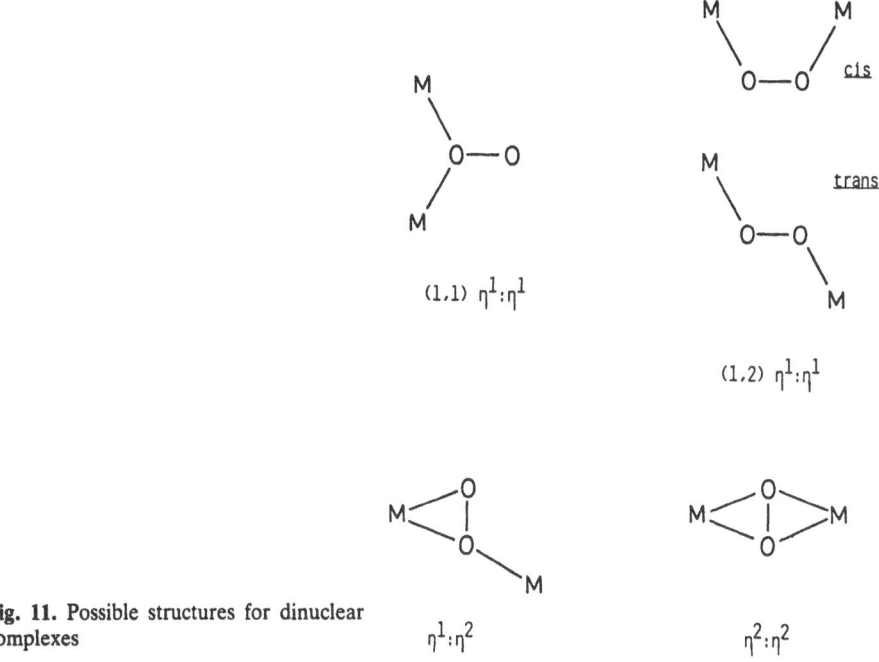

Fig. 11. Possible structures for dinuclear complexes

of the rarity of $(1, 1)$ $\eta^1:\eta^1$ complexes we shall not discuss them further and all $\eta^1:\eta^1$ complexes mentioned will be assumed of $(1, 2)$ type. In $\eta^2:\eta^2$ geometry the second metal overlaps with the two lobes of π^* that are free in the initial η^2 complex, while in $\eta^1:\eta^2$ geometry the second metal ion overlaps with one of these lobes.

It is however possible that the second metal ion will interact rather with the $\pi_g^*(\perp)$ orbital of the first complex which is either non-bonding or antibonding with respect to the first metal ion, and thus lies at higher energy. The resulting dinuclear complex would show the two MO_2 planes at right angles. This is indeed the case for the only $\eta^1:\eta^2$ complex known (Fig. 2)[56]. For $\eta^2:\eta^2$ complexes, on the other hand, the two crystal structures available show a planar M-O_2-M unit.

The $\eta^1:\eta^1$ complexes show varying values of the dihedral angle θ between the two MO_2 planes (Table 5). For the μ-superoxo (Vaska type I b) dicobalt complexes, the dihedral angle is close to $0°$ or $180°$ and the in-plane π^* orbital is doubly occupied and functions as a σ donor to both metals, while the $\pi_g^*(\perp)$ orbital is singly occupied and acts as a π acceptor from both metals. The μ-peroxo (Vaska type II b) complexes show a much greater variation in θ. A molecular orbital treatment shows no particular preference for planar or non-planar geometry, and we may notice that for hydrogen peroxide (which is formally isologous with these complexes) the dihedral angle is $119.8°$ but the barrier to rotation is very low (~ 15 kJ/mol)[38]. In such circumstances steric interactions between other ligands bonded to the two metal ions or the formation of a cyclic bridged system in the presence of a second bridging ligand may have a determining effect on the value of the dihedral angle. For $[(H_3N)_5CoO_2Co(NH_3)_5]^{4+}$ the counter anion appears to influence the value of θ (Table 5[80–82]).

Consideration of d-π_g^* overlap therefore allows the rationalisation of the structures of dinuclear complexes. Other features of the model presented for mononuclear complexes are equally applicable to the dinuclear complexes. Thus, for oxyhaemocyanin (Fig. 4) the coordination of the copper ions may be understood in terms of the necessity, for a metal at the extreme right of the transition metal series, to destabilise very strongly one of the metal d-orbitals to allow electron transfer to dioxygen. A correlation between the donor strength of the other ligands present and the stability of the dioxygen complex has been found for dinuclear cobalt species[186]. The absence or instability of μ-dioxygen complexes of the early transition metals may be explained by the readiness with which they may undergo d_π-$3\sigma_u^*$ electron transfer to give two M=O groups (Fig. 12). Although this reaction is unknown for the dicobalt species, it has been proposed for $\eta^1:\eta^1$ dioxygen complexes of iron[163, 260].

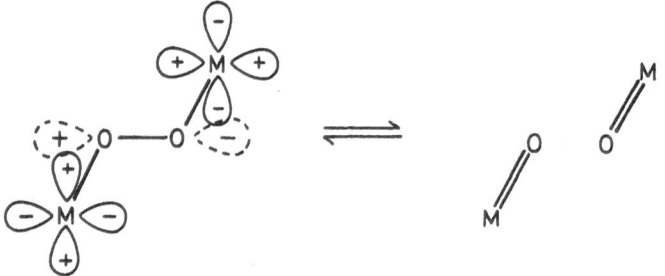

Fig. 12. $d_\pi \rightarrow 3\sigma_u^*$ transfer leading to O-O bond fission

Although the model we have presented here overlaps in many respects with the descriptions of electronic structure already given for various compounds, we feel that insufficient attention has been paid hitherto to the relative energies of metal d-orbitals and the dioxygen π_g^* orbitals. Consideration of this question rationalises the stability of the complexes and explains the role of the other ligands present. In the remainder of this review we shall discuss the results of calculations, spectroscopic measurements and studies of the reactivity of these complexes in terms of this model.

II. Calculations on Dioxygen Complexes

A large number of electronic structure calculations have been undertaken for these complexes, especially for those more or less comparable with oxyhaemoglobin. We give here a selective review of the more recent work in this field, and we shall concentrate on those calculations whose results give a qualitative explanation of the bonding.

Titanium and Manganese Compounds. Veillard and co-workers have published preliminary results for Ti(porph)O$_2$[261] and a more detailed discussion in a paper reviewing the results of their series of ab initio LCAO MO SCF calculations[262]. Using the experimental geometry for the metal porphyrin unit[59] in which the titanium atom lies out of the porphyrin plane, they find an η^2 structure to be appreciably more stable than an η^1

complex. Apart from the type (iii) overlap expected in this system they also find a weak δ interaction between $\pi_g^*(\perp)$ and the d_{xy} orbital (Fig. 13). This interaction is more favourable than that with $d_{x^2-y^2}$ which lies at higher energy (it is Ti-porph antibonding) and accounts for the barrier to rotation about the TiO_2 bond axis (Fig. 3 b)[59], and the eclipsed TiO and Ti-N bonds. The same argument may be used to explain the eclipsed bonds in $Mo(porph)(O_2)_2$[263]. Veillard et al.[262] also draw attention to the increasing stabilisation of the d-orbitals on crossing the series.

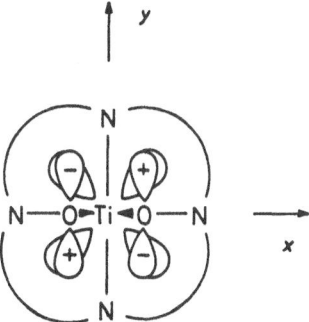

Fig. 13. The δ interaction between $\pi_g^*(\perp)$ (occupied) and d_{xy} (empty) in $Ti(porph)\eta^2(O_2)$

For $Mn(porph)O_2$ Veillard et al.[262, 264] propose an η^1 structure which they formulate as $Mn(III)O_2^-$. As mentioned before such a configuration has been proposed for the manganese phthalocyanine oxygen adduct[141] but the porphyrin oxygen adduct is thought to have η^2 geometry[143, 144]. Extended Hückel calculations on the porphyrin system support an η^2 structure, but the results depend strongly on the distance of the manganese atom from the porphyrin plane[251]. Hanson and Hoffman[251] claim that only the Griffith geometry and a $Mn(IV)O_2^{2-}$ formalism can explain the EPR and charge transfer spectra. They also find a $\pi_g^*(\perp)$ δ type interaction as for the titanium complex, but in this case the d_{xy} orbital of manganese is singly occupied and a more favourable interaction is obtained if $\pi_g^*(\perp)$ overlaps with the empty $d_{x^2-y^2}$ orbital leading to a staggered conformation.

Chromium and Molybdenum Compounds. The results of calculations on these complexes are particularly interesting as they are the only cases studied in which the metal is in a high formal oxidation state. The high symmetry (D_{2d}) of the MO_8^{n-} species makes them particularly attractive subjects for calculation. Two ab initio calculations on CrO_8^{3-} have been published.

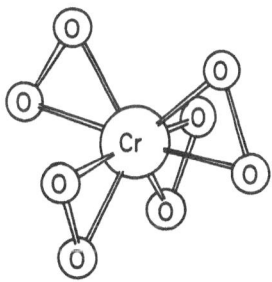

Fig. 14. The structure of $[CrO_8]^{3-}$

Dacre and Elder[264] find the unpaired electron to be localised on the chromium atom in the $d_{x^2-y^2}$ ($3b_1$) orbital, corresponding to a $Cr(V)4(O_2^{2-})$ formalism, although the chromium dioxygen bonding is essentially covalent. Analysis of the form of the wavefunctions shows that the bonding involves essentially those orbitals of dioxygen lying in the plane of the η^2CrO_2 unit.

The $3\sigma_g$ and π_u orbitals are mixed and act as donors to the metal ion. There is an energy gap of some $3\,eV$ between the M.O. formed from the $3\sigma_g$ and π_u and those formed from overlap of the π_g^* with the chromium. The results of an ab initio calculation of Fischer, Veillard and Weiss[265] agree on the general nature of the CrO_2 interaction but find rather surprisingly that the unpaired electron is localised on the dioxygen ligands. This is in disagreement with Dacre and Elder, with an EHMO calculation by Nösch and Hoffman[266] and with experiment[267].

A recent X-α calculation agrees well with the results of Dacre and Elder, although it finds less participation of π_u and $3\sigma_g$ orbitals. The π_g^* orbitals in the CrO_2 plane are strongly involved in bonding whereas the $\pi_g^*(\perp)$ are essentially non-bonding. The unpaired electron is in the $3b_1(d_{x^2-y^2})$ orbital which lies at substantially lower energy than the other Cr d-orbitals[268].

Brown and Perkins have published results of semiempirical calculations for $[MoO(C_2O_4)(O_2)_2]^{2-}$ and MoO_8^{2-}[269]. They find a very large gap between the HOMOs (localised on the oxygen atoms) and the LUMOs (mainly localised on the molybdenum). This does not agree with the X-α calculations on MoO_8^{2-}[268] which show the lowest empty orbital to be the $3b_1(d_{x^2-y^2})$ which is singly occupied in CrO_8^{3-} and which lies within $2\,eV$ of the HOMOs. The $\pi_g^*(\perp)$ orbitals are still non-bonding but there is no longer a sharp gap between π_g^* and π_u based M.O. The higher oxidation state of the metal in this complex appears to cause a greater perturbation of the dioxygen electronic structure in MoO_8^{2-} than in CrO_8^{3-}, and this is also shown by the appearance of some $3\sigma_u^*$ dioxygen character in the unoccupied orbitals[268].

Iron Compounds. Virtually all the calculations published have treated dioxygen adducts of iron porphyrin systems taken as models for haemoglobin, myoglobin, and cytochrome P 450. The limited number of compounds studied is, however, more than compensated by the variety of methods employed, and a complete coverage would require a review in itself. We shall therefore give only a brief discussion of the more recent results using a variety of methods, and compare these results to our qualitative model. References to work not discussed here may be found in the papers cited. We shall also neglect calculations on geometries other than the η^1 (Pauling) structure as this is now generally accepted.

Kirchner and Loew have published iterative EHMO calculations on oxyhaemoglobin[270, 271] and cytochrome P 450[271]. In agreement with the qualitative model, and with all the other calculations on these compounds, they find donation to the iron by the doubly occupied π_g^* orbital in the plane of the MO_2 unit, and π acceptance by the $\pi_g^*(\perp)$ orbital from the d_{xz} orbital of iron. Rotation of the dioxygen about the Fe-O bond axis (Fig. 3 a) reduces this π acceptance from d_{xz} and increases π acceptance from d_{yz} with a concomitant reduction in the electric field gradient at the iron atom, in agreement with the observed temperature dependence of the Mössbauer quadrupole splitting. Replacement of the axial imidazole ligand *trans* to dioxygen by CH_3S^- to give a model for cytochrome P 450[271] is accompanied by the observation of π donation from sulphur to iron.

Veillard and co-workers have published ab initio LCAO SCF calculations on dioxygen iron porphyrins with varying axial ligands[262, 272, 273]. They find little change on substituting ammonia by imidazole as the axial ligand. The interaction of the in-plane π_g^* orbital is essentially with the d_{yz} orbital (a type (ii) overlap, Fig. 5) rather than with the d_{z^2} orbital (type (i) overlap) as the d_{z^2} orbital lies at higher energy. They find the lowest energy state to be a $Fe^{3+}O_2^-$ triplet, but a $Fe^{2+}O_2$ singlet state (corresponding to the qualitative model) lies close in energy, and the authors estimate that a correction for the error in correlation energy would make this the ground state, although the triplet $Fe^{3+}O_2^-$ state would be a low-lying excited state. The $Fe^{2+}O_2$ singlet state shows considerable π acceptance by $\pi_g^*(\perp)$ from d_{xz} which is absent in the analogous cobalt complex. They calculate a low barrier for rotation about the Fe-O bond axis (~ 25 kJ/mol) and find the η^2 geometry to be much less stable.

Case, Huynh and Karplus have performed Parriser-Parr-Pople SCF calculations with configuration interaction and X-α calculations on oxy- and carboxy-haemoglobin, and have compared their results with those obtained from EHMO and ab initio calculations[274]. They find configuration interaction to be very important, and suggest that the ground state may conveniently be described as a mixture of $\{Fe^{2+}$(low spin d^6) $-$ $O_2(^1\Delta_g)\}$ and $\{Fe^{2+}(S = 1, t_{2g}^5 e_g^1) - O_2(^3\Sigma_g^-)\}$. The results all show an interaction of the in-plane π_g^* orbital with d_{yz} and of the out-of-plane orbital $\pi_g^*(\perp)$ with d_{xz}. The authors discuss in detail the spectroscopic data available, and their agreement with theoretical calculations.

Loew and co-workers have applied configuration interaction in conjunction with an INDO SCF method to oxyhaemoglobin and oxycytochrome P 450. The results for cytochrome P 450 showed that a theoretical parametrization gave close agreement with ab initio LCAO SCF calculations, but that the use of empirical parameters gave close agreement with the experimental Mössbauer spectrum[275]. Their results show an important d_{xz}-$\pi_g^*(\perp)$ interaction with charge transfer to the dioxygen. This charge is drawn from the axial ligand (CH_3S^-) and the porphyrin as well as the iron atom, and this is a fairly general conclusion from the calculations reviewed here. Ab initio methods appear to give less charge transfer to the dioxygen. The empirical parametrization used for oxy cytochrome P 450 was applied to oxyhaemoglobin[276] and after extensive configuration interaction gave a singlet ground state with a low lying triplet state 130 cm^{-1} above the ground state. Such a state had previously been proposed from measurements of the magnetic susceptibility[277] which showed slight temperature dependent paramagnetism of oxyhaemoglobin although this phenomenon is not universally accepted[278]. A barrier to rotation about the Fe-O axis similar to that of Veillard et al. was calculated, and good agreement with the experimental Mössbauer spectrum and its temperature dependence was obtained. In these calculations there is appreciable donation into the $\pi_g^*(\perp)$ orbital and the in-plane π_g^* orbital of dioxygen interacts with the metal d_{xz} and d_{z^2}. This work was extended to calculate the optical spectra of the model oxyhaemoglobin and cytochrome P 450 complex[279] and compared with experimental data.

The valence bond method has also been applied to oxyhaemoglobin. Semi-empirical valence bond calculations give the ground state as a mixture of the configurations $\{Fe^{2+}(d_{xz}^1 d_{z^2}^1) - O_2(^3\Sigma_g^-)\}$, $\{Fe^{2+}(d_{xy}^2 d_{xz}^2 d_{yz}^2) - O_2(^1\Delta_g)\}$ and $\{Fe^{3+}(d_{xy}^2 d_{yz}^2 d_{xz}^1) - O_2^-(^2\pi_g)\}$ implying a certain amount of charge transfer to dioxygen[280]. Ab initio valence bond calculations by Goddard et al.[281, 282] re-emphasize the problems of correlation in these systems and give the configuration $\{Fe^{2+}(d_{xy}^2 d_{xz}^2 d_{yz}^1 d_{z^2}^1) - O_2(^3\Sigma_g^-)\}$ as dominant.

Makinen has reviewed the results of theoretical investigations of oxyhaemoglobin complexes before 1978[283]. In general there is agreement on the η^1 dioxygen structure and on the importance of π back donation from the metal to the empty $\pi_g^*(\perp)$ orbital. The importance of including electron correlation in any calculation is now generally accepted. Agreement with experimental spectroscopic data is now reasonable. There is however much less agreement on the amount of charge accepted by the dioxygen and this problem has unfortunately become entangled with the argument over the oxidation state. The terms charge density, configuration and oxidation state have different meanings although they are sometimes confused. The dioxygen in the complexes discussed above is generally agreed to accept a certain amount of electron density which comes partly from the iron atom and partly from the other ligands present. The configuration assigned to the complex depends, in a molecular orbital model, on the arbitrary localisation of molecular orbitals and, in a valence bond model, on the assumption of a predominant configuration (which appears to be unlikely). The oxidation state is generally a means of classification according to chemical or spectroscopic properties[243], and in these complexes the classification as neutral dioxygen acting as a strong π acceptor from low spin iron(II) seems now to be generally accepted.

In a recent calculation of a non-biological system a MINDO method was used to calculate the structure of FeO_2 trapped in a matrix[284] in order to compare the results with experimental observations[175].

Cobalt Complexes. A fair number of complexes of cobalt have been studied theoretically. Veillard and co-workers have studied the Schiff base adduct $Co(acacen)LO_2$ (L = none, H_2O, CO, CN^-, imidazole)[285] and the porphyrin complex $Co(porph)(NH_3)O_2$[262] using ab initio LCAO SCF methods. The calculations show the η^1 structure to be more stable than the η^2; a linear η^1 structure is also found to be unstable[285]. The most important interaction is that between the cobalt d_{z^2} orbital and the in-plane π_g^* orbital. The interactions of the in-plane π_g^* orbital with the d_{yz} orbital and $\pi_g^*(\perp)$ with d_{xz}, although present, are much less than in the analogous iron complexes as a result of the tighter binding of the d-orbitals in cobalt. The unpaired electron is localised essentially in the $\pi_g^*(\perp)$ orbital of dioxygen.

The results show appreciable charge transfer to the dioxygen, but a considerable part of this charge density comes from the ligands bound to cobalt. The role of the axial ligand *trans* to dioxygen is to raise the energy of the d_{z^2} orbital sufficiently to interact with the in-plane π_g^* orbital and this is effected by σ donation from the ligand, π bonding effects with the ligand being relatively unimportant. Fantucci and Valenti have used an INDO UHF method to calculate the structure of Co(acacen), Co(acacen)NH$_3$ and their dioxygen adducts[286]. They obtain similar results to those of Veillard et al. but find that the unpaired electron in Co(acacen) is in the plane of the ligand, in agreement with experimental results[287] and the effect of the axial ligand is to move the electron into the d_{z^2} orbital.

Boca has carried out a series of CNDO UHF calculations on the systems $ML_4L'O_2$ (M=Co L=Cl$^-$, L'=none, NH$_3$, py, imidazole; M=Mn, Fe, Co, Ni, Cu, L=Cl$^-$, NH$_3$, L'=NH$_3$) assuming an η^1 geometry[288-291]. For the complex $[CoCl_4LO_2]^{2-}$ he finds a distinct stabilisation if the proximal oxygen atom is moved slightly away from the normal through cobalt to the CoCl$_4$ plane (this corresponds to the "kinked" structure shown in Fig. 3 a). A similar result was found for EHMO calculations of FeO_2 complexes[270] and has been found in crystal structure determinations[58] although the disorder frequently found in these complexes makes it difficult to measure this distortion accurately.

Teo and Li[292] have used the non-empirical Fenske-Hall method to calculate the Walsh diagram for the interaction of dioxygen with the fragments $[Co(CN)_5]^{3-}$ (square pyramidal) and $[Co(PH_3)_4]^+$ (pyramidal, as in $[Co(2=phos)_2O_2]^{+ 112}$). They find energy minima at the correct geometries (η^1 and η^2 respectively). In $[Co(CN)_5O_2]^{3-}$ the most important interaction is that between $3 d_{z^2}$ of Co and the in-plane π_g^* of dioxygen. Interaction with $\pi_g^*(\perp)$ is relatively unimportant. In the second case, $[Co(PH_3)_4O_2]^+$, there is a strong type (iii) (Fig. 5) overlap with d_{xz}. They also find a fair amount of mixing of the $3 \sigma_g$ and π_u orbitals of dioxygen with the Co-CN and Co-P bonding orbitals, and appreciable charge transfer to dioxygen from Co and the other ligands. This charge transfer is greater for $[Co(PH_3)_4O_2]^+$ than for $[Co(CN)_5O_2]^-$ as suggested by infra-red and crystal structure data. Recent SCF-Xα-SW calculations on $[M(PH_3)_4O_2]^+$, M=Co, Rh, Ir, by Norman and Ryan[429] agree generally with those of Teo and Li, and find the HOMO to be $\pi_g^*(\perp)$. They argue however that d-p hybridisation is important in the binding of dioxygen to the metal.

Hyla-Krispin, Natakaniec and Jezowska-Trzebiatowska have carried out EHMO type calculations on the ions $[(H_3N)_5CoO_2Co(NH_3)_5]^{4,5+}$ and $[(NC)_5CoO_2Co(CN)_5]^{6- 293}$. The $3 \sigma_g$ and π_u orbitals of dioxygen lie well below the metal d-orbitals, the in-plane π_g^* orbital just below the t_{2g} metal d-orbitals and the $\pi_g^*(\perp)$ orbitals (the HOMO) between the t_{2g} and e_g^* metal d-orbitals. In agreement with the EPR spectrum[240] the unpaired electron in the superoxo complex is localised on dioxygen.

Nickel, Palladium and Platinum Complexes. These metals all form planar η^2 complexes of composition L_2MO_2 (Fig. 15). Their relatively simple geometry and the possibility of comparison with the isostructural ethylene complexes $L_2M(C_2H_4)$ have made them attractive subjects for calculations. All the calculations show that the degree of electron transfer to the ligand is much greater for dioxygen than for ethylene. Norman[256] has reported SCF X-α calculations for $(PH_3)_2PtO_2$. The in-plane π_g^* orbital interacts with the d_{xz} orbital (Fig. 15) and is stabilised while the doubly occupied $\pi_g^*(\perp)$ orbital is essentially non-bonding and is the HOMO.

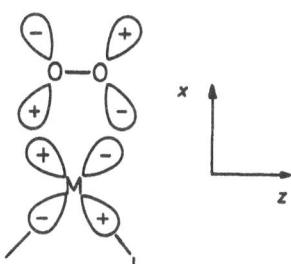

Fig. 15. Overlap of π_g^* with d_{xz} in $L_2M\eta^2(O_2)$

There is relatively little mixing between the in-plane π_g^* and the d_{xz} which is the LUMO and lies at much higher energy. There is however appreciable mixing of the other, lower-lying d-orbitals with the π_u and $3 \sigma_g$ of dioxygen, although as all these orbitals are occupied, this interaction will not have an overall bonding effect. This result is to be contrasted with the X-α results for CrO_8^{3-} and $MoO_8^{2- 268}$ which show extensive metal d-

π_g^* mixing, but little metal d-π_u, $3\sigma_g$ mixing. Norman suggests that much of the stability of $(Ph_3P)_2PtO_2$ arises from the ionic attraction of the positively charged $(Ph_3P)_2Pt$ moiety and the negatively charged O_2.

INDO calculations on $(HNC)_2NiO_2$[294] agree well with the results of Norman. Increasing the donor power of the *trans* ligands (by decreasing the ionisation potential of hydrogen) gives an increase in charge transfer, and a drop in the NiO_2 overlap population. This work was followed by INDO calculations on $(HNC)_2NiA$ (A=O_2, HNNH, C_2H_4) and by a study of the effect of changing the P-M-P angle in $(H_3P)_2MO_2$ (M=Ni, Pd)[295]. Binding of O_2 is favoured by a P-M-P angle of 90° giving a maximum destabilisation of the d_{xz} orbital. If this angle is increased the d_{xz} orbital is stabilised and the M-O_2 interaction is weakened. This agrees nicely with the observation that oxygenation of $(Ph(t\text{-}Bu)_2P)_2MO_2$ is reversible for M = Pd (P-Pd-P = 115.4°), but irreversible for M = Pt (P-Pt-P = 113.1°)[123].

Sakaki, Hori and Ohyoshi have published CNDO calculations on $(H_3P)_2MO_2$ (M = Ni, Pd, Pt) and $MX(CO)(PH_3)_2O_2$ (M = Rh, Ir; L = Cl, I)[296]. Their results show a much greater charge transfer to the in-plane π_g^* orbital than found by other workers. They also find a certain amount of π_u and $3\sigma_g$ interaction with the metal s and p orbitals.

III. Spectroscopic Studies

We have already discussed the use of spectroscopic data for the structural classification of dioxygen complexes in section C and in this sub-section we shall be concerned with the information on electronic structure to be obtained from spectra.

Electronic Spectra. Systematic studies of the electronic spectra of dioxygen complexes have been limited by problems such as the interference from absorptions due to other ligands present and the instability of many of the complexes in solution. For this reason most published work has been concerned with complexes of cobalt, and with biological compounds such as oxyhaemoglobin where the incentives to overcome the difficulties appear to be greater. The electronic spectra have been reviewed recently by Lever and Gray[298] and their reviews contain references to most of the published spectra.

Lever and Gray discuss the spectra in terms of charge transfer to and from the π_g^* orbitals of dioxygen, and the molecular orbital description given in Sect. D. I may be used to discuss the spectra. If we consider the M.O. diagram for an η^2 complex (Fig. 7b), two ligand to metal charge transfer (LMCT) bands are possible. That due to transfer from $\pi_g^*(\perp)$ is generally weak, while that from the bonding, in-plane π_g^* orbital is frequently found at high energy[298]. For the η^1 complex (Fig. 7a) the d_{xz}-$\pi_g^*(\perp)$ orbital is empty or partially occupied and it is possible to observe metal to ligand charge transfer (MLCT) into this orbital. The band corresponding to excitation from the in-plane π_g^* bonding orbital to d_{xz}-$\pi_g^*(\perp)$ is sometimes observed in the near infra-red[299]. π_u-π_g^* transitions are not normally observed for dioxygen complexes, but may occasionally be observed at high energy (~ 45 kK). They are blue shifted in comparison with the free superoxide anion[300]. The assumption of a considerable π_u-π_g^* separation is therefore justified.

The treatment of the charge transfer bands in $\eta^1 : \eta^1$ complexes is essentially identical and the reader is referred to the reviews for further details[298]. The variation of the energy of the LMCT transitions in η^2 complexes on change of metal ion is broadly in agreement with that expected. Lever, Ozin and Gray have recently studied the optical spectra of metal dioxygen complexes trapped in an inert matrix[301]. The energy of the LMCT band correlates extremely well with the electron affinity of the M^{2+} ion. They also find a good correlation between the energy of the charge transfer transition and the dioxygen stretching frequency, the stretching frequency falling (implying greater charge transfer to dioxygen in the ground state) as the energy of the charge transfer band increases. Pickens and Martell have shown for a series of $\eta^1 : \eta^1$ peroxo cobalt complexes that the energy of the charge transfer (LMCT) band correlates well with the cobalt reduction potential and the logarithm of the stability constant of the dioxygen complex[302].

The general conclusion to be drawn from electronic spectroscopic studies is that the model considering only the π_g^*-d interactions is satisfactory. The absence of reliable data for many transition metals does not allow us to study the effects of variation of metal d-orbital energy by this technique.

Apart from the references given above and those in the reviews, we may also mention the studies on spectra of doubly bridged $\eta^1 : \eta^1$ dioxygen complexes of cobalt by Fallab and his group[94, 303, 435, 438] and the detailed studies of biological complexes such as oxy-haemoglobin[304], oxymyoglobin[305] and haemerythrin[306].

Vibrational Spectra. In his review[4] Vaska discussed the dioxygen stretching frequencies measured in dioxygen complexes, and showed how the values then known fell into two groups. For η^1 and $\eta^1 : \eta^1$ superoxo complexes the frequencies lay in the range 1075–1195 cm^{-1} and for η^2 and $\eta^1 : \eta^1$ peroxo complexes in the range 790–932 cm^{-1}. The frequencies of ionic superoxide and peroxide respectively are close to the middle of these two ranges (Table 1) and this was invoked frequently in the discussion of oxidation states. As more compounds were prepared and studied it was found that some complexes (especially those trapped in O_2/Ar matrices) showed dioxygen stretching frequencies outside these limits, and in a recent article Nakamoto[307] has shown a more or less continuous range of values from 742 cm^{-1} for haemocyanin[308] to 1278 cm^{-1} for Co(TPP)O_2[196]. Nonetheless, the more stable dioxygen complexes normally show stretching frequencies within one or other of the two ranges noted by Vaska. If one accepts that the donation of charge to dioxygen is accompanied by a fall in the stretching frequency, one may deduce that, depending on the geometry adopted, the dioxygen in the more stable complexes accepts a more or less constant charge density. Furthermore, this charge density is greater for the η^2 and $\eta^1 : \eta^1$ peroxo complexes where both M.O. derived from the π_g^* orbitals are doubly occupied, than for η^1 and $\eta^1 : \eta^1$ superoxo where one of these orbitals is empty or singly occupied.

Since the appearance of Vaska's review, the increasing availability of Raman spectra has greatly widened the scope of vibrational spectroscopy in the study of dioxygen complexes, and has notably allowed the determination of O-O stretching frequencies in complexes where this vibration is infra-red inactive[310–311]. The use of resonance Raman spectroscopy not only increases the sensitivity of the technique[312, 313], but is of use in identifying the LMCT bands in the absorption spectrum which are coupled with the dioxygen stretching. It has also become evident that some of the early assignments of dioxygen stretching frequencies were incorrect, and comparison of $^{16}O_2$ and $^{18}O_2$ fre-quencies should be made whenever possible in order to confirm any assignment made.

The recent work of greatest significance to the electronic structure of these complexes is that carried out by Nakamoto and co-workers. The complexes $Mn(TPP)O_2$, $Fe(TPP)O_2$ and $Co(TPP)O_2$ show stretching frequencies of 983 cm^{-1}[314], 1195 and 1106 cm^{-1}[174], and 1278 cm^{-1}[196] respectively in inert gas matrices. The steady increase in the frequency is in complete agreement with the expected decrease in donor strength of the metal as the d-orbitals are stabilised on crossing the series. The compound $Co(OEP)O_2$ shows a stretching frequency almost identical with $Co(TPP)O_2$[199]. If, however, the porphyrin is replaced by a Schiff base ligand (acacen) the frequency falls to 1146 cm^{-1}[197]. This difference was attributed to the electron withdrawing properties of the porphyrin ligand, and it was shown that substitution of the acetylacetone in acacen by an acetyl group on the C3 carbon to give the Schiff base J-en has essentially the same effect, the O_2 stretching frequency of $Co(J\text{-}en)O_2$ being found at 1261 cm^{-1}[198].

The same workers have investigated the effect of various ligands added *trans* to the η^1 dioxygen in these complexes. In solution $Co(TPP)(1\text{-}MeIm)O_2$ and $Co(acacen)(1\text{-}MeIm)O_2$ show stretching frequencies roughly 130 cm^{-1} below those of the corresponding "base-free" species[197], and a similar lowering of frequency has been found for $Co(J\text{-}en)LO_2$ with a variety of axial bases L[198]. The $\eta^1:\eta^1$ peroxo complexes of cobalt $[Co(acacen)L]_2O_2$ and $[Co(J\text{-}en)L]_2O_2$ have also been studied and show a correlation between base strength of L and $\nu_{O\text{-}O}$, the frequency decreasing as the base strength increases. In solution at low temperatures no η^1 complex is observed if the axial base is too weak. Finally the complexes $[Co(salen)L]_2O_2$ (L = py, DMF, pyO, none) have been investigated in the solid state[315, 316] and the presence of the axial base has been found to lower the frequency from 1011 cm^{-1} (L = none) to around 900 cm^{-1}. Their results for $\nu_{O\text{-}O}$ agree well with those obtained by Hester and Nour[317] who have published a number of papers on resonance Raman spectroscopy of peroxy compounds[173, 318].

The measurement of the dioxygen stretching frequency is currently the best method for measuring charge transfer to dioxygen. The results of Nakamoto and co-workers show very clearly the variation of the quantity of charge donated to dioxygen as the metal changes, and the importance of the other ligands present in the destabilisation of the metal d-orbitals and in acting themselves as donors to the dioxygen.

Electron Spin Resonance Spectroscopy. Practically all the ESR studies on dioxygen complexes have been concerned with η^1 cobalt dioxygen complexes, and the results have been summarised in the review by Basolo et al.[19]. We limit our discussion here to the essential points relating to the electronic structure, and the reader is invited to consult the references given for fuller details.

In the cobalt(II) complexes that react with dioxygen the hyperfine coupling between the cobalt nucleus and the unpaired electron is large, but on addition of dioxygen the hyperfine coupling decreases very considerably, and this was generally accepted as indicating the localisation of the unpaired electron on the dioxygen. This was later confirmed by studies with $^{17}O_2$[319] which showed slightly greater spin density on the terminal oxygen. There has been a considerable debate, however, over the origin of the hyperfine coupling of the unpaired electron with the cobalt nucleus. One school of thought considers the coupling to arise from the partial mixing of d_{xz} and $\pi_g^*(\perp)$ orbitals resulting in the unpaired electron residing in an orbital with some Co d_{xz} character (Fig. 7a). Others have considered that this explanation is unsatisfactory, and that the hyperfine coupling arises from the spin polarisation of the pair of electrons in the molecular orbital arising from the in-plane π_g^*-cobalt $3\,d_{z^2}$ interaction (Fig. 5(i)). Drago and co-workers are the principal supporters of the second point of view, and have derived estimates of the degree of charge transfer to dioxygen from their analysis of the spectra[244, 320]. Their results do not agree particularly well with those obtained by vibrational spectroscopy. Recently, Smith and co-workers have analysed in some detail the ESR spectra of a series of dioxygen adducts and have concluded that both effects contribute to the hyperfine coupling and that electron transfer cannot be reliably estimated at present[321]. Haas and von Zelewsky[340] have investigated the *trans* influence of the axial base in a series of cobalt-Schiff base dioxygen complexes, and have found the cobalt hyperfine coupling constant to vary considerably, increasing with the ionisation potential of the axial base.

There has been much less work on other types of dioxygen complex. We have already mentioned the use of ESR to show the localisation of the unpaired electron on the dioxygen bridge in μ-superoxo dicobalt complexes[240] and on the chromium atom in K_3CrO_8[267]. Studies on the ESR spectra of the dioxygen adducts of manganese porphyrin complexes show no spin density on the dioxygen, and favour a $Mn(IV)O_2^{2-}$ formalism[143(a), 322]. Rhodium analogues of η^1 and $\eta^1 : \eta^1$ superoxo complexes show no hyperfine coupling of the unpaired electron with rhodium nuclei[210, 211].

Mössbauer Spectroscopy. Mössbauer spectroscopy has been used extensively in the study of iron-dioxygen complexes, and in particular for the study of oxymyoglobin and oxyhaemoglobin. A review of this subject has appeared recently to which the reader is referred for full details[323]. The Mössbauer spectra of oxymyoglobin, oxyhaemoglobin (and the α and β sub units[324]), and oxycytochrome P 450[325] are very similar, and consist of a quadrupole split doublet with an isomer shift typical for low spin Fe(II) or Fe(III) and with a large quadrupole splitting which decreases at high temperatures. Recent measurements suggest that the principal axis of the electric field gradient in oxymyoglobin lies in the plane of the porphyrin[326], and this is consistent with strong π acceptance from an essentially t_{2g}^6 Fe(II) ion by $\pi_g^*(\perp)$ of dioxygen. As mentioned before, the temperature dependence of the quadrupole splitting may be explained by the rotation of the dioxygen about the Fe-O axis[278]. The Mössbauer spectra of model compounds agree quite closely with those of natural oxygen carriers[323].

The Mössbauer spectrum of oxyhaemerythrin shows the presence of two non-equivalent high spin Fe(III) ions[327]. Recently a study of the Mössbauer spectrum of the complex $(TPP)FeO_2Fe(TPP)$ mentioned earlier[163] has confirmed its description as a μ-peroxo complex of two high spin Fe(III) ions, and its dissociation into the Fe(IV) complex $(TPP)FeO$[328].

The only non-iron complexes to have been studied are those of iridium. Addition of dioxygen to Vaska's compound produces a slight increase in isomer shift corresponding to a partial oxidation of the iridium, with an accompanying increase in the carbonyl stretching frequency[329]. Oxygenation of $[Ir(dppe)_2]^+$ is also accompanied by a relatively small increase in isomer shift, but a considerable change in the ^{31}P nmr shift of the ligands[330]. Both these observations are consistent with those calculations which suggest that an appreciable part of the charge donated to dioxygen comes from the other ligands present.

IV. Electronic Structure – Conclusions

Theoretical treatments (both qualitative and quantitative) and spectroscopic data now give a fairly consistent picture of the nature of the bonding in transition metal dioxygen complexes. The bonding arises essentially from the interaction of the dioxygen π_g^* orbitals with the d-orbitals of the metal. The dioxygen ligand acts essentially as an electron acceptor. The weakly bound d electrons of the early transition metals are readily donated to dioxygen, but, as one crosses the series, the other ligands play an increasingly important role in stabilising the dioxygen complex. The ligands may favour electron transfer by destabilising one or more metal d-orbitals in such a way as to favour interaction with dioxygen or may act themselves as electron donors, their charge being delocalised onto the dioxygen via the metal.

E. Reactivity of Dioxygen Complexes

Our intention in this section is to discuss the reactivity of dioxygen complexes in terms of their structure, and we shall therefore limit our coverage to the reactions of well characterised complexes. We shall not discuss the many metal ion catalysed auto-oxidation reactions where the intervention of MO_2 complexes has frequently been postulated but rarely established unambiguously. The fact that a metal complex is capable of forming a dioxygen adduct, and that it catalyses an auto-oxidation reaction does not imply that the dioxygen complex is the catalytically active species. In a recent book on homogeneous catalysis[332], none of the practical auto-oxidation catalysts discussed involves a dioxygen complex. Metal complexes may equally catalyse auto-oxidation of substrates by acting as electron transfer catalysts, by activation of the substrate to attack by dioxygen, by activation of hydroperoxides[333, 334], or by Wacker type processes. In a recent study one of the classic dioxygen complex "catalysed" reactions (the oxidation of tertiary phosphines by $(Ph_3P)_2PtO_2$) was in fact shown to involve hydrogen peroxide liberated from the dioxygen complex by catalytic amounts of acid[335]. Even in biological oxidations, current evidence would suggest that metal dioxygen complexes may be precursors of catalytically active species, but that they are not themselves active[17, 25, 26]. The metal ion catalysis of oxidation reactions is thus a very complex subject and the reader is referred to one of the many comprehensive reviews on this subject[13–16, 336–339].

There is considerable variation of reactivity between different structural types and within a given structural type, but one general feature of the reactivity of dioxygen complexes may usefully be mentioned before discussing each structural type. The binding of dioxygen to a transition metal is accompanied (i) by a reduction of the number of unpaired electrons on the oxygen (to one or zero) and (ii) by the transfer of charge density to the dioxygen. As a result of the first effect the kinetic barrier to spin changes in reactions of dioxygen is eliminated, and as a result of the second, the dioxygen ligand acquires basic or nucleophilic character. Dioxygen complexes typically react readily with diamagnetic electrophiles which are inert to molecular oxygen. The lack of catalytic processes involving these complexes may be related to the fact that the bonds that one would most like to oxidise, carbon-carbon double bonds and C-H single bonds, are more generally susceptible to oxidation by electrophiles.

I. Reactions of η^1 Complexes

Most of the published work on these complexes has been concerned with cobalt complexes, but there is a certain amount of data available for iron and rhodium complexes.

Nucleophilic Behaviour. The best known reaction of η^1 complexes is the dimerisation reaction with another cation to give an $\eta^1 : \eta^1$ MOOM complex. Although we discuss this below as an electron transfer reaction, it should be noted that the ready attack of the dioxygen on a metal cation is typical of nucleophilic dioxygen. Unambiguous data on the basicity of η^1 dioxygen is lacking, as the addition of acid to η^1 complexes generally produces a rapid decomposition of the complex, especially for anionic species such as $[Co(CN)_5O_2]^{3-}$ [341]. Cationic species are however appreciably more stable with respect to protonation; and Endicott and Kumar[342] have shown the macrocyclic complex $[Co(cyclam)(OH_2)O_2]^{2+}$ to undergo inner or outer sphere reduction more rapidly than protonation. Nishinaga has found $[Co(CN)_5O_2]^{3-}$ to act as a base in the auto-oxidation of 2,6 di-*t*-butylphenols in dichloromethane[343, 344]; the neutral dioxygen complexes do not act as bases (vide infra). The slightly basic character of η^1 dioxygen is also shown by the formation of hydrogen bonds between a neutral η^1 cobalt-Schiff base dioxygen adduct and 2,2,2-trifluoroethanol in methylene chloride[345], and hydrogen bonding by η^1 dioxygen bound to iron in oxygen carriers has also been postulated[71].

Electron Transfer Reactions. The reactions in which the oxidation state of the metal dioxygen system is changed may be discussed either in terms of outer or inner sphere electron transfer or as abstraction reactions.

Endicott and Kumar[342] have shown that $[Co(cyclam)(OH_2)O_2]^{2+}$ reacts with typical outer sphere reducing agents to give $[Co(cyclam)(OH_2)O_2]^+$. The oxidised complex has a lower intrinsic barrier to electron transfer than other η^1 cobalt dioxygen complexes, and outer sphere electron transfer may therefore compete successfully with inner sphere electron transfer or dimerisation. Inner sphere electron transfer may also be prevented when the other ligands present prevent the formation of $\eta^1 : \eta^1$ dimeric complexes. Oxygenation of $Co(CN)_2(PMe_2Ph)_3$ leads to an η^2 dioxygen complex $[(NC)_2(PhMe_2P)_3Co(\mu\text{-}CN)Co(CN)_2(PPhMe_2)_2O_2]$ in which the initial η^1 dioxygen complex has apparently been reduced by an inner sphere mechanism with a cyanide acting as a bridging ligand[113]. Oxygenation of $[Co(2 = phos)_2]^{2+}$ leads to 50% yield of η^2 $[Co(2 = phos)_2O_2]^+$ in which the rest of the cobalt has apparently acted as an outer sphere reducing agent[253]. In this case the phenyl groups of the chelating phosphine ligand prevent inner sphere attack by a cobalt(II) complex[254].

Electrochemical reduction of η^1 iron porphyrin dioxygen complexes leads to a species identified as $[Fe(III)(porph)O_2^{2-}]^-$ [346]. This species has also been reported to be formed on addition of superoxide to $[Fe(porph)]$ [170, 309].

The inner sphere reduction of η^1 cobalt dioxygen complexes by a second cobalt(II) ion is extremely well established in the oxygenation reactions of cobalt(II) solutions[347], and is discussed in many reviews[9, 10, 19]. Endicott[348] has recently investigated the kinetics of the two step reaction

$$Co(II) + O_2 \rightarrow \eta^1 CoO_2 \tag{1}$$

$$\eta^1 CoO_2 + Co(II) \rightarrow \eta^1 : \eta^1 CoO_2Co \tag{2}$$

for a series of macrocyclic complexes of Co(II). The speed of the second step varies considerably, and is fastest for the most reducing Co(II) complexes.

For the complex $[Co(cyclam)(OH_2)O_2]^{2+}$ reaction (2) is sufficiently slow for competing inner sphere reductions to be observed on a stopped flow time scale, and the spectrum of the transient CoOOFe species formed on inner sphere reduction by Fe^{2+} has been reported[342]. $[Co(CN)_5O_2]^{3-}$ reacts with $MoCl_5$ in the presence of excess cyanide to give a bridged CoOOMo species[349] which decomposes to an η^2 MoO_2 complex. Isotopic labelling studies of this complex show however that the oxygen in the η^2 complex does not originate from $[Co(CN)_5O_2]^{3-}$ [350]. There is some evidence that η^1 cobalt dioxygen complexes react via an inner sphere mechanism with phenoxy radicals[344(a), 351, 356].

The formation of $\eta^1 : \eta^1$ complexes has also been reported for rhodium(II) porphyrin complexes on oxygenation[211], and the η^1 iron dioxygen complexes will also undergo inner sphere reduction to give Fe(III)OOFe(III)[163, 495].

The final type of electron transfer reaction, radical abstraction, is specific to η^1 cobalt dioxygen complexes. These complexes are radicals (Fig. 7 a) and there is now abundant evidence of their participation in abstraction reactions. Pratt and co-workers[352] showed that a cobalt Schiff base dioxygen complex and the dioxygen complex of vitamin B_{12r} could abstract hydrogen from hydroquinone and N,N'-tetramethyl-p-phenylene-diamine. Nishinaga[344, 353] has published a large number of papers dealing with the catalytic auto-oxidation of phenols in the presence of cobalt Schiff base complexes in which the dioxygen complex activates the substrate by hydrogen abstraction. The abstraction by the neutral Schiff base complexes is to be contrasted with the basic behaviour of the anionic complex $[Co(CN)_5O_2]^{3-}$ (see above). Drago[351, 354] has also studied the oxidation of phenols by η^1 cobalt Schiff base dioxygen complexes, and has shown that they retain their catalytic activity when bound to a polymer support[354].

Little work on the photochemistry of η^1 dioxygen complexes has been published, although the photodissociation of oxyhaemoglobin and oxymyoglobin, which is notably less efficient than the photodissociation of CO from these complexes, has attracted some attention[355]. γ-irradiation of frozen glasses of oxymyoglobin and oxyhaemoglobin leads to electron capture by the FeO_2 system. As might be expected for a system iso-electronic with η^1 cobalt systems, the η^1 geometry is retained, and the unpaired electron is extensively delocalised on the dioxygen[357].

II. Reactions of η^2 Complexes

These complexes have been studied intensively as possible catalysts for organic oxidations, and their chemistry has been reviewed frequently[13, 16, 336–339, 358]. In the great majority of reactions studied the coordinated dioxygen behaves as a nucleophile. If the description of the electronic structure of η^2 complexes is recalled (Fig. 7 b) it will be seen that the HOMO is the $\pi_g^*(\perp)$ orbital lying perpendicular to the MO_2 plane. Norman[256] suggested that the structure of the $\eta^1 : \eta^2$ complex $[(Ph_3P)_2ClRhO_2]_2$ (Fig. 2) could be considered as two pyramidal $(Ph_3P)_2ClRhO_2$ molecules in which the $\pi_g^*(\perp)$ of one dioxygen is attacking the Rh of the other pyramid. If we accept this eminently reasonable proposition two points of interest may be noted: (i) an electrophile will attack the coordinated dioxygen on one oxygen atom only (the $\pi_g^*(\perp)$ orbital has a nodal plane between the two oxygen atoms) and the direction of attack will be perpendicular to the

MO_2 plane and (ii) the changes in bond length in comparison with the simple η^2 complex $(Ph_3P)_3ClRhO_2$[56] show the oxygen atom bound to both rhodium atoms to move away from the first rhodium, whereas the other oxygen atom moves closer to the rhodium:

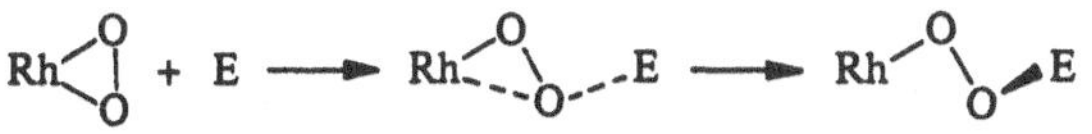

$$Rh\overset{O}{\underset{O}{\big|}} + E \longrightarrow Rh \overset{O}{\underset{O \cdots O}{\cdots}} E \longrightarrow Rh \overset{O}{\underset{O}{\diagdown}} E \tag{3}$$

In $[(Ph_3P)_2ClRhO_2]_2$ the Rh-O bond lengths in the RhO_2 plane are 1.980(7) and 2.198(7) Å compared with an average of 2.043 Å in $(Ph_3P)_3ClRhO_2$[56]. There is every reason to believe that the stereochemistry of electrophilic attack on η^2 dioxygen shown in this example is followed in other cases.

Protonation Reactions. The reaction of η^2 coordinated dioxygen with acids has been known for many years. If the anion of the acid is a good ligand this can even be a useful route to the *cis* disubstituted complex:

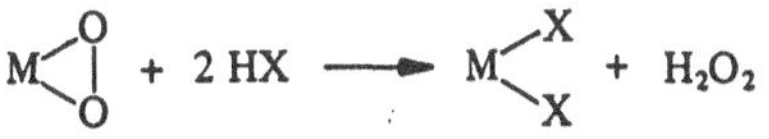

$$M\overset{O}{\underset{O}{\big|}} + 2\,HX \longrightarrow M\overset{X}{\underset{X}{\diagdown}} + H_2O_2 \tag{4}$$

Even fairly weak acids will react in this way, and recent reports include H_2S and thiols[359], *ortho*phenylene diamines and *ortho*catechols[360, 361], *ortho*amino phenols[361], and carboxylic acids[362] reacting with $\eta^2(Ph_3P)_2MO_2$ (M = Pd, Pt). The dibasic aromatic amines and phenols chelate the metal after loss of H_2O_2. Similar formation of chelates is observed with aroyl hydroxlamines and aroyl hydrazines[363]. The reaction of $(Ph_3P)_2MO_2$ (M = Pd, Pt) with carboxylic acids in the presence of triphenyl methyl bromide results in formation of triphenyl methyl hydroperoxide and $(Ph_3P)_2MBr(OCOR)$[364].

Platinum(II) forms a number of stable hydroperoxo compounds[365] and these may also be formed by reaction of η^2 MO_2 with one equivalent of acid[364]. Similar behaviour has been observed for $(Ph_3P)_2PdO_2$ reacting with non-coordinating acids in dichloromethane; the resulting hydroperoxide can oxidise terminal olefins[366]. Reaction of $Ph_3P)_3RhClO_2$ with acetylacetone in benzene gives $(Ph_3P)_2RhCl(OOH)(acetylacetonate)$[367].

In the presence of an alcohol however the treatment of $(Ph_3P)_2PtO_2$ with a non-coordinating acid[96] or simply the addition of $NaBPh_4$ to an alcoholic solution[368] leads to the formation of the $\eta^1:\eta^1$ complex $[(Ph_3P)_2Pt(O_2)(OH)Pt(PPh_3)_2]^+$.

Nucleophilic Attack on Polar Bonds. The η^2 dioxygen complexes of the group VIII transition metals react readily with polar bonds as shown in reactions 5 A and 5 B:

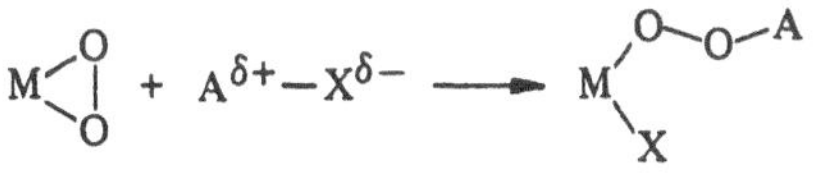

$$M\overset{O}{\underset{O}{\big|}} + A^{\delta+}\!-\!X^{\delta-} \longrightarrow M\overset{O-O-A}{\underset{X}{\diagdown}} \tag{5 A}$$

$$M\overset{O}{\underset{O}{\big|}} + A^{\delta+}\!=\!X^{\delta-} \longrightarrow M\overset{O-O}{\underset{X-A}{\diagup\big|}} \tag{5 B}$$

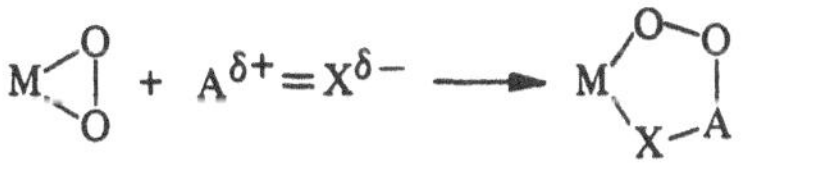

The cyclic product of reaction 5 B (which Mimoun calls a dioxo-metallacycle[358]) shows varying stability, and it is probably an intermediate in the reaction of η^2 MO_2 with many substrates (vide infra).

Otsuka and co-workers[369] have recently shown that attack of $(Ph_3P)_2PtO_2$ on PhEtCHBr to give $(Ph_3P)_2Pt(Br)OOCHPhEt$ produces an inversion of configuration at the carbon atom suggesting the nucleophilic attack to proceed via an S_{N2} mechanism. Reaction of this and similar platinum complexes with RBr (R = Ph_3C, PhCO) produces the peroxy species $L_2Pt(Br)OOR$. Similar reactions have been reported for $(tBuNC)_2$-NiO_2 with PhCOBr and Ph_3CX (X = Br, BF_4) where two equivalents of the organic substrate react to give cis MX_2 complexes and the organic peroxide[370]. The formation of an acyl peroxide was also reported for the reaction of $(Ph_3P)_2PtO_2$ with acyl chlorides where the organic peroxide was detected by the epoxidation of alkenes added to the reaction mixture[371]. The cationic iridium complex $[Ir(dppe)_2O_2]^+$ also reacts with acetyl chloride in the presence of cyclohexene to give trans $[Ir(dppe)_2Cl_2]^+$, cyclohexene oxide and acetic anhydride[341]. In what is practically the only paper comparing the reactivity of group VI and group VIII η^2 dioxygen complexes, Regen and Whitesides found that n-BuLi reacts with $(Ph_3P)_2PtO_2$ at low temperatures to give 88% yield of $(Ph_3P)_2Pt(n\text{-Bu})_2$[372]. The iridium complex $(Ph_3P)_2Ir(CO)IO_2$ does not react in this way. The group VI complexes $MoO(HMPT)(O_2)_2$ and $CrO(py)(O_2)_2$ gave good yields of butoxide, implying a more electrophilic character for dioxygen in these compounds.

Nucleophilic attack on polar double bonds has attracted more interest and the reactions of η^2 dioxygen complexes with ketones have been studied in some detail. Recrystallisation of $(Ph_3P)_2PtO_2$ in acetone gives the dioxo-metalla cycle (6)

$$\tag{6}$$

whose crystal structure has been determined[373]. The kinetics of the addition of ketones to $(Ph_3P)_2PtO_2$ has been studied in detail by Ugo and his co-workers[374]. They find two reaction pathways, both showed by the presence of coordinating solvents. The major pathway, first order in complex and ketone, is thought to involve coordination of the ketone to an axial site of square planar $(Ph_3P)_2PtO_2$ followed by intramolecular nucleophilic attack on the carbonyl by dioxygen. The second pathway, first order in complex only, is suggested to be an activation of dioxygen, possibly by an "opening out" to η^1 geometry. The rate of reaction is greater for aldehydes and fluoroketones than for alkyl ketones. Isotopic labelling has confirmed that the peroxidic oxygens of (6) originate from $(Ph_3P)_2PtO_2$[369]. Other reports of reactions with carbonyl compounds and their derivatives with $(Ph_3P)_2PtO_2$ include ketones, aldehydes and oximes[375] and oximes[376]. α-diketones can add two $(Ph_3P)PtO_2$ molecules to give a light-sensitive complex containing two five membered cycles[377]. $(Ph_3P)_2PtO_2$ reacts with aromatic diketones of the type ArCOCOAr to give acid anhydrides ArCOOCOAr, and with acid anhydrides to give acid peroxides ArCOOOCOAr[378]. 1,1,1-trifluoroacetone reacts with $(Ph_3)_2PtO_2$ similarly to acetone, but somewhat faster. In presence of an excess of hexafluoroacetone however a 2 : 1 adduct (7) is formed[379]

$$\text{Ph}_3\text{P} \diagdown \!\!\!\!\!\!\! \overset{\displaystyle \text{O}-\text{O}}{\underset{\displaystyle \text{O}-\text{C}-\text{O}}{\text{Pt}}} \!\!\!\!\! \diagup \overset{\displaystyle \text{C}(\text{CF}_3)_2}{\underset{\displaystyle (\text{CF}_3)_2}{}}$$

(7)

On recrystallisation this gives a normal five membered cyclic species which may be treated with triphenylphosphine to give a four membered cyclic species (8)

$$\text{Ph}_3\text{P} \diagdown \!\!\!\! \underset{\text{Ph}_3\text{P}}{\overset{\text{O}}{\text{Pt}}} \!\!\!\! \diagup \overset{\text{O}}{\underset{\text{O}}{}} \text{C}(\text{CF}_3)_2$$

(8)

Iridium complexes appear to be less reactive towards carbonyl compounds, although they react with aldehydes[341] and hexafluoroacetone[380]. This last reaction has been studied kinetically, and the authors find it to be first order in dioxygen complex and in hexafluoroacetone. They conclude the reaction takes place by a direct electrophilic attack on the coordinated dioxygen. This contrasts sharply with the results of Ugo et al. for $(\text{Ph}_3\text{P})_2\text{PtO}_2$: the slow substitution reactions of Ir(III) complexes and the absence of vacant coordination sites exclude the coordination of the substrate prior to the attack on coordinated dioxygen, and this may account for the much slower reactions of the iridium complexes.

Reactions of η^2 coordinated dioxygen complexes with many simple polar species have been reported and are summarised in Table 8. In very few cases have detailed studies of the mechanism been made, but it seems probable that the greater part take place via a first step similar to reaction 5 B.

The reaction of $(\text{Ph}_3\text{P})_2\text{PtO}_2$ with CO_2 leads to a peroxy carbonate intermediate which decomposes on recrystallisation in presence of Ph_3P to the carbonate[375]

$$\text{Pt} \overset{\text{O}}{\underset{\text{O}}{\diagdown}} \text{O} \xrightarrow{\text{CO}_2} \text{Pt} \overset{\text{O}-\text{O}}{\underset{\text{O}-\text{C}}{\diagup}}\!\!\!\!\! {}_{\text{O}} \xrightarrow{\text{PPh}_3} \text{Pt} \overset{\text{O}}{\underset{\text{O}}{\diagdown}} \text{CO}$$

(9)

The crystal structure of the carbonate has been reported[381] and isotopic labelling of the dioxygen has confirmed the five membered cyclic intermediate[369].

The reaction of SO_2 with η^2 dioxygen complexes gives a chelated sulphate complex. Isotopic labelling studies with $^{18}\text{O}_2$ show that one of the terminal oxygen atoms of the sulphate originates from the dioxygen complex and the other from the SO_2. This has been interpreted in terms of the rearrangement of a five membered cyclic intermediate[382].

One or two minor points may be added to Table 8 above: the adducts of aldehydes and ketones to palladium and platinum complexes react with SO_2, CO_2 and NO_2 with displacement of the organic moiety to give the products expected for addition of these molecules to a dioxygen complex; diphenylacetylene displaces both the carbonyl and the dioxygen ligand[375]. Adducts of CS_2 and thiourea have also been reported[375]. The products obtained with NO_2 depend on the reaction conditions, especially the solvent[387]; reaction of NO_2 with $(t\text{-BuNC})_2\text{MO}_2$ gives a *trans* dinitrate for M = Ni and a *cis* dinitrate for M = Pd[370].

Table 8. Addition of small molecules to η^2MO_2

Molecule	Product	Metal	Ref.
SO_2	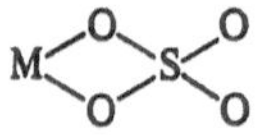 $M\diagdown\!\!\overset{O}{\underset{O}{}}\!\!S\!\!\overset{O}{\underset{O}{}}$	Ru	383
		Rh	117, 384–7
		Ir	382, 388–393
		Ni	370
		Pd	388
		Pt	382, 389, 394
CO_2	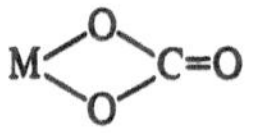 $M\diagdown\!\!\overset{O}{\underset{O}{}}\!\!C{=}O$	Ir	395
		Ni	370
		Pd	375, 396
		Pt	375, 381, 396
CO	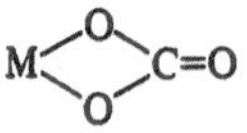 $M\diagdown\!\!\overset{O}{\underset{O}{}}\!\!C{=}O$	Rh	384–6
		Ni	370
		Pd	370, 375
		Pt	375
NO_2	$M(NO_3)_2$	Rh	384, 387
		Ir	388
		Ni	370
		Pd	370, 388
		Pt	388, 394
NO	$M(NO_2)_2$	Ni	370
		Pd	388
		Pt	388, 397
NO^+	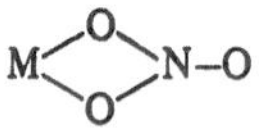 $M\diagdown\!\!\overset{O}{\underset{O}{}}\!\!N{-}O$	Pt	398
C_3O_2	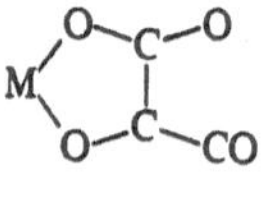 $M\diagdown\!\!\overset{O{-}C{=}O}{\underset{O{-}C{-}CO}{}}$	Pt	399

The complexes of rhodium and iridium are generally less reactive than those of palladium and platinum: no reaction was observed between Rh and Ir dioxygen complexes and CO_2, CS_2, aldehydes and ketones[375] or between an iridium complex and CO, CO_2 and NO[388]. The iridium complexes are not totally inert however, and reactions with aldehydes[341] and CO_2[395] have been observed in certain cases.

Atom Transfer Reactions. The apparent catalysis of the auto-oxidation of triphenylphosphine by $(Ph_3P)_2PtO_2$ aroused considerable interest as a possible model for enzymatic mono-oxygenase reactions. This reaction was recently reinvestigated by Halpern and Sen[335] who showed that intramolecular oxygen transfer does not occur. The reaction takes place *via* coordination of free phosphine to the platinum followed by loss of H_2O_2 or HO_2^- which can react with free phosphine. The Pt(II) species can be reduced to Pt(O) and may react with molecular oxygen. The reaction of triphenylphosphine with dioxo-metalla cycles[375, 379)] or with free peroxides[371)] formed after an addition reaction offers another possible route for phosphine oxidation. Similar reactions may be responsible for the oxidation of isocyanides observed in certain reactions of $(RNC)_2NiO_2^{370)}$.

Reactions with Carbon-Carbon Multiple Bonds. The nucleophilic group VIII η^2 dioxygen complexes will not react with carbon-carbon multiple bonds unless these latter are rendered electrophilic by the presence of electron-withdrawing substituents. Tetracyanoethylene reacts with rhodium[400)], iridium[341)] and platinum[401)] complexes to give a five membered ring as in reaction 5 B. The most detailed study, by Sheldon and van Doorn[402)], shows that one of the carbon atoms of an alkene must carry two electron withdrawing groups, and nucleophilic attack by O_2 will then occur on the other carbon atom. The resulting five membered ring is not cleaved by H_2 or PPh_3, but is decomposed by acid or by pyrolysis (to two ketones).

Clark and co-workers have studied the reactions of $(R_3P)_2PtO_2$ with acetylenes[403)]. When the acetylene carries two electron withdrawing groups (carboxylates or CF_3), the final product is a 1,2 addition of the dioxygen across the acetylinic bond

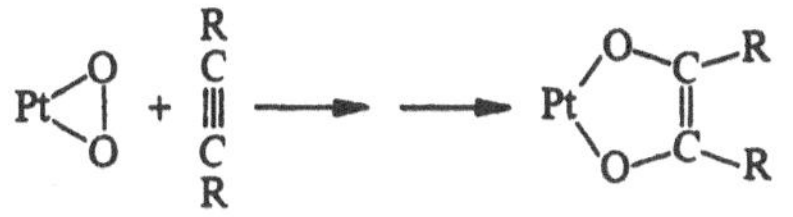

$$\tag{10}$$

Two intermediates were observed by low temperature nmr, and the possibility of the reaction proceeding *via* a five membered cyclic intermediate (5 B) cannot be discounted. PhCCPh, $MeCCCO_2Me$ and diethyl maleate do not react with $PtO_2^{403)}$. Isotopic labelling studies have shown that both oxygen atoms in the product originate in the PtO_2 complex[369)].

Group VIII η^2 dioxygen complexes do not in general react readily with unsaturated organic compounds that are not susceptible to nucleophilic attack, and the auto-oxidation of alkenes catalysed by these compounds has been shown to take place *via* radical reactions[404, 405)] although there are a few exceptions (vide infra). Group VI η^2 dioxygen complexes, on the other hand, react readily with carbon-carbon double bonds. The best studied compound is $MoO(O_2)_2HMPT$ where the HMPT ligand occupies the fifth coordination site of the pentagon in Fig. 8. Mimoun[406)] has shown that this compound reacts readily with olefins to form an epoxide. In the presence of H_2O_2 the peroxymolybdenum complex is reformed and the process is catalytic.

On the basis of kinetic data Mimoun proposes that the reaction takes place in two steps: (i) coordination of the olefin in the vacant site *trans* to the oxo ligand (Fig. 8) followed by (ii) an intramolecular nucleophilic attack by dioxygen on the coordinated olefin. The coordination of the olefin activates it towards nucleophilic attack. The five membered dioxo-metalla cycle decomposes to give epoxide and Mo = O:

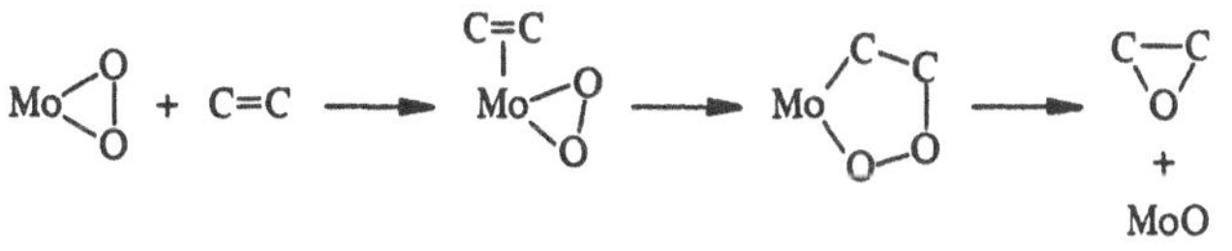

$$\tag{11}$$

The presence of donors which can occupy the alkene coordination site slows down the reaction. A kinetic analysis of the epoxidation of a number of alkenes has shown that the coordinating power of the olefin, and its tendency to undergo the intramolecular attack, increase with substitution of the alkene with electron donor groups provided they are not too bulky[407].

Although Mimoun's mechanism explains many features of the reaction it is open to some criticism; in a paper which showed by isotopic labelling that the oxygen of the epoxide is derived from the dioxygen and not the oxy ligand, Sharpless analysed the structure reactivity ratios for various alkenes, and concluded that a direct nucleophilic attack on one oxygen of the dioxygen ligand could equally be considered[408]. Mimoun's mechanism would however be consistent with structure-reactivity patterns if coordination to the metal was the rate-determining step.

Two other criticisms of the mechanism (11) may be made. The complexes $LL'MoO(O_2)_2$ where LL' is a bidentate ligand have no vacant coordination site but an example is known (111) where the complex is an efficient reagent for epoxidation and will even allow asymmetric epoxidation[409]. Secondly, the dioxygen bound to Mo(VI) is not notably nucleophilic: there is evidence from the reaction with n-BuLi[372] and from M.O. calculations[268] that it has some electrophilic character. Reaction of tetracyanoethylene with $MoO(O_2)_2$HMPT does not give a dioxo-metalla cycle as with the group VIII complexes, but a product where the nitrogen is apparently bonded to the metal[406]. A complex containing both an alkene and a nucleophilic dioxygen, $Ir(PPh_3)_2Cl(O_2)(C_2H_4)$, has been isolated and is relatively stable[410], although it might be argued that the metal-alkene π donation expected for a d^6 system would deactivate the alkene for nucleophilic attack.

Whatever the exact mechanism, the reactivity of these group VI η^2 dioxygen complexes is well established, and they react with alkenes to give epoxides[406–409, 411, 412], with enolates to give α-hydroxyketones[413], with allylalcohols to give epoxides[110], and with cyclic ketones to give lactones[414]. They are also efficient oxidants for alcohols[110, 415], although these reactions may involve hydroperoxides. The use of molybdenum and vanadium catalysts for epoxidation using alkylhydroperoxides (the Halcon process) is extremely important industrially but does not appear to involve dioxygen complexes[416].

We have concentrated on the molybdenum dioxygen complexes, but tungsten and chromium complexes generally behave in an analogous manner. Mimoun[101] has recently prepared a series of η^2 dioxygen titanium complexes which show properties similar to the group VIII complexes. The titanium complexes do not react with alkenes, allylic alcohols or cyclic ketones as the molybdenum compounds do, but react with tetracyanoethylene to give a dioxo-metalla cycle (reaction 5B) and slowly oxidise Ph_3P to triphenylphosphine oxide. The inertness of the titanium dioxygen complexes contrasts sharply with the effectiveness of Ti(IV) compounds in catalysing oxidations with t-butyl hydroperoxide[334]. Another case of oxidation of triphenylphosphine by an η^2 titanium complex to give an oxo-complex and triphenylphosphine oxide has recently been reported[127].

The possible catalytic oxidation of alkenes by group VIII complexes has been re-examined recently in conditions where free-radicals and Wacker type processes may be excluded. $RhCl(PPh_3)_3$ and $RhH(PPh_3)_3$ react with terminal olefins in the presence of oxygen to form methyl ketones and Ph_3PO[417]. Isotopic labelling studies have shown that the oxygen incorporated does not come from traces of water[418]. The complex $[RhCl(C_2H_4)_2]_2$ reacts with styrene and molecular oxygen at 110 °C in the presence of a radical inhibitor to form a mixture of acetophenone and benzaldehyde. The coordination of dioxygen and styrene is necessary[419]. Similarly [Rh(C$_8$H$_{14}$)$_2$Cl]$_2$ reacts with molecular oxygen to form a mixture of cyclooctanone and cyclo-1-ene-3-one[420].

These observations led Mimoun[421, 422] to investigate the reaction of [Rh(AsPh$_3$)$_4$O$_2$]$^+$ with terminal alkenes. In CH$_2$Cl$_2$ solution methyl ketones are formed together with a certain amount of Ph$_3$AsO. The oxidation of the arsine was prevented if the complex was suspended in pure alkene. The oxygen transferred to the alkene originates from the complex. Internal alkenes are unreactive. The complex reacts with tetracyanoethylene to give what is apparently a dioxo-metallacycle (reaction 5 B) but in which two triphenylarsines have been lost, [Rh(AsPh$_3$)$_2$O$_2$(C$_6$N$_4$)]$^+$. It appears to be necessary for the rhodium to have a vacant coordination site (or to be able to form one) for reaction to take place as the chelated complex [Rh(dppe)$_2$O$_2$]$^+$ is unreactive[422]. Mimoun and his co-workers have subsequently been able to obtain catalytic oxidation of alkenes using rhodium trichloride/Cu(II) mixtures[423].

Finally, a recent report suggests that the dioxo-metallacycle formed by addition of a ketone to (Ph$_3$P)$_2$PtO$_2$ may be reactive towards double bonds[424]. On adding an α-β unsaturated ketone to (Ph$_3$P)$_2$PtO$_2$, the initial product was the adduct across the C=O bond, but this could react slowly with an excess of the ketone in solution to give the product expected for addition across the C=C bond.

Miscellaneous Reactions. Coordinated dioxygen may be displaced from these complexes in certain cases. Partial dissociation of O$_2$ from (*t*-BuNC)$_2$NiO$_2$ by tetracyanoethylene has been reported[370, 425] and diphenylacetylene displaces O$_2$ from (Ph$_3$P)$_2$PtO$_2$[375]. Halogens will also displace O$_2$ from rhodium[386] and platinum complexes[426]. Tetrahydroborate reacts slowly with (Ph$_3$P)$_2$PtO$_2$ to give *trans* (Ph$_3$P)$_2$PtH$_2$; a better method of synthesis of these complexes uses BH$_4^-$ reduction of the CO or CO$_2$ adducts[427]. In the presence of ethylene, tetrahydroborate reduction gives (Ph$_3$P)$_2$Pt(C$_2$H$_4$)[426]. Electrochemical reduction of [Ir(dppe)$_2$O$_2$]$^+$ gives [Ir(dppe)$_2$]$^+$ and superoxide. This indicates the strongly M-O$_2$ antibonding nature of the LUMO in η^2 complexes (Fig. 7 b)[292].

Photochemistry. η^2 dioxygen complexes may generally be decomposed by irradiation with liberation of molecular oxygen. The complexes Ti(porph)O$_2$ are photolysed to give almost quantitative yields of the oxytitanium complex Ti(porph)O and less than 80% of the theoretical yield of O$_2$, the rest of the oxygen apparently reacting with the solvent (benzene)[430]. It was suggested that this might be due to formation of singlet oxygen. Mo(TPP)(O$_2$)$_2$ is photolysed cleanly to give a *cis* dioxo complex Mo(TPP)O$_2$[431].

Group VIII complexes show essentially similar behaviour. The oxygen adducts of IrCl(CO)(PPh$_3$)$_2$, IrI(CO)(PPh$_3$)$_2$, [Ir(dppe)$_2$]$^+$ and [Ir(2 = phos)$_2$]$^+$ are all deoxygenated on photolysis to give the parent complexes, but an inert gas purge is necessary to prevent the thermal back reaction[432]. The transition excited in this photolysis is thought to be iridium to phosphine charge transfer. No formation of singlet molecular oxygen could be detected.

(Ph$_3$P)$_2$PdO$_2$ is photolysed in dichloromethane solution to give *cis* (Ph$_3$P)PdCl$_2$; no singlet molecular oxygen could be detected. The transition excited is thought to have essentially dioxygen-metal charge transfer character[433]. The photolysis of (Ph$_3$P)PtO$_2$ in chloroform, however, does apparently lead to formation of singlet oxygen[434]. The transition excited is thought to be similar to that for $(Ph_3P)PdO_2$ and it is difficult to see why singlet oxygen is produced in one case and not in the other; this may conceivably be a result of the different methods used to detect its formation.

III. Reactions of $\eta^1 : \eta^1$ Complexes

a) $\eta^1 : \eta^1$ Complexes of Cobalt

The $\eta^1 : \eta^1$ complexes of cobalt were the first dioxygen complexes to be prepared and their reactivity has attracted considerable attention. Fewer dinuclear complexes are known for other transition metals, and information on their reactivity is correspondingly limited. We shall therefore discuss the cobalt complexes separately, and treat the other complexes at the end of this section. The principal types of reaction shown by these complexes are schematised in Fig. 16.

Much of the early work on these complexes has been reviewed [10] and we give here a summary of the more recent work. Before discussing the different types of reaction in detail, it is important to notice two general features of the reactivity: the existence of several equilibria in aqueous solution, and the high charges carried by some of the species studied (e.g. $[(H_3N)_5Co(O_2)Co(NH_3)_5]^{5+}$. In many cases the reaction observed is actually that of a protonated or partially dissociated species in equilibrium with the dinuclear dioxygen species, and the high charges of the reactants result in considerable variation of reactivity with ionic strength and in the frequent observation of catalysis by anions. These problems were not always recognised in some of the early work in this field, and care should be taken in comparing results from different studies.

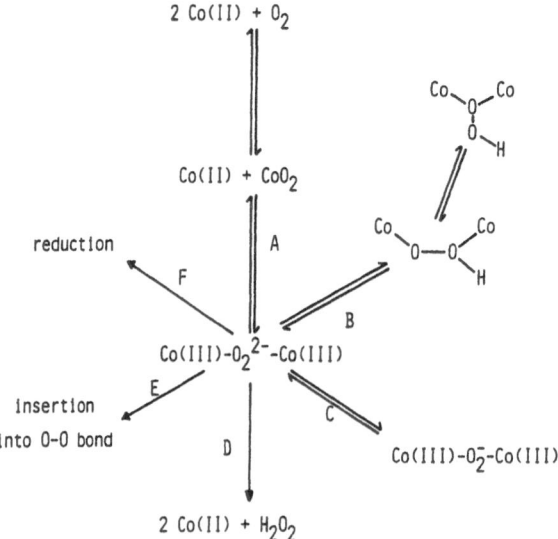

Fig. 16. The reactions of $\eta^1 : \eta^1$ dicobalt dioxygen complexes

Dissociation of Dioxygen. If one accepts the traditional formalism of Co(III)-O_2^{2-}-Co(III) for the $\eta^1:\eta^1$ peroxo complexes, this reaction (A in Fig. 16) corresponds to an intramolecular electron transfer which is the reverse of the reaction of formation of the complex:

$$\text{Co(III)}O_2^{2-}\text{Co(III)} \underset{k_2}{\overset{k_{-2}}{\rightleftharpoons}} \text{Co(III)-}O_2^- + \text{Co(II)} \underset{k_1}{\overset{k_{-1}}{\rightleftharpoons}} 2\,\text{Co(II)} + O_2 \qquad (12)$$

The various rate constants for this reaction were first determined by Wilkins et al.[347, 437)] who were able to follow the reaction by displacing the equilibrium to the right by addition of EDTA (to complex the Co(II)) or acid (to protonate the free ligands). Reaction 12 is sufficiently rapid to be studied as an equilibrium in aqueous solution, and a review of this subject has recently appeared[186)].

Recent work has included the determination of the rate constants of both steps of reaction 12 for a number of Co(II) macrocyclic complexes[348)], and a study of the kinetics of dioxygen uptake. This appears to involve dissociation of coordinated water from an octahedral Co(II) complex, and may be accelerated by a S_N1 CB mechanism[438)].

The rapid exchange of ligands at the formally Co(II) centre of the $\eta^1:\eta^1$ peroxo complexes is due to the lability of the Co(II) species produced *via* the equilibria 12. Thus the water molecules in $[((H_2O(cyclam)Co)_2O_2]^{4+}$ may readily be exchanged[439)]; the decomposition of this complex in acid solution is thought to take place *via* $[(H_2O)(cyclam)CoO_2]^{2+}$ and $[(H_2O)_2(cyclam)Co]^{2+}$ [440)]. The ammonia of $[((NH_3)en_2Co)_2O_2]^{4+}$ is quite labile[441)]; the rate of substitution is the same as the rate of decomposition of the complex in acid solution[442)]. The rate of exchange of dioxygen in $[tren(MeNH_2)-CoO_2Co(MeNH_2)tren]^{4+}$ has been found to be equal to the rate of decomposition of the complex in acid solution[443)]. It has been shown that the rate of reaction of Fe^{2+} with a μ-peroxo complex is independent of the concentration of Fe^{2+} but is the same as the rate of decomposition of the complex in acid solution[444)]. The Fe^{2+} itself does not appear to react with the peroxo complex.

The examples given above show the importance of reaction 12 in the chemistry of these complexes. The dissociation can however be prevented by the introduction of a bridging ligand or by the protonation of the peroxy ligand. In strongly basic media an aquo or ammino ligand *cis* to the peroxo group may be deprotonated and forms a dibridged μ-O_2 μ-X(X = OH, NH$_2$) species. The μ-amido complexes are even quite stable in acid solution[10)]. The rate of formation of μ-hydroxo bridges has been studied for $[(Co(tren)(NH_3))_2O_2]^{4+}$ [84)], $[(Co(en)_2(NH_3))O_2]^{4+}$ [442)] and $[(Co(tren)(MeNH_2))O_2]^{4+}$ [443)] and has been found to be the same as the rate of decomposition in acid solution (i.e. the rate of reaction[12)]; the formation of the bridge thus occurs *via* a Co(II) intermediate. This has recently been confirmed by showing that the rate of bridge formation is equal to the rate of oxygen exchange[443)]. The μ-OH μ-O$_2$ complex is kinetically much more stable than the single bridged species.

The μ-hydroxo bridge is hydrolysed in acid solution to give a single bridged species which decomposes rapidly[445-447)]. The complex $[(Co(tren)(NH_3))O_2]^{4+}$ reacts with excess tren to give a doubly bridged species $[(tren)Co(\mu\text{-}O_2)(\mu\text{-}tren)Co(tren)]^{4+}$ which also decomposes after acid hydrolysis of the bridging tren group[436)].

Apart from those cases where the addition of acid hydrolyses a second bridging group, the kinetics of decomposition of the μ-peroxo complex are generally independent of pH; the instability of these complexes in acid solution arises from the dissociation and protonation of the ligands bound to the labile Co(II) product of reaction 12. In very strongly acidic solution however the μ-peroxo group itself may be protonated (vide infra) and this stabilises the complex. A recent study of the decomposition of $[(Co(NH_3)_5)_2O_2]^{4+}$ formed in acid solution by one electron reduction of the superoxo complex shows the complex to be stabilised by protonation, by chloride ion and, more effectively, by sulphate (or rather by HSO_4^- in the conditions used)[448]. A similar stabilisation of the μ-peroxo complex was observed in the second step of the reduction of $[((NH_3)_5Co)_2O_2]^{5+}$ by V^{2+}, Cr^{2+} and Eu^{2+} [449].

Electrophilic Attack on Dioxygen. The nucleophilic character noted for coordinated dioxygen in mononuclear complexes is also found for $\eta^1:\eta^1$ dioxygen in the peroxo complexes, but there is no evidence for nucleophilic character of $\eta^1:\eta^1$ dioxygen in the superoxo complexes. The best known example is the protonation of the μ-amido μ-peroxo complexes $[(NH_3)_4Co(\mu-NH_2)(\mu-O_2)Co(NH_3)_4]^{3+}$ [450] and $[(en)_2Co(\mu-NH_2)(\mu-O_2)Co(en)_2]^{3+}$ [451] (Fig. 16 reaction B).

The reaction apparently takes place in two steps: protonation followed by isomerisation;

$$(13)$$

There is no firm evidence for the structure of the initial protonated species, but, bearing in mind that the HOMO of a μ-peroxo complex is essentially a π_g^* orbital of dioxygen (Sect. D), and on the basis of other structural evidence, it seems probable that the proton is bound to one oxygen only. The structure of the final isomerised product has been determined by X-ray crystallography[75]. Protonation of a μ-peroxo complex of cobalt with tertiary arsine ligands has also been reported[452].

The nucleophilic nature of the μ-peroxo linkage is also shown in the solid state. The crystal structure of $K_5[(NC)_5Co(O_2)Co(CN)_5] \cdot 2 KNO_3 \cdot 4 H_2O$ [89] shows the four molecules of water to be hydrogen bonded to the μ-peroxo group. Crystallisation of $[(en)_2Co(NH_2)(O_2)Co(en)_2]^{3+}$ from silver nitrate solution gives the non-stoichiometric compound $[(en)_2Co(NH_2)(O_2)Co(en)_2](NO_3)_3 \cdot 15/8\, AgNO_3 \cdot H_2O$ in which the silver cations are bound to the bridging dioxygen unit. In aqueous solution however the Ag^+ appears to be only very weakly bound to the peroxide linkage[453]. The nucleophilic character of the $\eta^1:\eta^1$ peroxo group is also illustrated by a number of insertion reactions into the oxygen-oxygen bond as observed for mononuclear η^2 complexes (see below).

$\eta^1:\eta^1$ Peroxide-$\eta^1:\eta^1$ Superoxide Electron Transfer. The peroxide-superoxide redox reaction (Fig. 16 reaction C) was one of the earliest reactions of these systems to be discovered, and has been one of the most popular for kinetic studies. Sykes has given a short review of some of the work in this field up to 1974[454]. The oxidation of μ-peroxo complexes to μ-superoxo complexes by strong oxidising agents such as persulphate is well known in preparative chemistry, but the majority of kinetic studies have been concerned

with reduction of μ-superoxo complexes as most μ-superoxo complexes are reasonably strong oxidising agents (the complex $[(NC)_5CoO_2Co(CN)_5]^{5-}$ is a notable exception).

There have been a number of attempts to establish the redox potentials of these couples[455-458], but these have not always allowed for possible dissociation or protonation of the μ-peroxo complex (reactions A and B, Fig. 16) and do not always agree. The most recent values[458] obtained by rapid cyclic voltammetry allow for these effects and show that protonation of the μ-peroxo complex stabilises it to oxidation.

The reaction of μ-superoxo complexes with reducing metal ions generally follows an outer sphere mechanism, and kinetic data have been reported for reduction by Fe^{2+} [459], V^{2+} and Cr^{2+} [449, 460], Eu^{2+} [449], cobalt(II) chelates[455], $Mo(V)$[461], $[Ru(NH_3)_6]^{2+}$ [462] and the excited state of $[Ru(bipy)_3]^{2+}$ [463]. The pressure effect of the Mo(V) reduction has been studied[464] and the stereoselectivity has been shown to depend on the ionic strength of the reaction medium[465]. The quenching of fluorescence of the excited state of $[Ru(bipy)_3]^{2+}$ by a μ-superoxo complex leads to formation of $[Ru(bipy)_3]^{3+}$ and a μ-peroxo species[463] which reacts rapidly to give a μ-superoxo complex and $[Ru(bipy)_3]^{2+}$ [457].

The outer sphere character of these reactions has encouraged some workers to apply Marcus theory to the rate constants obtained[342, 455, 457]. Given the uncertainty in the values of the electrode potentials and the considerable electrostatic work function involved in the formation of the precursor complex, the significance of the intrinsic rate parameters obtained is not clear.

Inner sphere reduction has been postulated for the reduction of μ-superoxo complexes by $[Ti(OH_2)_6]^{3+}$ [466] and Cd_{aq}^+ produced by pulse radiolysis[467]. The exchange reaction between $[(en)_2Co(\mu\text{-}NH_2)(\mu\text{-}O_2)Co(en)_2]^{3+}$ and a series of μ-superoxo complexes has been studied[468], and the unprotonated form of the μ-peroxo complex has been shown to act as the reductant.

Reduction of the μ-superoxo complexes by neutral and ionic species has also been studied, and the initial product is that expected for outer sphere electron transfer, the μ-peroxo complex. Reducing agents studied include I^- [459(b), 469], NO[470], SO_3^{2-} [471-474], HNO_2[472-475], $As(III)$[473] and $Se(IV)$[472]. The sulphite ion SO_3^{2-} reduces the μ-superoxo to the μ-peroxo complex but does not reduce it further; in most other cases further reduction of the μ-peroxo group occurs. Reduction by O_2^-, by the solvated electron, and by organic radicals generated by pulse radiolysis has been studied[476, 477]. Positively charged μ-superoxo complexes react much more rapidly with the formate radical and the solvated electron than the negatively charged $[(NC)_5CoO_2Co(CN)_5]^{5-}$ [477], and this charge effect is also shown by the rapid reaction of the cationic complexes with O_2^- whereas the anionic cyanide complex reacts with HO_2[476]. The complex $[(H_3N)_5Co(\mu\text{-}O_2)Co(NH_3)_5]^{5+}$ has recently been reported to react with the cyclic nitrone spin trap 5,5-dimethyl-1-pyrroline-1-oxide[478].

The last three reactions of Fig. 16 (D, E, F) involve the decomposition of a μ-peroxo dicobalt complex. The interpretation of the results is complicated by the presence of equilibria 12 and 13, and it is not always clear whether the reaction studied involves the μ-peroxo complex or one of the complexes formed by equilibria 12 and 13.

Formation of Hydrogen Peroxide. There have been several reports of the decomposition of μ-peroxo complexes in aqueous solution to give Co(III) complexes and H_2O_2, although the hydrogen peroxide is frequently found in less than stoichiometric quantities, and in some cases hydroperoxides have been observed. Systems showing this behaviour include cobalt peptide dioxygen complexes[479], $[((NO_2)(en)_2Co)_2O_2]^{2+}$ [480], $[((NC)_5Co)_2O_2]^{6-}$ [481] and trans $[(X(cyclam)Co)_2O_2]^{2+}$ [482]. In the last case it seems very probable that the reaction is due to an outer sphere electron transfer between the oxidising CoO_2 and the reducing Co(II) complexes produced in the first step of reaction 12[440]. The decomposition in acid solution of $[(trien)Co(\mu\text{-}OH)(\mu\text{-}O_2)Co(trien)]^{3+}$ follows path A between pH 6 and pH 1.8 but below pH 1.8, or in the presence of sulphate, path D is followed[445] and hydrogen peroxide is formed. The complex [(py(acacen)-Co)$_2$O$_2$] decomposes in the presence of acetic acid in pyridine solution to give $[Co(acacen)(py)_2]^+$ and H_2O_2 [483].

The reaction of cobalt(III) complexes with H_2O_2, the reverse of reaction D, is also known: reaction of cis$[Co(en)_2(H_2O)_2]^{3+}$ with H_2O_2 gives $[Co(en)_2(OOH)(OH_2)]^{2+}$ at pH 4, and on increasing the pH the μ-hydroxo-μ-peroxo complex $[(en)_2Co(\mu\text{-}O_2)(\mu\text{-}OH)Co(en)_2]^{3+}$ is formed[484]. The reverse of this mechanism would seem to be the most probable mechanism for the formation of H_2O_2 from a μ-peroxo complex. It is perhaps significant that there are no reports of formation of hydrogen peroxide from the μ-amido μ-peroxo complexes where dissociation in acid solution via reaction 12 does not occur.

Insertion into the O-O Bond (Fig. 16, reaction E). The addition of small electrophilic molecules across the O-O bond is a typical reaction of the mononuclear η^2 dioxygen complexes, and is also observed for binuclear $\eta^1:\eta^1$ complexes. In most cases the reaction has been observed as a second step after the reduction of the μ-superoxo species. The most straightforward examples are the reactions of the μ-amido μ-peroxo complexes which do not dissociate in the acid solutions used by most workers. Sulphur dioxide adds across the μ-peroxo bridge to give a μ-SO$_4$ dicobalt(III) complex[470, 472–474], and SeO$_2$ gives a μ-selenato complex[472]. The pH dependence of the reaction with SO$_2$ shows that it is SO$_2$ and not the sulphite ion that participates in the reaction. Nitrite ion reacts with doubly-bridged complexes to give μ-nitrito complexes[473, 474] and NO also forms a μ-nitrito complex[470].

Reaction of SO$_2$ or HNO$_2$ with the singly bridged μ-peroxo complexes gives a variety of mononuclear complexes which have not always been characterised[470]. Treatment of $[(NH_3)_5Co(O_2)Co(NH_3)_5]^{5+}$ with acid sulphite solutions gives Co(II), free sulphate and a sulphato complex of Co(III). Isotopic labelling of the μ-superoxo group has shown the oxygen to be transferred to the sulphate[471]. Reaction of the same complex with a considerable excess of nitrous acid gives $[Co(NH_3)_5NO_3]^{2+}$, but with a smaller excess of HNO$_2$ or in presence of bromide as a catalyst, $[Co(NH_3)_5(OH_2)]^{3+}$ is also formed[475]. The μ-peroxo group does not react with acetaldehyde[473].

Reduction of the O-O Bond (Fig. 16, reaction F). In discussing the other reductions of the μ-peroxo complexes it is convenient to begin with the reactions of the doubly bridged μ-amido-μ-peroxo species where competition from the dissociation reaction (A in Fig. 16) can be neglected. Iodide ion reacts rapidly with the isomerised protonated complex AH

$$(14)$$

to give the μ-amido-μ-hydroxo dicobalt(III) complex as product[469, 474, 485]. The non-isomerised form BH reacts much more slowly, and the non-protonated form is apparently unreactive. It is assumed that the iodide ion attacks the non-bridging oxygen atom of AH to form HOI which itself reacts with an excess of I^- to form I_2[485]. A similar mechanism is thought to be involved in the Cl^- and Br^- catalysed disproportionations of $[(en)_2Co(\mu\text{-}NH_2)(\mu\text{-}O_2)Co(en)_2]^{3+}$

$$3\,[(en)_2Co(\mu\text{-}NH_2)(\mu\text{-}O_2)Co(en)_2]^{3+} \;\; + 3\,H^+ \xrightarrow{\;Cl^- \text{ or } Br^-\;}$$

$$2\,[(en)_2Co(\mu\text{-}NH_2)(\mu\text{-}O_2)Co(en)_2]^{4+} \; +$$

$$[(en)_2Co(\mu\text{-}NH_2)(\mu\text{-}OH)Co(en)_2]^{4+} \; + H_2O$$

The μ-amido μ-peroxo complex reacts with the halide to form the μ-amido μ-hydroxo complex and HOX which oxides two μ-amido μ-peroxo complexes to μ-amido μ-superoxo complexes[486].

The reduction of the μ-amido μ-peroxo complex with arsenite also gives a μ-NH$_2$ μ-OH species and may follow a mechanism similar to the iodide reduction[473]. A detailed study by Sykes[460] of the Cr^{2+} reduction of $[(en)_2Co(\mu\text{-}NH_2)(\mu\text{-}O_2)Co(en)_2]^{3+}$ has shown the initial step to involve inner sphere attack of Cr^{2+} on the peroxo bridge followed by loss of Co^{2+} to give a $Co(III)\text{-}O_2^{2-}\text{-}Cr(III)$ complex. The succeeding steps involve protonation, isomerisation and reduction of the peroxo bridge, and only in the last step is the second Co(III) group reduced.

The reaction of iodide with $[(en)_2Co(\mu\text{-}OH)(\mu\text{-}O_2)Co(en)_2]^{3+}$ in acid solution gives $[Co(en)_2(OH_2)_2]^{3+}$ and iodine. The rate has been shown to be the same as that of dissociation in acid solution (reaction 12) and the initial step is hydrolysis of the μ-hydroxo group followed by I^- attack on the monobridged species[447]. The monobridged complex $[(H_3N)_5Co(O_2)Co(NH_3)_5]^{4+}$ produced by iodide reduction of the superoxo complex does not react with iodide but decomposes to Co(II) and ammonia[469]. The reduction of the mononuclear μ-peroxo complex $[(H_3N)_5Co(O_2)Co(NH_3)_5]^{4+}$ by V^{2+}, Cr^{2+} or Eu^{2+} has been studied and found to be inversely dependent on the concentration of H^+ suggesting that the protonation of the peroxo unit stabilises it towards reduction as well as towards dissociation[449]. It is not always easy to distinguish these two rections and the apparent reduction of a series of μ-peroxo complexes by Fe^{2+} [487] has been shown to proceed at the same rate as the decomposition in acid solution in the absence of Fe^{2+} [444].

If the strongly reducing complex $[Co(CN)_5]^{3-}$ is treated with less than the stoichiometric quantity of dioxygen, the complex $[Co(CN)_5(OH_2)]^{2-}$ is formed, presumably as a result of the reduction of $[((NC)_5Co)_2O_2]^{6-}$ by excess $[Co(CN)_5]^{3-}$ [481]. An investigation of the electrochemical reduction of dioxygen bound to a dicobalt "face-to-face" diporphyrin complex has shown that H_2O_2 or water may be produced, the product distribution depending on the geometry of the porphyrin complex and the pH[194].

The oxidation of organic compounds by μ-peroxo complexes has not been studied in any detail. There have been several reports of the decomposition of μ-peroxo complexes containing dipeptide ligands[479, 488] in which the peptide is oxidised. A recent investigation has shown that this involves oxidation of the N-terminal position of the peptide, leading to a coordinated imine[489]. The complex $[(bipy)_2Co(\mu\text{-}OH)(\mu\text{-}O_2)Co(bipy)_2]^{3+}$ has recently been reported to catalyse the oxidation of 2,6-di-tert-butylphenol in methanol solution[490], but it is not clear if the oxidation is catalysed by the $\eta^1:\eta^1$ complex or the η^1 mononuclear complex in equilibrium with it. As discussed earlier η^1 cobalt dioxygen complexes have been shown by Nishinaga to catalyse this reaction.

Many of the reactions of $\eta^1 : \eta^1$ complexes discussed above show some photo-sensitivity, but only the μ-superoxo complexes have been studied in any detail. Valentine and Valentine showed that irradiation of the ligand to metal charge transfer band of $[(H_3N)_5Co(O_2)Co(NH_3)_5]^{5+}$ and $[(H_3N)_4Co(\mu\text{-}NH_2)(\mu\text{-}O_2)Co(NH_3)_4]^{4+}$ gave $[Co(NH_3)_5\text{-}(OH_2)]^{3+}$, Co^{2+} and oxygen[491]. In the presence of chloride $[Co(NH_3)_5Cl]^{2+}$ was also formed, suggesting the presence of a $[Co(NH_3)_5]^{3+}$ intermediate. Quantum yields fell sharply if the terminal ammine ligands were replaced by chelating amines[492]. The complex $[(NC)_5Co(O_2)Co(CN)_5]^{5-}$ is photolysed to $[Co(CN)_5(OH_2)]^{2-}$, O_2, and H_2O_2[491, 492], and a reinvestigation of this photolysis has shown that the quantum yield varies according to the nature of the charge transfer band excited, and according to the pH of the solution[494]. The primary photo product of this photolysis will reduce oxidised cytochrome c.

b) $\eta^1 : \eta^1$ Complexes Containing Metals Other Than Cobalt

Only a few of these complexes have been well characterised, and the data concerning their reactivity are correspondingly limited. The results available do, however, show considerable differences from the reactions of the cobalt complexes, suggesting that the latter cannot be taken as reliable models for other $\eta^1 : \eta^1$ complexes.

Nucleophilic character of a μ-peroxo group has been observed for $[(Ph_3P)_2Pt(\mu\text{-}O_2)(\mu\text{-}OH)Pt(PPh_3)_2]^{+}$ [96]: non-coordinating acids cleave the peroxo linkage to form a di-μ-hydroxo complex, and coordinating acids such as HCl form *cis* $[(Ph_3P)_2PtX_2]$. In dichloromethane, SO_2 adds across the peroxo group to give a μ-sulphato linkage[96], but in methanol the complex *trans* $[(Ph_3P)_2Pt(SO_3Me)_2]$ is formed[368]. No reaction is observed with aldehydes or ketones, but CO cleaves both bridges to give $[(Ph_3P)_2Pt(\mu\text{-}CO)_2Pt(PPh_3)_2]$[96].

The remaining results are concerned with redox reactions: μ-superoxo-μ-peroxo redox reactions have been reported for $\eta^1 : \eta^1$ rhodium complexes[208]. The complex $[(NC)_5Co(\mu\text{-}O_2)Mo(O)(OH_2)(CN)_5]^{5-}$ decomposes on standing in oxygen to $[(\eta^2O_2)Mo(O)(CN)_4]^{2-}$ [349], and isotopic labelling studies show that the dioxygen in this complex does not originate from $[(NC)_5Co(\mu\text{-}O_2)Mo(O)(OH_2)(CN)_5]^{5-}$ [350]. The details of this reaction are not clear but the final product could conceivably arise from the formation of a Mo(IV) complex on breakdown of the $\eta^1 : \eta^1$ complex followed by addition of dioxygen.

The most detailed and the most interesting studies concern the $\eta^1 : \eta^1$ complexes of iron. The complex $[(porph)Fe(\mu\text{-}O_2)Fe(porph)]$ has been shown to be formed (and to be stable at low temperatures) on addition of dioxygen to the complex $[Fe(porph)]$ at low temperatures[163, 495]. On addition of a base such as imidazole[260], or on warming the solution, cleavage of the O-O bond occurs to form $[(porph)FeO]$. This complex can react with free $[Fe(porph)]$ to form the μ-oxo Fe(III) dimer $[(porph)FeOFe(porph)]$. The complex $[(porph)FeO]$ has frequently been postulated as a reactive intermediate[51] and does in fact react with triphenylphosphine in the presence of a nitrogen containing base B such as imidazole, piperidine, or pyridine to give Ph_3PO and $[(porph)FeB_2]$[496].

No such O-O bond cleavage has been observed for $\eta^1:\eta^1$ dicobalt complexes but a similar reaction may occur on oxygenation of certain manganese complexes where the formation of μ-oxo complexes is also observed[141, 147]. As discussed earlier (Sect. D.I., Fig. 12), the cleavage may arise from electron transfer from the metal d-orbitals to the $3\sigma_u^*$ orbital of the dioxygen. The greater d-orbital binding energy of cobalt may render this reaction impossible. In support of this we may recall that one electron reduction of a dicobalt μ-peroxo complex involves reduction of a cobalt centre in the first step[460], whereas reduction of an η^1 iron-dioxygen complex involves electron transfer to the dioxygen moiety[346, 357], followed, in one case, by electron transfer from the iron to the partially reduced dioxygen to give a formal $Fe(III)$-O_2^{2-} complex[346]. This change in the site of reduction is yet another example of the importance of the relative energies of metal d-orbitals and dioxygen orbitals.

IV. Reactions of $\eta^1:\eta^2$ and $\eta^2:\eta^2$ Complexes

In the only known $\eta^1:\eta^2$ complex, $[Rh(PPh_3)_2Cl(O_2)]_2$ (Fig. 2), the $\pi_g^*(\perp)$ orbital normally responsible for the nucleophilic character of an η^2 dioxygen complex is acting as a donor to a second rhodium atom, and its nucleophilicity would therefore be expected to be reduced. Bennett and Donaldson report that the complex is indeed inert to strong acid and strong base, and its very low solubility has hindered any further investigation of its chemical properties[56].

Discussion of the reactivity of the $\eta^2:\eta^2$ complexes is hindered by the lack of structural data, but, if we assume that the planar M_2O_2 structure found for uranium and lanthanum complexes is adopted, then the doubly occupied π_g^* orbital perpendicular to the M_2O_2 plane will be non-bonding with respect to the metal ions, and would be expected to show nucleophilic character. All the reports concerning these complexes show the dioxygen ligand to be basic. The palladium complexes[214, 497] react with monobasic acids HX such as HCl, ROH, RSH and RNH$_2$ to form H_2O_2 and dimeric complexes $Pd(\mu$-X$)_2$Pd. Monobasic acids which may act as chelating ligands HL such as 8-hydroxyquinoline give monomeric complexes PdL. The rhodium complexes[53] show very similar behaviour but are, if anything, even more nucleophilic: cyclopentadiene is a sufficiently strong acid to displace H_2O_2 and give a monomeric complex $[(diene)Rh(C_5H_5)]$. Refluxing the rhodium complex with cyclohexanone gives cyclohexenone and phenol.

F. Conclusions

Our knowledge of dioxygen complexes has increased considerably since Vaska's review of 1976. Theoretical and spectroscopic studies have given a description of the bonding in these complexes which rationalises the structures observed experimentally, and which provides a useful basis for the discussion of the reactivity. Very few theoretical studies on dinuclear complexes have appeared however, and we know of only one investigation of the relative merits of the $\eta^1:\eta^1$ and $\eta^2:\eta^2$ bridging peroxo structures[498].

In the field of structure determination the many X-ray crystallographic studies in recent years have shown the basic validity of Vaska's classification, with the exception of the $\eta^1:\eta^2$ and $\eta^2:\eta^2$ bridged species. If the d-π_g^* interaction is indeed dominant in the bonding of dioxygen to transition metals, it seems improbable that any radically new structural types will be discovered. The current absence of any crystal structure data on manganese and copper complexes is a challenge to the synthetic chemist, especially since these metals are known to play important roles in biological dioxygen chemistry. We feel that an understanding of the nature of the M-O$_2$ bond, and, in particular, the requirements on the stereochemistry of the other ligands present should enable a more systematic approach to the synthesis of new complexes.

The reactivity of coordinated dioxygen is probably the area in which the historical divisions mentioned in the introduction have proved the most durable. The recent studies of the $\eta^1:\eta^1$ dioxygen complexes of iron have shown the very considerable changes produced by moving one element away from the well-known cobalt complexes. Dioxygen is exceptional in the extent to which its reactivity is changed by the metal to which it is coordinated, and there remains considerable scope for experimental investigation.

Note added in proof. A number of important papers have appeared or been brought to our attention since the completion of the review, and we mention below those that we consider the most important.

Crystal structures have been published for $[TaF_5(O_2)]^{2-}$ (redetermination)[499], $VO(O_2)pic.2\,H_2O$[500], $MoO(O_2)Cl(HMPT)\ pic$[501], and a triply bridged $\eta^1:\eta^1$ dicobalt peroxo-complex[502]. A careful reinvestigation of the manganese/phosphine/halide/dioxygen system has shown the extreme sensitivity of this system to water. Reversible dioxygen uptake does apparently take place, but phosphine oxidation is a competitive process[503]. EXAFS measurement suggest that dioxygen in oxyhaemerythrin is bound to only one of the iron atoms[504]. The reversible formation of $\eta^1:\eta^1$ peroxo-complexes of Ru(IV) has been postulated[505]. Synthetic work on cobalt complexes has included Schiff bases[506], pentadentate ligands[507], the effect of macrocycle ring size on the formation of $\eta^1:\eta^1$ peroxo-compounds[508], and a mercaptide cobalt porphyrin complex analogous to cobalt cytochrome P 450[509]. The crystal structure of the $\eta^1:\eta^1$ superoxo-dirhodium complex[208] shows it to be a mixed-valence oxo-bridged species, but the Basle group have prepared $\eta^1:\eta^1$ dirhodium complexes by other methods[510]. A nickel dioxygen complex is reported to be formed when the nickel is complexed by a macrocyclic ligand which stabilises Ni(III)[511]. Superoxide anion forms weak complexes with M(porph) (M = Zn, Mg, Cd)[512]. Some thorium complexes showing epoxidation activity have been prepared[513].

Calculations have been carried out on the following systems: Fe(porph) (O$_2$) and Fe(Pc)(O$_2$)[514]; the reaction between η^2-O$_2$ and ethylene in a Mo complex[515]; Fe(porph) (NH$_3$)(O$_2$)[516]. The last calculation disagrees with a previous ab initio calculation in suggesting a $Fe^{3+}O_2^-$ configuration for the groundstate; this disagreement is attributed to a different choice of basis functions. Boca has published a review of his calculations, and discusses other calculations and crystal structure data[517].

Resonance Raman spectroscopy has been applied to the study of hydrogen bonding by coordinated dioxygen in coboglobins[518] and to the prophyrin complex Co(porph) (py)(O$_2$)[519]. A detailed single crystal EPR study of the dioxygen adduct of vitamin B$_{12r}$ has been pusblished[520]. ^{59}Co nmr has been used to estimate the ligand field strength of dioxygen in $\eta:\eta^1$ peroxo complexes as close to that of iodide[521]. A recent article discusses the use of ^{17}O nmr in the study of dioxygen complexes[522].

Mimoun has published two important papers on the oxidation of organic substrates by η^2 dioxygen bound to a d^0 metal ion. The molybdenum catalysed epoxidation of alkenes proceeds by different mechanisms if a peroxide or an alkyl hydroperoxide is used as the source of oxygen[501]. Epoxidation catalysed by vanadium complexes probably proceeds *via* a hydroperoxide intermediate[500]. In the field of reactivity we may note papers concerning oxidation of cyclo-octa-1,5-diene by a rhodium complex[523], an iridium catalysed oxidation of cyclooctene involving a hydroperoxide[524], and two papers concerning the electro-chemical reduction of $\eta^1:\eta^1$ peroxo-dicobalt complexes[525, 526]. A review of organic oxidations involving dioxygen complexes has appeared[527]. In the oxidation of indoles catalysed by cobalt[528] and in the epoxidation of ethylene on a silver metal surface[529], however, dioxygen complexes are not involved.

G. References

1. Frémy, E.: Ann. Chim. Phys. *35*, 257 (1852)
2. Werner, A.: Ann. Chem. Liebigs *375*, 1 (1910)
3. Pauling, L., Coryell, C. D.: Proc. Nat. Acad. Sci. U.S.A. *22*, 210 (1936)
4. Vaska, L.: Acc. Chem. Res. *9*, 175 (1976)
5. O'Connor, J. A., Ebsworth, E. A. V.: Adv. Inorg. Chem. Radiochem. *6*, 279 (1964)
6. Collman, J. P., Chong, A. O., Jameson, G. B., Oakley, R. T., Rose, E., Schmitton, E. R., Ibers, J. A.: J. Amer. Chem. Soc. *103*, 516 (1981)
7. Henrici-Olivé, G., Olivé, S.: Angew. Chem. Int. Ed. Engl. *13*, 29 (1974)
8. Fallab, S.: ibid. *6*, 496 (1967)
9. McLendon, G., Martell, A. E.: Coord. Chem. Rev. *19*, 1 (1976)
10. Sykes, A. G., Weil, J. A.: Prog. Inorg. Chem. *13*, 1 (1970)
11. Valentine, J. S.: Chem. Rev. *73*, 235 (1973)
12. Collman, J. P.: Acc. Chem. Res. *1*, 136 (1963)
13. The first internat. symp. on activation of molecular oxygen and selective oxidation catalysed by transition metal complexes, Bendor (France) 1979, J. Mol. Catal. *7*, 1–320 (1980)
14. Davidson, J. M.: Catalysis, The Chemical Society, London *2*, 198 (1978)
15. Ugo, R.: Colloq. Int. CNRS *281*, 133 (1978)
16. Lyons, J. E.: Homogeneous catalytic activation of oxygen for selective oxidations. In: Fundamental Research in Homogeneous Catalysis. Tsutsui, M., Ugo, R. (eds.), pp. 1–52 New York, Plenum Press 1977
17. Spiro, T. G. (ed.): Metal ion activation of dioxygen. New York, Chichester, Brisbane, Toronto: J. Wiley 1980
18. Caughey, W. S. (ed.): Biochemical and clinical aspects of oxygen. New York, London, Sydney, San Francisco: Academic Press 1979
19. Jones, R. D., Summerville, D. A., Basolo, F.: Chem. Rev. *79*, 139 (1979)
20. Erskine, R. W., Field, B. O.: Structure and Bonding *28*, 1 (1976)
21. Basolo, F., Hoffmann, B. M., Ibers, J. A.: Acc. Chem. Res. *8*, 384 (1975)
22. Que, L. (Jr.): Structure and Bonding *40*, 39 (1980)
23. von Jagow, G., Engel, W. D.: Angew. Chem. Int. Ed. Engl. *19*, 659 (1980)
24. Nozaki, M.: Topics in current chemistry *78*, 145 (1979)
25. Williams, R. J. P., Moore, G. R., Wright, P. E.: Oxidation-reduction of cytochromes and peroxidases. In: Biological aspects of inorganic chemistry. Addison, A. W., Cullen, W. R., Dolphin, D., James, B. R. (eds.), pp. 369–401 New York: John Wiley 1977

26. Cadby, P. A., Jefford, C. W.: Molecular mechanisms of enzyme-catalysed dioxygenation, Prog. Chem. Org. Natural Products *40*, 191 (1981)
27. (a) Arnold, S. J., Browne, R. J., Ogryzio, E. A.: Photochem. Photobiol. *4*, 963 (1965); (b) Hoytink, G. H.: Acc. Chem. Res. *2*, 114 (1969)
28. Kearns, D. R.: Chem. Rev. *71*, 395 (1971)
29. Kasha, M., Khan, A. U.: Ann. N. Y. Acad. Sci. *171*, 5 (1970)
30. Abrahams, J. C.: Q. Rev. Chem. Soc. *10*, 407 (1956)
31. Shamir, J., Beneboym, J., Classen, H. H.: J. Amer. Chem. Soc. *90*, 6223 (1968)
32. Herzberg, G.: Molecular spectra and molecular structure. 2nd Ed. New York: Van Nostrand 1950
33. Herzberg, L., Herzberg, G.: Astrophys. J. *105*, 353 (1947)
34. (a) Templeton, D. H., Dauben, C. H.: J. Amer. Chem. Soc. *72*, 2251 (1950) (b) Zhdanov, G. S., Zvonkova, Z. V.: Dokl. Akad. Nauk. SSSR *82*, 743 (1952)
35. Halverson, F.: Phys. Chem. Solids *23*, 207 (1962)
36. Celotta, R. J., Bennett, R. A., Hall, J. L., Siegel, M. W., Levine, J.: Phys. Rev. A *6*, 631 (1972)
37. Blunt, F. J., Hendra, P. J., Mackenzie, J. R.: J. Chem. Soc. Chem. Comm. *1969*, 278
38. Readington, R. L., Olson, W. B., Cross, P. C.: J. Chem. Phys. *36*, 1311 (1962)
39. Busing, W. R., Levy, H. A.: ibid. *42*, 3054 (1965)
40. Föppl, H.: Z. Anorg. Allg. Chem. *291*, 12 (1957)
41. Tallman, R., Margrave, J. L., Bailey, S. W.: J. Amer. Chem. Soc. *79*, 2979 (1957)
42. Evans, J. C.: J. Chem. Soc. Chem. Comm. *1969*, 682
43. Abrahams, S. C., Kalnajs, J.: Acta Cryst. *7*, 838 (1954)
44. Hoffman, C. W. W., Ropp, R. C., Mooney, R. W.: J. Amer. Chem. Soc. *81*, 3830 (1959)
45. (a) Wilshire, J., Sawyer, D. T.: Acc. Chem. Res. *12*, 105 (1979); (b) Sawyer, D. T., Richens, D. T., Nanni, E. J. Jr., Stallings, M. D.: Redox reaction chemistry of superoxide ion. In: Chemical and Biochemical Aspects of Superoxide and Superoxide Dismutase. Bannister, J. V., Hill, H. A. O. (eds.), pp. 1–26. New York, Amsterdam, Oxford: Elsevier, North-Holland 1980
46. Hayashi, Y., Yamazaki, I.: ref. 18, pp. 157–165
47. Fridovich, I.: Chemical aspects of superoxide radical and of superoxide dismutases. In: Biochemical and medical aspects of active oxygen. Hayaishi, O., Asada, K. (eds.), pp. 3–12. Tokyo: Univ. of Tokyo Press 1977
48. Fridovich, I.: Acc. Chem. Res. *5*, 321 (1977)
49. Fee, J. A.: ref. 17, pp. 209–239
50. Walling, C.: Acc. Chem. Res. *8*, 125 (1975)
51. Groves, J. T.: ref. 17, pp. 125–162
52. Landolt Börnstein, Zahlenwerte und Funktionen aus Physik, Chemie, Astronomie, Geophysik und Technik, Band II 2 b, Berlin: Springer 1962
53. Sakurai, F., Suzuki, H., Moro-oka, Y., Ikawa, T.: J. Amer. Chem. Soc. *102*, 1749 (1980)
54. Boeyens, J. C. A., Haegele, R.: J. Chem. Soc. (Dalton) *1977*, 648
55. Bradley, D. C., Ghota, J. S., Hart, F. A., Hursthouse, M. B., Raithby, P. R.: ibid. *1977*, 1166
56. Bennett, M. J., Donaldson, P. B.: Inorg. Chem. *16*, 1585 (1977)
57. Nolte, M. J., Singleton, E., Laing, M.: J. Amer. Chem. Soc. *97*, 6396 (1975)
58. Brown, L. D., Raymond, K. N.: Inorg. Chem. *14*, 2595 (1975)
59. Guilard, R., Latour, J. M., Le Comte, C., Marchon, J.-C., Protas, J., Ripoll, D.: ibid. *17*, 1228 (1978)
60. Rodley, G. A., Robinson, W. T.: Nature (London) *235*, 438 (1972)
61. Calligaris, M., Nardin, G., Randaccio, L.: J. Chem. Soc. Chem. Comm. *1973*, 419
62. Gall, R. S., Rogers, J. F., Schaefer, W. P., Christoph, G. C.: J. Amer. Chem. Soc. *98*, 5135 (1976)
63. Avdeef, A., Schaefer, W. P.: ibid. *98*, 5153 (1976)
64. Gall, R. S., Schaefer, W. P.: Inorg. Chem. *15*, 2758 (1976)
65. Petsko, G. A., Rose, D., Tsernoglou, D., Ikeda-Saito, M., Yonetani, T.: Frontiers of Biological Energetics (Dutton, P. L., Scarpa, A., Leigh, J. S. eds.), p. 1011, Academic Press, New York 1978

66. Jameson, G. B., Rodley, G. A., Robinson, W. T., Gagné, R. R., Reed, C. A., Collman, J. P.: Inorg. Chem. *17*, 850 (1978)
67. Phillips, S. E. V.: Nature *273*, 247 (1978); J. Molec. Biol. *142*, 531 (1980)
68. Weber, E., Steigemann, W., Jones, T. A., Huber, R.: J. Mol. Biol. *120*, 327 (1978)
69. Jameson, G. B., Molinaro, F. S., Ibers, J. A., Collman, J. P., Brauman, J. F., Rose, E., Suslick, K.: J. Amer. Chem. Soc. *100*, 6769 (1978)
70. Jameson, G. B., Molinaro, F. S., Ibers, J. A., Collman, J. P., Brauman, J. F., Rose, E., Suslick, K. S.: ibid. *102*, 3224 (1980)
71. Shaanan, B.: Nature *296*, 683 (1982)
72. Schaefer, W. P., Marsh, R. E.: a) J. Amer. Chem. Soc. *88*, 178 (1966); b) Acta Cryst. *21*, 735 (1966)
73. Schaefer, W. P., Ealick, S. E., Marsh, R. E.: Acta Cryst. *B37*, 34 (1981)
74. Marsh, R. E., Schaefer, W. P.: ibid. *B24*, 246 (1968)
75. Thewalt, U., Marsh, R. E.: J. Amer. Chem. Soc. *89*, 6364 (1967)
76. Thewalt, U., Marsh, R. E.: Inorg. Chem. *11*, 351 (1972)
77. Christoph, G. G., Marsh, R. E., Schaefer, W. P.: ibid. *8*, 291 (1969)
78. Thewalt, U., Struckmeier, G.: Z. Anorg. Allg. Chem. *419*, 163 (1976)
79. Fronczek, F. R., Schaefer, W. P., Marsh, R. E.: Inorg. Chem. *14*, 611 (1975)
80. Schaefer, W. P.: ibid. *7*, 725 (1968)
81. Fronczek, F. R., Schaefer, W. P., Marsh, R. E.: Acta Cryst. *B30*, 117 (1974)
82. Thewalt, U.: Z. Anorg. Allg. Chem. *485*, 122 (1982)
83. Fritch, J. R., Christoph, G. G., Schaefer, W. P.: Inorg. Chem. *12*, 2170 (1973)
84. Thewalt, U., Zehnder, M., Fallab, S.: Helv. Chim. Acta *60*, 867 (1977)
85. Shibahara, T., Koda, S., Mori, M.: Bull. Chem. Soc. Japan *46*, 2070 (1973)
86. Timmons, J. H., Clearfield, A., Martell, A. E., Niswander, R. H.: Inorg. Chem. *18*, 1042 (1979)
87. Timmons, J. H., Niswander, R. H., Clearfield, A., Martell, A. E.: ibid. *18*, 2977 (1979)
88. Zehnder, M., Thewalt, U.: Z. Anorg. Allg. Chem. *461*, 53 (1980)
89. Fronczek, F. R., Schaefer, W. P.: Inorg. Chim. Acta *9*, 143 (1974)
90. Wang, B. C., Schaefer, W. P.: Science *166*, 1404 (1969)
91. a) Calligaris, M., Nardin, G., Randaccio, L., Ripamonti, A.: J. Chem. Soc. A *1970*, 1069
    b) Calligaris, M., Nardin, G., Randaccio, L.: J. Chem. Soc. Chem. Comm. *1969*, 763
92. Lindblom, L. A., Schaefer, W. P., Marsh, R. E.: Acta Cryst. *B27*, 1461 (1971)
93. Avdeef, A., Schaefer, W. P.: Inorg. Chem. *15*, 1432 (1976)
94. Fallab, S., Zehnder, M., Thewalt, U.: Helv. Chim. Acta *63*, 1491 (1980)
95. Zehnder, M., Thewalt, U.: ibid. *59*, 2290 (1976)
96. Bhaduri, S., Casella, L., Ugo, R., Raithby, P. R., Zuccaro, C., Hursthouse, M. B.: J. Chem. Soc. (Dalton) *1979*, 1624
97. Guilard, R., Fontesse, M., Fournari, P., Le Comte, C., Protas, J.: J. Chem. Soc. Chem. Comm. *1976*, 161
98. Schwarzenbach, D.: Z. Krist. *143*, 429 (1976)
99. Manohar, H., Schwarzenbach, D.: Helv. Chim. Acta *57*, 1086 (1974)
100. Schwarzenbach, D., Girgis, K.: ibid. *58*, 2391 (1975)
101. Mimoun, H., Postel, M., Casabianca, F., Fischer, J., Mitschler, A.: Inorg. Chem. *21*, 1303 (1982)
102. Ruzic-Toros, Z., Kojic-Prodic, B., Gabela, F., Sljukic, M.: Acta Cryst. *B33*, 692 (1977)
103. Bkonche-Waksman, I., Bois, C., Sala-Pala, J., Guerchais, J. E.: J. Organometal. Chem. *195*, 307 (1980)
104. Massa, W., Pausewang, G.: Z. Anorg. Allg. Chem. *456*, 169 (1979)
105. Dewan, J. C., Edwards, A. J., Calves, J. Y., Guerchais, J. E.: J. Chem. Soc. (Dalton) *1977*, 981
106. Edwards, A. J., Slim, D. R., Guerchais, J. E., Kergoat, R.: ibid. *1977*, 1966
107. Chevrier, B., Dibold, T., Weiss, R.: Inorg. Chim. Acta *19*, L57 (1976)
108. Jacobson, S. E., Tang, R., Mares, F.: Inorg. Chem. *17*, 3055 (1978)
109. Edwards, A. J., Slim, D. R., Guerchais, J. E., Kergoat, J. R.: J. Chem. Soc. (Dalton) *1980*, 289
110. Tomioka, H., Takai, K., Oshima, K., Nozaki, H.: Tetrahedron Lett. *21*, 4843 (1980)

111. Winter, W., Mark, C., Schurig, V.: Inorg. Chem. *19*, 2045 (1980)
112. Terry, N . W., Amma, E. L., Vaska, L.: J. Amer. Chem. Soc. *94*, 653 (1972)
113. Halpern, J., Goodall, B. L., Khare, G. P., Lim, H. S., Pluth, J. J.: ibid. *97*, 2301 (1975)
114. Crump, D. B., Stepaniak, R. F., Payne, N. C.: Can. J. Chem. *55*, 438 (1977)
115. Nolte, M., Singleton, E.: Acta Cryst. *B32*, 1410 (1976)
116. Bennett, M. J., Donaldson, P. S.: Inorg. Chem. *16*, 1581 (1977)
117. Ellerman, J., Hohenberger, E. F., Kehr, W., Pürzer, A., Thiele, G.: Z. Anorg. Allg. Chem. *464*, 45 (1980)
118. Gash, A. G., Terry, N. W., Amma, E. L.: Amer. Cryst. Assocn (Winter Meeting) *1973*, 40
119. Nolte, M. J., Singleton, E.: Acta Cryst. *B31*, 2223 (1975)
120. Nolte, M. J., Singleton, E.: ibid. *B32*, 1838 (1976)

121. Nolte, M. J., Singleton, E., Laing, M.: J. Chem. Soc. (Dalton) *1976*, 1979
122. Weiniger, M. S., Griffith, E. A. H., Sears, C. T., Amma, E. L.: Inorg. Chim. Acta *60*, 67 (1982)
123. Yoshida, T., Tatsumi, K., Matsuomoto, M., Nakatsu, K., Nakamura, A., Fueno, T., Otsuka, S.: Nouv. J. Chim. *3*, 761 (1979)
124. Latour, J. M., Marchon, J. C., Nakajima, M.: J. Amer. Chem. Soc. *101*, 3974 (1979)
125. Catton, P. R., Premovic, P., Stavdal, L., West, P.: J. Chem. Soc. Chem. Comm. *1983*, 863
126. Chimura, Y., Beppu, M., Yoshida, S., Tarama, K.: Bull. Chem. Soc. Japan *50*, 691 (1977)
127. Goedken, V. L., Ladd, J. A.: J. Chem. Soc. Chem. Comm. *1982*, 142
128. Schwendt, P., Petrovic, P., Uskert, D.: Z. Anorg. Allg. Chem. *466*, 232 (1980)
129. Howarth, O. W., Hunt, J. R.: Chem. Soc. (Dalton) *1979*, 1388
130. Cooper, S. R., Koh, Y. B., Raymond, K. N.: J. Amer. Chem. Soc. *104*, 5092 (1982)

131. Cheung, S. K., Grimes, C. J., Wong, J., Reed, C. A.: ibid. *98*, 5028 (1976)
132. Carlton, L., Lindsell, W. E., Preston, P. N.: J. Chem. Soc. Chem. Comm. *1981*, 531
133. Poliakoff, M., Smith, K. P., Turner, J. J., Wilkinson, A. J.: J. Chem. Soc. (Dalton) *1982*, 651
134. Kellerman, R., Hutta, P. J., Klier, K.: J. Amer. Chem. Soc. *96*, 5946 (1974)
135. Imamura, T., Terui, M., Takahashi, Y., Numatatsu, T., Fujimoto, M.: Chem. Lett. *1979*, 89
136. Westland, A. D., Haque, F., Bouchard, J. M.: Inorg. Chem. *19*, 2255 (1980)
137. Etcheverry, S. B., Baran, E. J.: Z. Anorg. Allg. Chem. *465*, 153 (1980)
138. Hocks, L., Durbab, P., Teyssie, P.: J. Mol. Catal. *7*, 75 (1980)
139. Coleman, W. M., Taylor, L. T.: Coord. Chem. Rev. *32*, 1 (1980)
140. Elvidge, J. A., Lever, A. B. P.: Proc. Chem. Soc. London *1959*, 195

141. Lever, A. B. P., Wilshire, J. P., Quan, S. K.: Inorg. Chem. *20*, 761 (1981)
142. Moxon, N. T., Fielding, P. E., Gregson, A. K.: J. Chem. Soc. Chem. Comm. *1981*, 98
143. a) Hoffman, B. M., Szymanski, T., Brown, T. G., Basolo, F.: J. Amer. Chem. Soc. *100*, 7253 (1978);
     b) Jones, R. D., Summerville, D. A., Basolo, F.: ibid. *100*, 4416 (1978)
144. Jones, R. D., Budge, J. R., Ellis, P. E., Linard, J. E., Summerville, D. A., Basolo, F.: J. Organomet. Chem. *181*, 151 (1979)
145. Yarino, T., Matsushita, T., Masuda, I., Shinra, K.: J. Chem. Soc. Chem. Comm. *1970*, 1317
146. Coleman, W. M., Taylor, L. T.: Inorg. Chim. Acta *30*, L291 (1978)
147. Matsushita, T., Shono, T.: Bull. Chem. Soc. Japan *54*, 3743 (1981)
148. Coleman, W. M., Taylor, L. T.: Inorg. Chim. Acta *61*, 13 (1982)
149. Lindsell, W. E., Preston, P. N.: J. Chem. Soc. (Dalton) *1979*, 1105
150. Magers, K. D., Smith, C. G., Sawyer, D. T.: Inorg. Chem. *19*, 492 (1980)

151. Lynch, M. W., Hendrickson, D. N., Fitzgerald, B. J., Pierpoint, C. G.: J. Amer. Chem. Soc. *103*, 3961 (1981)
152. Lauffer, R. B., Heistand, R. H., Que, L.: ibid. *103*, 3947 (1981)
153. a) Hosseiny, A., McAuliffe, C. A., Minten, K., Parrott, M. J., Pritchard, R., Tames, J.: Inorg. Chim. Acta *39*, 227 (1980);
     b) McAuliffe, C. A., Al-Khateeb, H.: ibid. *45*, L195 (1980);
     c) Barber, M., Bordoli, R. S., Hosseiny, A., Minten, K., Perkin, C. R., Sedgwick, R. D., McAuliffe, C. A.: ibid. *45*, L89 (1980);
     d) McAuliffe, C. A., Al-Khateeb, H., Jones, M. H., Levason, W., Minten, K.: J. Chem. Soc. Chem. Comm. *1979*, 736

154. a) Brown, R. M., Bull, R. E., Green, M. L. H., Grebenik, P. D., Martin-Polo, J. J., Mingos, D. M. P.: J. Organometal. Chem. *201*, 437 (1980);
     b) McAuliffe, C. A.: ibid. *228*, 255 (1982)
155. Wang, J. H.: J. Amer. Chem. Soc. *80*, 3168 (1958)
156. Ledon, H., Brigandat, Y.: J. Organomet. Chem. *190*, L 87 (1980)
157. Battersby, A. R., Hamilton, A. D.: J. Chem. Soc. Chem. Comm. *1980*, 117
158. Momenteau, M., Loock, B.: J. Mol. Catal. *7*, 315 (1980)
159. Herron, N., Busch, D. H.: J. Amer. Chem. Soc. *103*, 1236 (1981)
160. Traylor, T. G.: Acc. Chem. Res. *14*, 102 (1981)
161. Budge, J. R., Ellis, P. E., Jones, R. D., Linard, J. E., Basolo, F., Baldwin, J. E., Dyer, R. L.: J. Amer. Chem. Soc. *101*, 4760 (1979)
162. Schappacher, M., Ricard, L., Weiss, R., Montiel-Montoya, R., Bill, E., Gonser, U., Trautwein, A.: ibid. *103*, 7646 (1981)
163. Chin, D. H., La Mar, G. N., Balch, A. L.: ibid. *102*, 4344 (1980); ibid. *99*, 5486 (1977)
164. Ercolani, C., Rossi, G., Monacelli, F.: Inorg. Chim. Acta *44*, L 215 (1980)
165. Kimura, E., Kodama, M., Machida, R., Ishizu, K.: Inorg. Chem. *21*, 595 (1982)
166. Marini, P. J., Murray, K. S., West, B. O.: J. Chem. Soc. Chem. Comm. *1981*, 726
167. See, for example, Cohen, I. A.: Structure and Bonding *40*, 1 (1980)
168. Stenkamp, R. E., Sieker, L. C., Jensen, L. H.: Proc. Nat. Acad. Sci. USA *73*, 349 (1976)
169. Loehr, J. S., Loehr, T. M., Mauk, A. G., Gray, H. B.: J. Amer. Chem. Soc. *102*, 6992 (1980)
170. McCandlish, E., Mikstal, A. R., Nappa, M., Sprenger, A. Q., Valentine, J. S., Strong, J. D., Spiro, T. G.: ibid. *102*, 4268 (1980)
171. Walling, C., Kurz, M., Schugar, H. J.: Inorg. Chem. *9*, 931 (1970)
172. McClune, G. J., Fee, J. A., McCluskey, G. A., Groves, J. T.: J. Amer. Chem. Soc. *99*, 5220 (1977)
173. Hester, R. E., Nour, E. M.: J. Raman Spectroscopy *11*, 35 (1981)
174. Nakamoto, K., Watanabe, T., Ama, T., Urban, M. W.: J. Amer. Chem. Soc. *104*, 3744 (1982)
175. Chang, S., Blyholder, G., Fernandez, J.: Inorg. Chem. *20*, 2813 (1981)
176. Farrell, N., Dolphin, D., James, B. R.: J. Amer. Chem. Soc. *100*, 324 (1978)
177. Laing, K. R., Roper, W. R.: J. Chem. Soc. Chem. Comm. *1968*, 1556
178. Christian, D. F., Roper, W. R.: ibid. *1971*, 1271
179. Personal communication from J. A. Ibers, cited in Graham, B. W., Laing, K. R., O'Connor, C. J., Roper, W. R.: J. Chem. Soc. (Dalton) *1972*, 1237
180. James, B., Markham, L., Rattray, A., Wang, D.: Inorg. Chim. Acta *20*, L 25 (1976)
181. Taqui Khan, M. M., Ramachandraiah, G.: Inorg. Chem. *21*, 2109 (1982)
182. Gustafson, B. L., Lin, M.-J., Lunsford, J. H.: J. Phys. Chem. *84*, 3211 (1980)
183. Chen, L. S., Koehler, M. E., Pestel, B. C., Cummings, B. C.: J. Amer. Chem. Soc. *100*, 7243 (1978)
184. Hay, R. W., Norman, P. R., McLaren, F.: Inorg. Chim. Acta *44*, L 125 (1980)
185. Bedell, S. A., Timmons, J.-H., Martell, A. E., Murase, I.: Inorg. Chem. *21*, 874 (1978)
186. Martell, A. E.: Acc. Chem. Res. *15*, 155 (1982)
187. Pickens, S. R., Martell, A. E., McLendon, G., Lever, A. B. P., Gray, H. B.: Inorg. Chem. *17*, 2190 (1978)
188. Kozuka, M., Suzuki, M., Nishida, Y., Kida, S., Nakamoto, K.: Inorg. Chim. Acta *45*, L 111 (1980)
189. Berry, K. J., Moya, F., Murray, K. S., van den Bergen, A. M. B., West, B. O.: J. Chem. Soc. (Dalton) *1982*, 109
190. Drago, R. S., Stahlbush, J. R., Kitko, D. J., Breese, J.: J. Amer. Chem. Soc. *102*, 1884 (1980)
191. Miskowski, V. M., Robbins, J. L., Hammond, G. S., Gray, H. B.: ibid. *98*, 2477 (1976)
192. Williams, A. F.: unpublished results
193. Ng, C. Y., Martell, A. E., Motekaitis, R. J.: J. Coord. Chem. *9*, 255 (1979)
194. Collman, J. P., Denisevich, P., Konai, Y., Marrocco, M., Koval, C., Anson, F. C.: J. Amer. Chem. Soc. *102*, 6027 (1980)
195. Suzuki, M., Kanatomi, H., Murase, I.: Chem. Lett. *1981*, 1745
196. Kozuka, M., Nakamoto, K.: J. Amer. Chem. Soc. *103*, 2162 (1981)
197. Urban, M. W., Nonaka, Y., Nakamoto, K.: Inorg. Chem. *20*, 1046 (1982)

198. Nakamoto, K., Nonaka, Y., Ishiguro, T., Urban, M. W., Suzuki, M., Kozuka, M., Nishida, Y., Kida, S.: J. Amer. Chem. Soc. *104*, 3386 (1982)
199. Urban, M. W., Nakamoto, K., Kincaid, J.: Inorg. Chim. Acta *61*, 77 (1982)
200. Simon, J., LeMoigne, J., Markovitsi, D., Dayantis, J.: J. Amer. Chem. Soc. *102*, 7247 (1980)
201. Schoonheydt, R. A., Pelgrims, J.: J. Chem. Soc. (Dalton) *1981*, 914
202. Andreev, A., Prahov, L., Shopov, D.: Coll. Czech. Chem. Comm. *45*, 1780 (1980)
203. Fieldhouse, S. A., Fullam, B. W., Neilson, G. W., Symons, M. C. R.: J. Chem. Soc. (Dalton) *1974*, 567
204. Ozin, G. A., Hanlan, A. J. L., Power, W. J.: Inorg. Chem. *18*, 2390 (1979)
205. James, B. R., Mahajan, D.: Can. J. Chem. *58*, 996 (1980)
206. Lumpkin, O., Dixon, W. T., Poser, J.: Inorg. Chem. *18*, 982 (1979)
207. Lawson, D. N., Mays, M. J., Wilkinson, G.: J. Chem. Soc. A *1966*, 52
208. Addison, A. W., Gillard, R. D.: ibid. *1970*, 2523
209. Endicott, J. F., Wong, C.-L., Inoue, T., Natarajan, P.: Inorg. Chem. *18*, 450 (1979)
210. Raynor, J. B., Gillard, R. D., Pedrosa de Jesus, J. D.: J. Chem. Soc. (Dalton) *1982*, 1165

211. Wayland, B. B., Newman, A. R.: Inorg. Chem. *20*, 3093 (1981)
212. Empsall, H. D., Heys, P. N., McDonald, W. S., Norton, M. C., Shaw, B. L.: J. Chem. Soc. (Dalton) *1978*, 1119
213. Hughes, G. R., Mingos, D. M. P.: Transition Metal Chem. (Weinheim, Germany) *3*, 381 (1978)
214. a) Chung, R. J., Suzuki, H., Moro-oka, Y.: Chem. Lett. *1980*, 63;
     b) Suzuki, M., Mizutani, K., Moro-oka, Y., Ikawa, T.: J. Amer. Chem. Soc. *101*, 748 (1978)
215. Chimura, Y., Beppu, M., Yoshida, S., Tarama, K.: Chem. Lett. *1976*, 375
216. Huber, H., Klotzbücher, N., Ozin, G. A., Vander Voet, A.: Can. J. Chem. *51*, 2722 (1973)
217. Ozin, G. A., Klotzbücher, W. E.: J. Amer. Chem. Soc. *97*, 3965 (1975)
218. Hanlan, A. J. L., Ozin, G. A.: Inorg. Chem. *16*, 2848, 2857 (1977)
219. van Holde, K. E., Miller, K. I.: Q. Rev. Biophysics *15*, 1 (1982)
220. a) Co, M. S., Hodgson, K. O., Eccles, T. K., Lontie, R.: J. Amer. Chem. Soc. *103*, 984 (1981);
     b) Brown, J. M., Powers, L., Kincaid, B., Larrabee, J. A., Spiro, T. G.: ibid. *102*, 4220 (1980)
221. Larrabee, J. A., Spiro, T. G.: ibid. *102*, 4217 (1980)
222. Himmelwright, E. S., Eickmann, N. C., Lubien, C., Lerch, K., Solomon, E. I.: ibid. *102*, 7339 (1980)
223. Richardson, J. S., Thomas, K. A., Rubin, B. H., Richardson, J. C.: Proc. Nat. Acad. Sci. USA *72*, 1349 (1975)
224. Simmons, M. G., Merrill, C. L., Wilson, L. J., Bottomley, L. A., Kadish, K. M.: J. Chem. Soc. (Dalton) *1980*, 1827
225. Nishida, Y., Takahashi, K., Kuramoto, H., Kida, S.: Inorg. Chim. Acta *54*, L103 (1981)
226. Karlin, K. D., Dahlstrom, P. L., Cozzette, S. N., Scensny, P. M., Zubieta, J.: J. Chem. Soc. Chem. Comm. *1981*, 881
227. Bulkowski, J. E., Burke, P. L., Ludmann, M. F., Osborn, J. A.: ibid. *1977*, 498
228. a) Zuberbühler, A. D.: Metal Ions in Biological Stems (H. Sigel, ed.), M. Dekker, New York 1976, Vol. 5, pp. 325–268;
     b) Gampp, H., Zuberbühler, A. O.: J. Mol. Catal. *7*, 81 (1980)
229. Nappa, M., Valentine, J. S., Miksztal, A. R., Schugar, H. J., Isied, S. S.: J. Amer. Chem. Soc. *101*, 7744 (1979)
230. Valentine, J. S.: ibid. *99*, 3522 (1977)

231. Djordjevic, C., Vuletic, I.: Inorg. Chem. *19*, 3049 (1980)
232. Westland, A. D., Tarafder, M. T. H.: ibid. *20*, 3992 (1981)
233. Pauling, L.: Haemoglobin (Roughton, F. J. W., Kendrew, J. C., eds.) Butterworths, London, 1949, p. 57;
     Pauling, L.: Nature *203*, 182 (1964)
234. Griffith, J. S.: Proc. Roy. Soc. A *235*, 23 (1956)
235. Weiss, J. J.: Nature *202*, 83 (1964)
236. Tuck, D. G., Walters, R. M.: Inorg. Chem. *2*, 428 (1963)

237. Griffith, W. P.: J. Chem. Soc. *1964*, 5248
238. Stomberg, R.: Ark. Kemi *24*, 283 (1965)
239. Swalen, J. D., Ibers, J. A.: J. Chem. Phys. *37*, 17 (1962)
240. Weil, J. A., Kinnaird, J. K.: J. Phys. Chem. *71*, 3341 (1967)
241. Mason, R.: Nature *217*, 543 (1968)
242. Summerville, D. A., Jones, R. D., Hoffman, B. M., Barolo, F.: J. Chem. Ed. *56*, 157 (1979)
243. Jørgensen, C. K.: Oxidation Numbers and Oxidation States, Berlin, Heidelberg, New York; Springer Verlag 1969
244. Drago, R. S., Corden, B. B.: Acc. Chem. Res. *13*, 353 (1980)
245. Mingos, D. M. P.: Nature Phys. Sci. *229*, 193 (1971); ibid. *230*, 154 (1971)
246. Hoffmann, R., Chen, M. M. L., Thorn, D. L.: Inorg. Chem. *16*, 503 (1977)
247. Williams, A. F.: to be published
248. Schaefer, W. P., Huie, B. T., Kurilla, M. G., Ealick, S. E.: Inorg. Chem. *19*, 340 (1980)
249. Turner, D. W., Baker, C., Baker, A. D., Brundle, C. R.: Molecular Photoelectron Spectroscopy, London, New York, Sydney, Toronto; J. Wiley 1970
250. Ellinger, Y., Latour, J. M., Marchon, J.-C., Subra, R.: Inorg. Chem. *17*, 2024 (1978)

251. Hanson, L. K., Hoffman, B. M.: J. Amer. Chem. Soc. *102*, 4602 (1980)
252. Goddard, W. D., Olafson, B. D.: ref. 18, pp. 87–124
253. Miskowski, V. M., Robbins, J. L., Hammond, G. S., Gray, H. B.: J. Amer. Chem. Soc. *98*, 2477 (1976)
254. Gubelmann, M. H., Williams, A. F.: unpublished observations
255. Herman, F., Skillman, S.: Atomic Structure Calculations, New Jersey; Prentice-Hall, Englewood Cliffs 1963
256. Norman, J. G.: Inorg. Chem. *16*, 1328 (1977)
257. Vaska, L., Chen, L. S., Senoff, C. V.: Science *174*, 587 (1971)
258. Thewalt, U., Marsh, R. E.: J. Amer. Chem. Soc. *89*, 6364 (1967)
259. Hrncir, D. C., Rogers, R. D., Atwood, J. L.: ibid. *103*, 4277 (1981)
260. Chin, D. H., Balch, A. L., La Mar, G. N.: ibid. *102*, 1446 (1980)

261. Rohmer, M.-M., Barry, M., Dedieu, A., Veillard, A.: Int. J. Quantum Chem. (Quantum Biology Symp.) *4*, 337 (1977)
262. Dedieu, A., Rohmer, M.-M., Veillard, H., Veillard, A.: Nouv. J. Chimie *3*, 653 (1979)
263. Bachmann, C., Demuynck, J., Veillard, A.: J. Amer. Chem. Soc. *100*, 2366 (1978)
264. Dacre, P. D., Elder, M.: J. Chem. Soc. (Dalton) *1972*, 1426
265. Fischer, J., Veillard, A., Weiss, R.: Theoret. Chim. Acta *24*, 317 (1972)
266. Rösch, N., Hoffmann, R.: Inorg. Chem. *13*, 2656 (1974)
267. Dalal, N. S., Millar, J. M., Jagadeesh, M. S., Seehra, M. S.: J. Chem. Phys. *74*, 1916 (1981)
268. Weber, J., Roch, M., Williams, A. F.: to be published
269. Brown, D. H., Perkins, P. G.: Inorg. Chim. Acta *8*, 285 (1974)
270. Kirchner, R. F., Loew, G. H.: J. Amer. Chem. Soc. *99*, 4639 (1977)

271. Loew, G. H., Kirchner, R. F.: ibid. *97*, 7388 (1975)
272. Dedieu, A., Rohmer, M.-M., Benard, M., Veillard, A.: ibid. *98*, 3717 (1976)
273. Dedieu, A., Rohmer, M.-M., Veillard, A.: Proc. 9th Jerusalem Conf. on Quantum Chemistry and Biochemistry (B. Pullman, N. Goldblum, eds.), Dordrecht, Boston: D. Reidel, 1976, part 2, p. 101
274. Case, D. A., Huynh, B. H., Karplus, M.: J. Amer. Chem. Soc. *101*, 4433 (1979)
275. Rohmer, M.-M., Loew, G. H.: Int. J. Quantum Chemistry (Quantum Biology Symp.) *6*, 93 (1979)
276. Herman, Z. S., Loew, G. H.: J. Amer. Chem. Soc. *102*, 1815 (1980)
277. (a) Cerdonio, M., Congin-Castellano, L., Calabrese, L., Morante, S., Pispisa, B., Vitale, S.: Proc. Natl. Acad. Sci. USA *75*, 4916 (1978);
(b) Cerdonio, M., Congin-Castellano, L., Mogno, F., Pispisa, B., Romani, G. L., Vitale, S.: ibid. *74*, 398 (1977)
278. Pauling, L.: ibid. *74*, 2612 (1977)
279. Loew, G. H., Herman, Z. S., Zerner, M. C.: Int. J. Quantum Chem. *18*, 481 (1980); Loew, G. H., Rohmer, M.-M.: J. Amer. Chem. Soc. *102*, 3655 (1980)

280. Otsuka, J., Seno, Y., Fuchikami, N., Matsuoka, O.: Proc. 9th Jerusalem Conf. on Quantum Chemistry and Biochemistry (B. Pullman, N. Goldblum, eds.), Dordrecht, Boston: D. Reidel, 1976, part 2, p. 60
281. Olafson, B. D., Goddard, W. A.: Proc. Natl. Acad. Sci. USA 74, 1315 (1977)
282. Olafson, B. D., Goddard, W. A.: ref. 18, p. 87
283. Makinen, M. W.: ref. 18, p. 143
284. Blyholder, G., Head, J., Ruette, F.: Inorg. Chem. 21, 1539 (1982)
285. Dedieu, A., Rohmer, M.-M., Veillard, A.: J. Amer. Chem. Soc. 98, 5789 (1976)
286. Fantucci, P., Valenti, V.: ibid. 98, 3832 (1976)
287. Daul, C., Schläpfer, C. W., von Zelewsky, A.: Structure and Bonding 36, 129 (1979)
288. Boca, R.: J. Mol. Structure 65, 173 (1980)
289. Boca, R., Pelikan, P.: Inorg. Chim. Acta 44, L65 (1980)
290. Boca, R.: J. Mol. Catal. 9, 275 (1980)
291. Boca, R.: ibid. 10, 187 (1981)
292. Teo, B.-K., Li, W.-K.: Inorg. Chem. 15, 2005 (1976)
293. Hyla-Krispin, I., Natakaniec, L., Jezowska-Trzebiatowska, B.: (a) Chem. Phys. Lett. 35, 311 (1975);
     (b) Bull. Acad. Polonaise des Sciences 25, 193 (1977)
294. Tatsumi, K., Fueno, T., Nakamura, A., Otsuka, S.: Bull. Chem. Soc. Japan 49, 2164 (1976)
295. Yoshida, T., Tatsumi, K., Otsuka, S.: Pure Appl. Chem. 52, 713 (1980)
296. Sakaki, S., Hori, K., Ohyoshi, A.: Inorg. Chem. 17, 3183 (1978)
297. Carlton, L., Lindsell, W. E., Preston, P. N.: J. Chem. Soc. (Dalton) 1982, 1483
298. Lever, A. B. P., Gray, H. B.: Acc. Chem. Res. 11, 348 (1978); Lever, A. B. P.: J. Mol. Struct. 59, 123 (1980)
299. Miskowski, V. M., Robbins, V. M., Treitel, I. M., Gray, H. B.: Inorg. Chem. 14, 2318 (1975)
300. Lever, A. B. P., Pickens, S. R.: Inorg. Chim. Acta 45, L185 (1980)
301. Lever, A. B. P., Ozin, G. A., Gray, H. B.: Inorg. Chem. 19, 1823 (1980)
302. Pickens, S. R., Martell, A. E.: ibid. 19, 15 (1980)
303. (a) Braun-Steinle, D., Mäcke, H., Fallab, S.: Helv. Chim. Acta 59, 2032 (1976);
     (b) Zehnder, M.: ibid. 62, 2854 (1979)
304. Eaton, W. A., Hanson, L. K., Stephens, P. J., Sutherland, J. C., Dunn, J. B. R.: J. Amer. Chem. Soc. 100, 4991 (1978)
305. (a) Churg, A. K., Makinen, M. W.: J. Chem. Phys. 68, 1913 (1978);
     (b) Makinen, M. W., Churg, A. K., Glick, H. A.: Proc. Natl. Acad. Sci. USA 75, 2291 (1978)
306. Loehr, J. S., Loehr, T. M., Mauk, A. G., Gray, H. B.: . Amer. Chem. Soc. 102, 6992 (1980)
307. Suzuki, M., Ishiguro, T., Kozuka, M., Nakamoto, K.: Inorg. Chem. 20, 1993 (1981)
308. Loehr, J. S., Freedman, T. B., Loehr, T. M.: Biochem. Biophys. Res. Commun. 56, 510 (1974)
309. Shirazi, A., Goff, H. M.: J. Amer. Chem. Soc. 104, 6318 (1982)
310. Shibahara, T., Mori, M.: Bull. Chem. Soc. Japan 51, 1374 (1978)
311. Barraclough, C. G., Lawrance, G. A., Lay, P. A.: Inorg. Chem. 17, 3317 (1978)
312. Szymanski, T., Cape, T. W., van Duyne, R. P., Basolo, F.: J. Chem. Soc. Chem. Comm. 1979, 5
313. Brunner, H.: Naturwiss. 61, 129 (1974)
314. Urban, M. W., Nakamoto, K., Basolo, F.: Inorg. Chem. 21, 3406 (1982)
315. Nakamoto, K., Suzuki, M., Ishiguro, T., Kozuka, M., Nishida, Y., Kida, S.: ibid. 19, 2822 (1980)
316. Suzuki, M., Ishiguro, T., Kozuka, M., Nakamoto, K.: ibid. 20, 1993 (1981)
317. Nour, E. M., Hester, R. E.: J. Mol. Struct. 62, 77 (1980)
318. Hester, R. E., Nour, E. M.: J. Raman Spectrosc. 11, 43, 59, 64 (1981)
319. Dickinson, L. C., Chien, J. C. W.: Proc. Natl. Acad. Sci. USA 77, 1235 (1980)
320. Tovrog, B. S., Kitko, D. J., Drago, R. S.: J. Amer. Chem. Soc. 98, 5144 (1976)
321. Smith, T. D., Ruzic, I. M., Tirant, S.: J. Chem. Soc. (Dalton) 1982, 363
322. Hoffman, B. M., Weschler, C. J., Basolo, F.: J. Amer. Chem.Soc. 98, 5473 (1976)
323. Maeda, Y.: J. Phys. Colloque C2, 40, C2–514 (1979)
324. Tsai, T. E., Groves, J. T., Wu, C. S.: J. Chem. Phys. 74, 4306 (1981)

325. Sharrock, M., Debrunner, P. G., Schulz, C., Lipscomb, J. D., Marshall, V., Gunsalus, I. C.: Biochim. Biophys. Acta *420*, 8 (1976)
326. Maeda, Y., Harami, T., Morita, Y., Trautwein, A., Gonser, U.: J. Phys. Colloque C2, *40*, C2–500 (1979)
327. (a)Okamura, M. Y., Klotz, I. M., Johnson, C. E., Winter, M. R. C., Williams, R. J. P.: Biochem. *8*, 1951 (1969);
     (b) Garbett, K., Johnson, C. E., Klotz, I. M., Okamura, M. Y., Williams, R. J. P.: Arch. Biochem. Biophys. *142*, 574 (1971);
     (c) York, J. L., Bearden, A. J.: Biochem. *9*, 4549 (1970)
328. Simonneaux, G., Scholz, W. F., Reed, C. A., Lang, G. A.: Biochim. Biophys. Acta *716*, 1 (1982)
329. Wickman, H. H., Silverthorn, W. E.: Inorg. Chem. *10*, 2333 (1971)
330. Williams, A. F., Jones, G. C. H., Maddock, A. G.: J. Chem. Soc. (Dalton) *1975*, 1952
331. McIntosh, D., Ozin, G. A.: Inorg. Chem. *16*, 59 (1977);
     ibid. *15*, 2869 (1976)
332. Parshall, G. W.: Homogeneous Catalysis, New York, Chichester, Brisbane, Toronto: John Wiley 1980
333. Sheldon, R. A.: J. Mol. Catal. *7*, 107 (1980)
334. Katsuki, T., Sharpless, K. B.: J. Amer. Chem. Soc. *102*, 5974 (1980);
     Rossiter, B. E., Katsuki, T., Sharpless, K. B.: ibid. *103*, 464 (1981)
335. Sen, A., Halpern, J.: ibid. *99*, 8337 (1977)
336. Lyons, J. E.: Aspects Homogeneous Catalysis *3*, 1 (1977)
337. James, B. R.: Adv. Chem. Ser. *191*, 253 (1980)
338. Mimoun, H.: J. Mol. Catal. *7*, 1 (1980);
     Rev. Inst. Fr. Petr. *33*, 259 (1978)
339. Sheldon, R. A., Kochi, J. K.: Metal catalysed oxidations of organic compounds, Academic Press, New York 1981, chapter 4
340. Haas, O., von Zelewsky, A.: J. Chem. Research (S) *1980*, 78;
     J. Chem. Research (M) *1980*, 1228
341. Gubelmann, M. H., Williams, A. F.: unpublished results
342. Endicott, J. F., Kumar, K.: ACS Symp. Series *198*, 425 (1982)
343. Nishinaga, A., Tomita, H., Matsuura, T.: Tetrahedron Letters *21*, 2833 (1980)
344. (a) Nishinaga, A.: ref. 47, pp. 13–27;
     (b) Nishinaga, A.: ref. 13, pp. 179–199
345. Drago, R. S., Cannady, J. P., Leslie, K. A.: J. Amer. Chem. Soc. *102*, 6014 (1980)
346. Welborn, C. H., Dolphin, D., James, B. R.: ibid. *103*, 2869 (1981)
347. Wilkins, R. G.: Adv. Chem. Ser. *100*, 111 (1971)
348. Wong, C. L., Switzer, J. A., Endicott, J. F., Balakrishnan, K. P.: J. Amer. Chem. Soc. *102*, 5511 (1980)
349. Arzoumanian, H., Lai, R., Lopez Alvarez, R., Petrigniani, J. F., Metzger, J., Furhop, J.: ibid. *102*, 845 (1980)
350. Arzoumanian, H., Lai, R., Lopez Alvarez, R., Metzger, J., Petrigniani, J. F.: ref. 13, pp. 44–49
351. Zombeck, A., Drago, R. S., Corden, B. B., Gaul, J. H.: J. Amer. Chem. Soc. *103*, 7580 (1981)
352. Abel, E. W., Pratt, J. M., Whelan, R., Wilkinson, P. J.: ibid. *96*, 7119 (1974)
353. (a)Nishinaga, A., Tomita, H., Matsuura, T.: Tetrahedron Lett. *20*, 2893 (1979);
     (b) Nishinaga, A., Shimizu, T., Matsuura, T.: ibid. *21*, 4097 (1980);
     (c) Nishinaga, A., Tomita, H., Nishizawa, K., Matsuura, T., Ooi, S., Hirotsu, K.: J. Chem. Soc. (Dalton) *1981*, 1504;
     (d) Nishinaga, A., Tomita, H., Oda, M., Matsuura, T.: Tetrahedron Lett. *23*, 339 (1982);
     (e) Nishinaga, A., Shimizu, T., Toyoda, Y., Matsuura, T., Hirotsu, K.: J. Org. Chem. *47*, 2278 (1982)
354. Drago, R. S., Gaul, J., Zombeck, A., Straub, D. K.: J. Amer. Chem. Soc. *102*, 1033 (1980)
355. Martin, J. L., Migus, A., Poyart, C., Lecarpentier, Y., Antonetti, A., Orszag, A.: Biochem. Biophys. Res. Commun. *107*, 803 (1982) and references therein
356. Nishinaga, A., Tomita, H., Matsuura, T.: Tetrahedron Lett. *21*, 3407 (1980)

357. Symons, M. C. R., Petersen, R. L.: Proc. Roy. Soc. *B 201*, 285 (1978);
     Biochim. Biophys. Acta *535*, 241 (1978);
     ibid. *537*, 70 (1978)
358. Mimoun, H.: Angew. Chem. Int. Ed. *21*, 734 (1982)
359. Briant, C. E., Hughes, G. R., Minshall, P. C., Mingos, D. M. P.: J. Organomet. Chem. *202*, C18 (1980)
360. Pizzotti, M., Cenini, S., La Monica, G.: Inorg. Chim. Acta *33*, 161 (1978)
361. Muto, S., Tasaka, K., Kamiya, Y.: Bull. Chem. Soc. Japan *50*, 2493 (1977)
362. Muto, S., Ogata, H., Kamiya, Y.: Chem. Lett. *1975*, 809
363. Belloni, P. L., Cenini, S., Demartin, F., Manassero, M., Pizzotti, M., Porta, F.: J. Chem. Soc. (Dalton) *1980*, 2060
364. Muto, S., Kamiya, Y.: Bull. Chem. Soc. Japan *49*, 2587 (1976)
365. Michelin, R. A., Ros, R., Strukul, G.: Inorg. Chim. Acta *37*, L491 (1979)
366. Igersheim, F., Mimoun, H.: Nouv. J. Chimie *4*, 711 (1980)
367. Suzuki, H., Matsuura, S., Moro-oka, Y., Ikawa, T.: Chem. Lett. *1982*, 1011
368. Hughes, G. R., Mingos, D. M. P.: Transition Metal Chem. *3*, 381 (1978)
369. Tatsuno, Y., Otsuka, S.: J. Amer. Chem. Soc. *103*, 5832 (1981)
370. Otsuka, S., Nakamura, A., Tatsuno, Y., Miki, M.: ibid. *94*, 3761 (1972)
371. Chen, M. J. Y., Kochi, J. K.: J. Chem. Soc. Chem. Commun. *1977*, 204
372. Regen, S. L., Whitesides, G. M.: J. Organomet. Chem. *59*, 293 (1973)
373. Ugo, R., Conti, F., Cenini, S., Mason, R., Robertson, G. B.: J. Chem. Soc. Chem. Commun. *1968*, 1498
374. Ugo, R., Zanderighi, G. M., Fusi, A., Carrieri, D.: J. Amer. Chem. Soc. *102*, 3745 (1980)
375. Hayward, P. J., Blake, D. M., Wilkinson, G., Nyman, C. J.: ibid. *92*, 5873 (1970)
376. Maeda, M., Moritani, I., Hosokawa, T., Murahashi, S.: Tetrahedron Lett. *1974*, 797
377. Hayward, P. J., Safitch, S. J., Nyman, C. J.: Inorg. Chem. *10*, 1311 (1971)
378. Aida, S., Ohta, H., Kamiya, Y.: Chem. Lett. *1981*, 1639
379. Hayward, P. J., Nyman, C. J.: J. Amer. Chem. Soc. *93*, 617 (1971)
380. Beaulieu, W. B., Mercer, G. D., Roundhill, D. M.: ibid. *100*, 1147 (1978)
381. Cariati, G., Mason, R., Robertson, G. B., Ugo, R.: J. Chem. Soc. Chem. Commun. *1967*, 408
382. Horn, R. W., Weissenberger, E., Collman, J. P.: Inorg. Chem. *9*, 2367 (1970)
383. Stiddard, M. H. B., Townsend, R. E.: J. Chem. Soc. Chem. Commun. *1969*, 1372
384. Nappier, T. E., Meek, D. W., Kirchner, R. M., Ibers, J. A.: J. Amer. Chem. Soc. *95*, 4194 (1973)
385. Booth, B. L., McAuliffe, C. A., Stanley, G. L.: J. Organometal. Chem. *226*, 191 (1982)
386. Vasapollo, G., Giannoccaro, P., Nobile, C. F., Sacco, A.: Inorg. Chim. Acta *48*, 125 (1981)
387. Mague, J. T., Davis, E. T.: Inorg. Chem. *16*, 131 (1977)
388. Levinson, J. J., Robinson, S. D.: J. Chem. Soc. A *1971*, 762
389. Brown, C. K., Georgiou, D., Wilkinson, G.: J. Chem. Soc. A *1971*, 3120
390. Reed, C. A., Roper, W. R.: J. Chem. Soc. Chem. Commun. *1971*, 1556
391. Stiddard, M. H. B., Townsend, R. E.: J. Chem. Soc. A *1970*, 2719
392. Nash, D. H., Harris, R. O.: Can. J. Chem. *49*, 3821 (1971)
393. Valentine, J., Valentine, D., Collman, J. P.: Inorg. Chem. *10*, 219 (1971)
394. Cook, C. D., Jauhal, G. S.: J. Amer. Chem. Soc. *89*, 3066 (1967)
395. Mason, M. G., Ibers, J. A.: ibid. *104*, 5153 (1982)
396. Nyman, C. J., Wymore, C. J., Wilkinson, G.: J. Chem. Soc. A *1968*, 561
397. Collman, J. P., Kobuta, H., Hosking, J.: J. Amer. Chem. Soc. *89*, 4809 (1967)
398. Phillips, D. A., Kubota, M., Thomas, J.: Inorg. Chem. *15*, 118 (1978)
399. Paiaro, G., Pandolfo, L.: Angew. Chem. Int. Ed. Engl. *20*, 288 (1981)
400. Nakamura, A., Tatsuno, Y., Otsuka, S.: Inorg. Chem. *11*, 2053 (1972)
401. Ugo, R.: Engelhard Ind. Tech. Bull. *11*, 45 (1970)
402. Sheldon, R. A., van Doorn, J. A.: J. Organomet. Chem. *94*, 115 (1975)
403. Clark, H. C., Goel, A. B., Wong, C. S.: J. Amer. Chem. Soc. *100*, 6241 (1978)
404. Lyons, J. E., Turner, J. O.: J. Org. Chem. *37*, 2881 (1972);
     Tetrahedron Lett. *1972*, 2903
405. Mercer, G. D., Beaulieu, W. B., Roundhill, D. M.: J. Amer. Chem. Soc. *99*, 6551 (1977)
406. Mimoun, H., Sérée de Roch, I., Sajus, L.: Tetrahedron *26*, 37 (1970)

407. Arakawa, H., Moro-oka, Y., Ozaki, A.: Bull. Chem. Soc. Japan 47, 2958 (1974)
408. Sharpless, K. B., Townsend, J. M., Williams, D. R.: J. Amer. Chem. Soc. 94, 295 (1972)
409. Kagan, H. B., Mimoun, H., Mark, C., Schurig, V.: Angew. Chem. Int. Ed. Engl. 18, 485 (1979)
410. van Gaal, H., Cuppers, H. G. A. M., van der Ent, A.: J. Chem. Soc. Chem. Comm. 1970, 1694;
     van der Ent, A., Onderderlinden, A. L.: Inorg. Chim. Acta 7, 204 (1973)
411. Simandi, L. I., Zahonyi-Budo, E., Bodnar, J.: Inorg. Chim. Acta 65, L181 (1982)
412. Westland, A. D., Haque, F., Bouchard, J.-M.: Inorg. Chem. 19, 2255 (1980)
413. Vedejs, E.: J. Amer. Chem. Soc. 96, 5944 (1974)
414. Jacobson, S. E., Tang, R., Mares, F.: J. Chem. Soc. Chem. Comm. 1978, 888
415. Jacobson, S. E., Muccigrosso, D. A., Mares, F.: J. Org. Chem. 44, 921 (1979)
416. Chong, A. O., Sharpless, K. B.: ibid. 42, 1587 (1977)
417. Dudley, C. W., Read, G., Walker, P. J. C.: J. Chem. Soc. (Dalton) 1974, 1926;
     Read, G., Walker, P. J. C.: ibid. 1977, 883
418. Tang, R., Mares, F., Neary, N., Smith, D. E.: J. Chem. Soc. Chem. Comm. 1979, 274
419. Farrar, J., Holland, D., Milner, D. J.: J. Chem. Soc. (Dalton) 1975, 815
420. Holland, D., Milner, D. J.: ibid. 1975, 2440
421. Igersheim, F., Mimoun, H.: J. Chem. Soc. Chem. Comm. 1978, 559
422. Igersheim, F., Mimoun, H.: Nouv. J. Chimie 4, 161 (1980)
423. Mimoun, H., Perez Machirant, M. M., Sérée de Roch, I.: J. Amer. Chem. Soc. 100, 5437 (1978)
424. Broadhurst, M. J., Brown, J. M., John, R. A.: Angew. Chem. Suppl. 1983, 1
425. Otsuka, S., Nakamura, A., Tatsumo, Y.: J. Amer. Chem. Soc. 91, 6994 (1969)
426. Cook, C. D., Jauhal, G. S.: Inorg. Nucl. Chem. Lett. 3, 31 (1967)
427. Clark, H. C., Goel, A. B., Wong, C. S.: J. Organomet. Chem. 152, C45 (1978)
428. Teo, B.-K., Ginsberg, A. P., Calabrese, J. C.: J. Amer. Chem. Soc. 98, 3027 (1976)
429. Norman, J. G., Ryan, P. B.: Inorg. Chem. 21, 3555 (1982)
430. Boreham, C. J., Latour, J.-M., Marchon, J. C., Boisselier-Cocolios, B., Guilard, R.: Inorg. Chim. Acta 45, L69 (1980)
431. Ledon, H., Bonnet, M., Lallemand, J. Y.: J. Chem. Soc. Chem. Comm. 1980, 702
432. Geoffroy, G. L., Hammond, G. S., Gray, H. B.: J. Amer. Chem. Soc. 97, 3933 (1975)
433. Deal, D., Zink, J. I.: Inorg. Chem. 20, 3995 (1981)
434. Vogler, A., Kunkely, H.: J. Amer. Chem. Soc. 103, 6222 (1981)
435. Mäcke, H., Zehnder, M., Thewalt, U., Fallab, S.: Helv. Chim. Acta 62, 1804 (1979)
436. Zehnder, M., Thewalt, U., Fallab, S.: ibid. 62, 2099 (1979)
437. (a) Simplicio, J., Wilkins, R. G.: J. Amer. Chem. Soc. 89, 6092 (1967);
     (b) ibid. 91, 1325 (1969);
     (c) Miller, F., Simplicio, J., Wilkins, R. G.: ibid. 91, 1962 (1969)
438. (a) Mäcke, H.: Helv. Chim. Acta 64, 1579 (1981)
     (b) Mäcke H., Exnar, I.: ibid. 60, 2504 (1977)
439. Bosnich, B., Poon, C. K., Tobe, M. L.: Inorg. Chem. 5, 1514 (1966)
440. Wong, C. L., Endicott, J. F.: ibid. 20, 2233 (1981)
441. Shibahara, T., Mori, M.: Bull. Chem. Soc. Japan 45, 1433 (1972)
442. Sasaki, Y., Suzuki, K. Z., Matsumoto, A., Saito, K.: Inorg. Chem. 21, 1825 (1982)
443. Fallab, S., Hunold, H. P., Maeder, M., Mitchell, P. R.: J. Chem. Soc. Chem. Comm. 1981, 469
444. Al-Shatti, N., Ferrer, M., Sykes, A. G.: J. Chem. Soc. (Dalton) 1980, 2533
445. Fallab, S.: Chimia 24, 76 (1970)
446. Zehnder, M., Mäcke, H., Fallab, S.: Helv. Chim. Acta 58, 2306 (1975)
447. Brüstlein-Banks, P., Fallab, S.: ibid. 60, 1601 (1977)
448. Ferrer, M., Hand, T. D., Sykes, A. G.: J. Chem. Soc. (Dalton) 1980, 14
449. Hoffman, A. B., Taube, H.: Inorg. Chem. 7, 1971 (1968)
450. Mori, M., Weil, J. A., Ishiguro, M.: J. Amer. Chem. Soc. 90, 615 (1968)
451. Mori, M., Weil, J. A.: ibid. 89, 3732 (1967)
452. Bosnich, B., Jackson, W. G., Lo, S. T. D., McLaren, J. W.: Inorg. Chem. 13, 2605 (1974)
453. Shibahara, T., Mori, M., Matsumoto, K., Ooi, S.: Bull. Chem. Soc. Japan 54, 433 (1981)

454. Sykes, A. G.: Chemistry in Britain *10*, 170 (1974)
455. McLendon, G. L., Mooney, W. F.: Inorg. Chem. *19*, 12 (1980)
456. Harris, W. R., McLendon, G. L., Martell, A. E., Bess, R. C., Mason, M.: ibid. *19*, 21 (1980)
457. Chandrasekaran, K., Natarajan, P.: J. Chem. Soc. (Dalton) *1981*, 478
458. Richens, D. T., Sykes, A. G.: ibid. *1982*, 1621
459. (a) Davies, K. M., Sykes, A. G.: J. Chem. Soc. A *1971*, 1414;
     (b) Davies, R., Sykes, A. G.: ibid. *1968*, 2831
460. Hyde, M. R., Sykes, A. G.: J. Chem. Soc. (Dalton) *1974*, 1550
461. Sasaki, Y., Kawamura, R.: Bull. Chem. Soc. Japan *54*, 3379 (1981)
462. Hand, T. D., Hyde, M. R., Sykes, A. G.: Inorg. Chem. *14*, 1720 (1975)
463. Chandrasekaran, K., Natarajan, P.: J. Chem. Soc. Chem. Comm. *1977*, 774;
     Inorg. Chem. *19*, 1714 (1980)
464. Sasaki, Y., Ueno, F. B., Saito, K.: J. Chem. Soc. Chem. Comm. *1981*, 1135
465. Kondo, S., Sasaki, Y., Saito, K.: Inorg. Chem. *20*, 429 (1981)
466. Thompson, G. A. K., Sykes, A. G.: ibid. *15*, 639 (1976)
467. Natarajan, P., Raghavan, N. V.: J. Chem. Soc. Chem. Comm. *1980*, 268
468. Davies, K. M., Sykes, A. G.: J. Chem. Soc. A *1971*, 1418
469. Davies, R., Sykes, A. G.: ibid. *1968*, 2237
470. Yang, C. H., Keeton, D. P., Sykes, A. G.: J. Chem. Soc. (Dalton) *1974*, 1089
471. Davies, R., Hagopian, A. K. E., Sykes, A. G.: J. Chem. Soc. A *1969*, 623
472. Garbett, K., Gillard, R. D.: ibid. *1968*, 1725
473. Edwards, J. D., Yang, C. H., Sykes, A. G.: J. Chem. Soc. (Dalton) *1974*, 1561
474. Sykes, A. G., Mast, R. D.: J. Chem. Soc. A *1967*, 784
475. Green, M., Sykes, A. G.: ibid. *1970*, 3209
476. Natarajan, P., Raghavan, N. V.: J. Amer. Chem. Soc. *102*, 4518 (1980)
477. Natarajan, P., Raghavan, N. V.: J. Phys. Chem. *85*, 188 (1981)
478. Hill, H. A. O., Thornalley, P. J.: Inorg. Chim. Acta *67*, L35 (1982)
479. (a) Gillard, R. D., Phipps, D. A.: J. Chem. Soc. A *1971*, 1074;
     (b) Gillard, R. D., Spencer, A.: ibid. *1968*, 2718
480. Shibahara, T., Kuroya, H., Mori, M.: Bull. Chem. Soc. Japan *53*, 2834 (1980)
481. Haim, A., Wilmarth, W. K.: J. Amer. Chem. Soc. *83*, 509 (1961)
482. Bosnich, B., Poon, C. K., Tobe, M. L.: Inorg. Chem. *5*, 1514 (1966)
483. Pignatello, J. J., Jensen, F. R.: J. Amer. Chem. Soc. *101*, 5929 (1979)
484. Zehnder, M., Fallab, S.: Helv. Chim. Acta *58*, 2812 (1975)
485. Davies, R., Sykes, A. G.: J. Chem. Soc. A *1968*, 2237
486. Davies, R., Sykes, A. G.: ibid. *1968*, 2840
487. McLendon, G., Martell, A. E.: Inorg. Chem. *15*, 2662 (1976)
488. Harris, W. R., McLendon, G. L., Martell, A. E.: J. Amer. Chem. Soc. *98*, 8378 (1976)
489. Harris, W. R., Martell, A. E.: J. Mol. Catal. *7*, 99 (1980);
     J. Coord. Chem. *10*, 107 (1980)
490. Bedell, S. A., Martell, A. E.: Inorg. Chem. *22*, 364 (1983)
491. Valentine, J. S., Valentine, D. S.: J. Amer. Chem. Soc. *93*, 1111 (1971)
492. Valentine, J. S., Valentine, D. S.: Inorg. Chem. *10*, 393 (1971)
493. Miskowski, V. M., Robbins, J. L., Treitel, I. M., Gray, H. B.: ibid. *14*, 2318 (1975)
494. Hoshino, M., Nakajima, M., Takakubo, M., Imamura, M.: J. Phys. Chem. *86*, 221 (1982)
495. Latos-Grazynski, L., Cheng, R.-J., La Mar, G. N., Balch, A. L.: J. Amer. Chem. Soc. *104*,
     5992 (1982)
496. Chin, D.-H., La Mar, G. N., Balch, A. L.: ibid. *102*, 5945 (1980)
497. Sugimoto, R., Eikawa, H., Suzuki, H., Moro-oka, Y., Ikawa, T.: Bull. Chem. Soc. Japan *54*,
     2849 (1981)
498. Tatsumi, K., Hoffmann, R.: J. Amer. Chem. Soc. *103*, 3328 (1981)
499. Stomberg, R.: Acta Chem. Scand. A *36*, 423 (1982)
500. Mimoun, H., Saussine, L., Daire, E., Postel, M., Fischer, J., Weiss, R.: J. Amer. Chem. Soc.
     105, 3101 (1983)
501. Chaumette, P., Mimoun, H., Saussine, L., Fischer,J., Mitschler, A.: J. Organometal. Chem.
     250, 291 (1983)
502. Suzuki, M., Ueda, I., Kanatomi, H., Murase, I.: Chem. Lett. *1983* 185

503. Burkett, H. D., Newberry, V. F., Hill, W. E., Worley, S. D.: J. Amer. Chem. Soc. *105*, 4097 (1983)
504. Elam, W. T., Stern, E. A., McCallum, J. D., Sanders-Loehr, J.: J. Amer. Chem. Soc. *104*, 6369 (1982)
505. Taqui-Khan, M. M.: Pure Appl. Chem. *55*, 159 (1983)
506. Kasuga, K., Nagahara, T., Tsuge, A., Sogabe, K., Yakamoto, Y.: Bull. Chem. Soc. Jpn *56*, 95 (1983)
507. Braydich, M. D., Fortman, J. J., Cummings,S. C.: Inorg. Chem. *22*, 484 (1983)
508. Machida, R., Kimura, E., Kodama, M.: Inorg. Chem. *22*, 2055 (1983)
509. Doppelt, P., Weiss, R.: Nouveau J. Chimie *7*, 341 (1983)
510. S. Fallab, personal communication
511. Kimura, E., Sakonaka, A., Machida, R., Kodama, M.: J. Amer. Chem. Soc. *104*, 4255 (1982)
512. Ozawa, T., Hanaki, A.: Inorg. Chim. Acta *80*, 33 (1983)
513. Westland, A. D., Tarafder, M. T. H.: Inorg. Chem. *21*, 3228 (1982)
514. Sabelli, N. H., Melendres, C. A.: J. Phys. Chem. *86*, 4342 (1982)
515. Purcell, K. F.: J. Organometal. Chem. *252*, 181 (1983)
516. Nozawa, T., Hatano, M., Nagashima, U., Obara, S., Kashiwagi, H.: Bull. Chem. Soc. Jpn *56*, 1721 (1983)
517. Boca, R.: Coord. Chem. Rev. *50*, 1 (1983)
518. Kitagawa, T., Ondrias, M. R., Rousseau, D. L., Ikeda-Saito, M., Yonetani, T.: Nature *298*, 869 (1982)
519. Bajdor, K., Nakamoto, K.: J. Amer. Chem. Soc. *105*, 678 (1983)
520. Jörin, E., Schweiger, A., Günthard, Hs. H.: J. Amer. Chem. Soc. *105*, 4277 (1983)
521. Au-Yeung, S. C. F., Eaton, D. R.: Inorg. Chim. Acta *76*, L141 (1983)
522. Postel, M., Brevard, C., Arzoumanian, H., Riess, J. G.: J. Amer. Chem. Soc. *105*, 4922 (1983)
523. Carlton, L., Read, G., Urgelles, M.: J. Chem. Soc. Chem. Commun. *1983*, 586
524. Atlay, M. T., Preece, M., Strukul, G., James, B. R.: Can. J. Chem. *61*, 1332 (1983)
525. Geiger, T. C., Anson, F. C.: J. Amer. Chem. Soc. *103*, 7489 (1981)
526. Durand, R. C., Bencosme, C. S., Collman, J. P., Anson, F. C. : J. Amer. Chem. Soc. *105*, 2710 (1983)
527. Martell, A. E.: Pure Appl. Chem. *55*, 125 (1983)
528. Nishinaga, A., Ohara, H., Tomita, H., Matsuura, T.: Tetahedron Lett. *24*, 213 (1983)
529. Grant, R. B., Lambert, R. M.: J. Chem. Soc. Chem. Commun. *1983*, 662

The Role of Vibronic Coupling in the Interpretation of Spectroscopic and Structural Properties of Biomolecules

M. Bacci

Istituto di Ricerca sulle Onde Elettromagnetiche Firenze, Italy

The aim of the present article is to review several biomolecules, the spectroscopic and structural properties of which have been so far interpreted in the light of the Jahn-Teller effect. Indeed the importance of the vibronic coupling in the interpretation of physical properties of molecules and crystals is widely recognized now, but only a desultory attention has been paid to the role played by such a coupling in biological systems. After a brief outline of the Jahn-Teller effect, the whole problem is critically re-examined separating the systems for which the experimental evidence of the Jahn-Teller effect is clear, from those where its effectiveness is still debated. Furthermore some suggestions are given for a proper inclusion of the vibronic coupling in future analysis on biomolecules.

Structure and Bonding 55
© Springer-Verlag Berlin Heidelberg 1983

List of Symbols and Abbreviations

JT	Jahn-Teller	ODMR	Optically detected magnetic resonance
APES	Adiabatic potential energy surface		
ESR	Electron spin resonance	MIDP	Microwave induced delayed phosphorescence
MCD	Magnetic circular dichroism		
LCAO	Linear combination of atomic orbitals	NiEtio	Etioporphyrin Ni(II)
		CrTPPCl	Tetraphenylporphyrin Cr(III) chloride
MO	Molecular orbital		
AOM	Angular overlap model	CuTPP	Tetraphenylporphyrin Cu(II)
Hb	Hemoglobin	HOMO	Highest occupied molecular orbital
Mb	Myoglobin	HP	High potential iron protein
MeP	Metalloporphyrin	WKB	Wentzel-Kramers-Brillouin

1 Introduction

The term *vibronic coupling* is widely spread in literature nowadays, although it is rather vague and sometimes misleading. The adjective *vibronic* derives from the contraction of two adjectives, *vibrational* and *electronic*, and, in its more common meaning, indicates a dynamic interaction between vibrational and electronic motions. Another frequently used definition of vibronic coupling is "the breakdown of the Born-Oppenheimer approximation". It is outside the scope of the present article to give an accurate treatment of the different approaches and approximations inherent to the vibronic coupling problem, so that here only some fundamental concepts will be given to avoid any misunderstanding. The interested reader is referred to more extensive reviews on the argument[1-3].

The total Hamiltonian for a molecular system can be written as

$$H(r, Q) = T_{el}(r) + T_N(Q) + V(Q) + V(r, Q) \tag{1}$$

where r stands for the electronic coordinates and Q stands for the normal nuclear coordinates; $T_{el}(r)$ and $T_N(Q)$ are the electronic and nuclear kinetic energy operators, respectively, while $V(Q)$ is the potential energy of the nuclei and $V(r, Q)$ is the electron-nuclei interaction operator.

Since the Hamiltonian (1) does not allow a separation of variables, the Schrödinger equation

$$H(r, Q)\,\Psi(r, Q) = E\,\Psi(r, Q) \tag{2}$$

cannot be solved exactly. Usually the terms containing the electronic coordinate r are separated and the electronic part of Eq. (2) is solved assuming the nuclear coordinates Q as parameters:

$$[T_{el}(r) + V(r, Q)]\,\phi_i(r, Q) = \varepsilon_i(Q)\,\phi_i(r, Q) \tag{3}$$

Then, for a non-degenerate electronic term, the complete wavefunction (*vibronic state*) is expressed as a product of the electronic and nuclear wavefunctions

$$\Psi(r, Q) = \phi_i(r, Q)\,\chi_i(Q) \tag{4}$$

where $\chi_i(Q)$ is obtained from the following equation

$$[T_N(Q) + V(Q) + \varepsilon_i(Q)]\,\chi_i(Q) = E_i\,\chi_i(Q) \tag{5}$$

Equation (5) is only approximately correct, because it implies that the electronic wavefunction $\phi_i(r, Q)$ is insensitive to small nuclear displacements, i.e. the two integrals

$$\langle \phi_i(r, Q)\,|T_N(Q)|\,\phi_j(r, Q)\rangle \qquad \text{and}$$

$$\left\langle \phi_i(r, Q) \left| \frac{\partial}{\partial Q_n} \right| \phi_j(r, Q) \right\rangle$$

are vanishing. This is the so-called *Born-Oppenheimer adiabatic approximation*[2].

For degenerate (or quasi-degenerate) electronic states, such an approximation becomes invalid, because the kinetic energy terms may appreciably mix these states, and, instead of Eq. (4), the full wavefunction can be expressed as a superposition of single Born-Oppenheimer products

$$\Psi(r, Q) \quad = \sum_{i=1}^{f} \phi_i(r, Q) \chi_i(Q) \tag{6}$$

In Eq. (6) f is the degree of degeneracy and the vibrational wavefunctions $\chi_i(Q)$ obey a system of coupled equations

$$[T_N(Q) + V(Q) + \varepsilon_i(Q)] \chi_i(Q) + \sum_{\substack{j=1 \\ j \neq i}}^{f} A_{ij} \chi_j(Q) = E_i \chi_i(Q) \tag{7}$$

where A_{ij} are the electronic non-diagonal matrix elements of the Hamiltonian[3].

The Jahn-Teller (JT) effect[4] represents a classical example of vibronic admixture and, in the present review, we shall mainly deal with such an aspect, while the aspects connected with the wide field of tunnel effects will be treated only marginally.

Since the first formulation of the JT effect in 1937[4], many years elapsed before its importance in physics and chemistry was fully realized and much more it had to be waited for applications to biomolecules[5]. In fact, to our knowledge, the first interpretation of physical properties of a biological molecule in the light of the JT effect is due to Kamimura and Mizuashi in 1968[6], who attempted to explain the large anisotropy of g factors of the low spin ferrihemoglobin azide.

The main purpose of the present article is to review the experimental results, obtained on molecules of biological interest, which have been hitherto interpreted with the help of the JT effect. Though the review is not intended to be comprehensive, the author will try to cover as far as possible the principal problems.

2 Generalities on the Jahn-Teller Effect

The JT effect is based on the following theorem[4, 7]: If a non-linear molecule (or polyatomic ion) has a degenerate electronic level (apart from Kramers degeneracy) it is unstable with respect to displacements of the atoms.

The original theorem was limited to molecules, but in 1939 Van Vleck, who had been acquainted with this theorem at the American Physical Society meeting in April 1936[8], extended it to ions in crystals[9]. Really no mathematical proof was given by Jahn and Teller, since they limited themselves to present a proof by exhaustion enumerating every point group symmetry and verifying the validity of the theorem for different molecular structures: in every case there was always a non-totally symmetrical normal vibration suitable to distort the molecule. Considerably later satisfactory general proofs have been given[10, 11] by an analysis in multi-dimensional spaces and only some non-physically important violations of the theorem were found[11].

In their papers Jahn and Teller neglected the kinetic energy of the nuclei so giving a "static" picture of the problem, whereby equivalent stable distorted structures, belonging to the sub-groups of the original point group, were obtained owing to the generalized JT force acting on the nuclei in the undistorted configuration. In the stable minima the electronic degeneracy is removed, but, if a jumping among the different equivalent minima is allowed so that the starting symmetry is re-obtained, a *vibronic* degeneracy substitutes the *electronic* degeneracy. Indeed, the JT effect *cannot* reduce the overall degeneracy of a level, since the overall Hamiltonian (1) remains totally symmetric under the operations of the point group of the system[12].

Let us examine now the main theoretical aspects and experimental manifestations of the JT effect in order to fully realize the importance and also the limits in applying such an effect to the interpretation of physical properties of biomolecules.

2.1 The Linear Jahn-Teller Effect

Let us expand the potential $V(r, Q)$ of Eq. (1) in a series of small displacements from the undistorted configuration ($Q_\gamma^\Gamma = 0$):

$$V(r, Q) = V_0(r) + \sum_{\Gamma_\gamma} \left(\frac{\partial V}{\partial Q_\gamma^\Gamma}\right)_0 Q_\gamma^\Gamma + \frac{1}{2} \sum_{\Gamma_\gamma'} \sum_{\Gamma_\gamma''} \left(\frac{\partial^2 V}{\partial Q_\gamma^{\Gamma'} \partial Q_\gamma^{\Gamma''}}\right)_0 Q_\gamma^{\Gamma'} Q_\gamma^{\Gamma''} + \ldots \qquad (8)$$

where $V_0(r)$ is the static crystal field and γ is the component of the irreducible representation Γ.

If the relation:

$$\left\langle \phi_i \left| \left(\frac{\partial V}{\partial Q_\gamma^\Gamma}\right)_0 \right| \phi_i \right\rangle \neq 0 \qquad (9)$$

is satisfied, the system is subject to a force, which is the analog of the classical force $F_{\Gamma_\gamma} = -\partial V/\partial Q_\gamma^\Gamma$, and will be distorted (Fig. 1).

Since $V(r, Q)$ is totally symmetric, the derivative $\partial V/\partial Q_\gamma^\Gamma$ must transform in the same way as Q_γ^Γ. Therefore it is easily seen by elementary group theory that the relationship (9) is valid if the irreducible representation Γ is contained in the symmetric square $[\Gamma_i^2]_{sym}$, where Γ_i is the representation corresponding to the electronic level ϕ_i. For non-degenerate levels $[\Gamma_i^2] = A_1$ holds and no lowering of symmetry occurs. The situation is different for degenerate levels, however, because the system will slip in equivalent minima with a lower symmetry, which is determined by the JT active vibrations[1].

A good visualization of the problem is obtained by the adiabatic potential energy surfaces (APES's), whose shape is determined by the Hamiltonian

$$H_{e-l} = V(r, Q) + V(Q) \qquad (10)$$

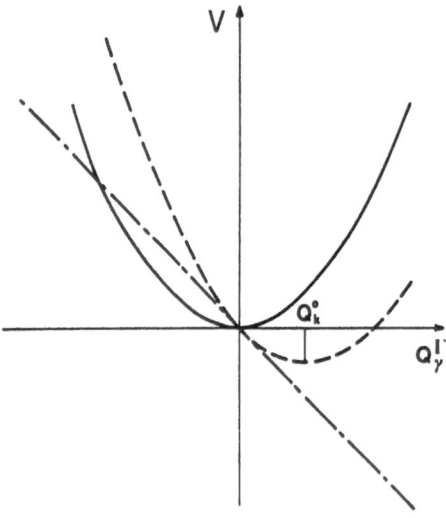

Fig. 1. Influence of linear terms of the electron-nuclei interaction operator $V(r, Q)$ $(-\cdot-\cdot)$ on the equilibrium configuration. In the harmonic approximation, the starting parabola (---) is shifted $(---)$ and the new minimum Q_k^0 is obtained. The constant crystal field $V_0(r)$ is neglected

According to the Hellmann-Feynman theorem[13, 14], the energies (ε_k^0) and the coordinates (Q_k^0) of the stationary points are given by the solutions of the following system:

$$\left\langle \phi \left| \frac{\partial V(r, Q)}{\partial Q_\gamma^\Gamma} \right| \phi \right\rangle + \frac{\partial V(Q)}{\partial Q_\gamma^\Gamma} = 0 \tag{11}$$

Then the stationary points obtained are minima if arbitrary increments q_k to the coordinates Q_k^0 cause a positive-definite increment of the energy ε_k^0:

$$U(q_k) = \varepsilon(Q_k^0 + q_k) - \varepsilon(Q_k^0) > 0 \tag{12}$$

The most extensively studied symmetry is the cubic one (O_h, T_d), where doubly and triply degenerate electronic levels (E and T_1 or T_2, respectively) occur. A doubly degenerate level E can be coupled by vibrational modes ε (stretching in O_h and bending in T_d symmetry)[1] and the resulting APES has rotational symmetry about the origin, the so-called "Mexican hat", with a continuum of equivalent minima[9, 12, 13, 15, 16]. An analogous situation is observed for the E $\otimes$ ε coupling in systems with trigonal symmetry.

   The problem for a triply degenerate level T_1 (or T_2) in O_h symmetry is more complicate because both tetragonal (ε) and trigonal (τ_2) vibrational modes are JT active. Öpik and Pryce showed[14] that in the linear approximation 3 tetragonal minima (D_{4h} symmetry) *or* 4 trigonal minima (D_{3d} symmetry) are stabilized, while 6 orthorhombic points (D_{2h} symmetry) are saddle points. Similar results are valid for T_d symmetry, though two τ_2 vibrations exist for tetrahedral entities MX_4. Commonly only the "softer" mode, which is essentially of the bending type, is taken into account, because for similar coupling constants it will give the larger JT stabilization energy.

---

1 Roman letters are used for representations of electronic states and greek letters for vibrational modes

Systems with only one four-fold symmetry axis C_4 can have at most a doubly-degenerate level E. In contrast to molecular geometries with trigonal symmetry, two modes (β_1 and β_2) are JT active in this case leading to rhombic or rectangular distortions[17-20].

If spin-orbit coupling is included, the shape of the APES's may be strongly affected, with the exception of ^2E states in cubic symmetry, where spin-orbit coupling has no influence in first-order. As a general rule, the spin-orbit interaction tends to cancel the JT effect shifting the minima towards the undistorted configuration[14].

Finally, it has to be recalled that internal strains or crystal packing forces have to be present, in order to stabilize one of several equivalent minima so that the system is "seen" as statically distorted. It has been stated that even small low-symmetry perturbations can be amplified by vibronic coupling[3]. A mechanism of this kind may be more effective in lifting electronic degeneracies than low-symmetry ligand field components. The important point is whether JT contributions are greater or comparable with the low-symmetry ones: if it is so, JT terms must be included in to the Hamiltonian and have to be considered before or together with the static low-symmetry crystal field.

2.2 Second Order Interactions

With the term "second order interactions" we mean two different kinds of interactions, which are profitably summarized by the relationships (13) and (14):

$$\left\langle \phi_i \left| \left(\frac{\partial V}{\partial Q_\gamma^\Gamma} \right)_0 \right| \phi_j \right\rangle \neq 0 ; \qquad i \neq j \tag{13}$$

$$\left\langle \phi_i \left| \left(\frac{\partial^2 V}{\partial Q_\gamma^\Gamma \partial Q_\gamma^\Gamma} \right)_0 \right| \phi_i \right\rangle \neq 0 ; \tag{14}$$

Relationship (13) corresponds to the pseudo-JT effect, that is a coupling between two different electronic levels by vibrations of suitable symmetry. Both levels may be non-degenerate.

In the absence of linear JT coupling, second order perturbation theory yields the energy of the ground state:

$$\varepsilon_i = \varepsilon_i^0 + \frac{1}{2} KQ^2 + Q^2 \sum_k \frac{\left| \left\langle \phi_i \left| \left(\frac{\partial V}{\partial Q} \right)_0 \right| \phi_k \right\rangle \right|^2}{\varepsilon_i^0 - \varepsilon_k^0} \tag{15}$$

Only one active coordinate Q is considered. K is the classical force constant and gives the restoring force due to the original electron distribution. It is always positive, whereas the third term in Eq. (15) is always negative, since $\varepsilon_k^0 > \varepsilon_i^0$. It is easily seen that the original configuration is unstable when

$$\left| \sum_k \frac{\left| \left\langle \phi_i \left| \left(\frac{\partial V}{\partial Q} \right)_0 \right| \phi_k \right\rangle \right|^2}{\varepsilon_i^0 - \varepsilon_k^0} \right| > \frac{1}{2} K \tag{16}$$

Rather curiously physicists have not paid much attention to the pseudo-JT effect (see, however, Englman[1] and Öpik and Pryce[14]), which has been widely applied, however, by chemists to interpret not only stereochemistry but also reaction mechanisms[21-25].

The pseudo-JT effect is often called also second-order JT effect. However, in order to avoid any misunderstanding, in this paper only the term pseudo-JT will be used. The term second-order JT effect will be reserved to describe second-order interactions within the same electronic level as in (Eq. 14). The quadratic JT effect is a particular case of second-order interactions, when $\Gamma' = \Gamma''$. Second-order contribution may strongly affect the shape of the APES's. Thus the "warping" of the "Mexican hat" leads to three equivalent tetragonal minima[1, 3, 12].

The influence of second-order JT terms on the APES's arising from the $T_{1(2)} \otimes (\varepsilon \times \tau_2)$ problem in cubic symmetry is perhaps more striking. In fact not only orthorhombic points may now become minima[26-29] but also different distortion symmetries (tetragonal + trigonal, tetragonal + orthorhombic, trigonal + orthorhombic, tetragonal + trigonal + orthorhombic) can coexist[28-30] (Fig. 2). Moreover the simultaneous presence of minima of different symmetry is not restricted to cubic symmetry. Even for tetragonal molecules rectangular and rhombic distortions can coexist, if the mixing of the normal vibrations β_1 and β_2 through totally symmetric vibrations is included[20].

2.3 Static and Dynamic Limit

In Sects. 2.1 and 2.2 we have shown that the vibronic coupling produces a displacement of the nuclei into equivalent potential wells. Indeed this is quite exact for isolated molecules or for an isotropic environment (gas, solution), while in crystals local forces may stabilize one of various distortions. On the other hand a fully *static* picture is true

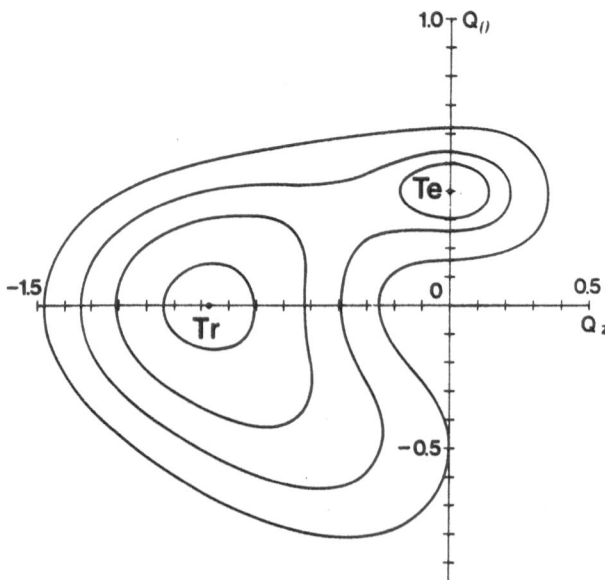

Fig. 2. Coexistence of tetragonal (Te) and trigonal (Tr) minima. Map in the Q_θ, Q_z ($\sqrt{3}\,Q_z = Q_\xi = Q_\eta = Q_\zeta$) for the $T_{1(2)} \times (\varepsilon + \tau_2)$ JT problem (adapted from Ref. 28)

only as a limiting case, since in every physical system, we may examine, there will be always terms which couples the different wells. In other words there is always a finite probability of tunneling among different distortions so that the real system is always dynamic.

A real system appears statically distorted or not depending on the time scale of the experimental method employed with respect to the transfer rate among the wells, so that, at the same temperature, a molecule may appear undistorted in a spectroscopic measurement, which is slow (i.e. ESR) or distorted, if it is fast.

The tunneling rate is affected by various parameters and may be suitably expressed in the semiclassical WKB formalism[19, 31, 32]

$$W = \left[1 - \exp\left(-\frac{\hbar\omega}{kT}\right)\right] \sum_n \nu_n D_n \exp\left(-n\frac{\hbar\omega}{kT}\right) \tag{17}$$

where ν_n is the vibrational frequency into the well for the n-th level and D_n, the transmission coefficient for the n-th level with energy E_n, is given by:

$$D_n = \left[1 + \exp\frac{2}{\hbar}\int_s \sqrt{2\mu(V(s) - E_n)}\, ds\right]^{-1} \tag{18}$$

Here the integral is computed along the trajectory connecting the minima; μ is the effective mass of the particle and $V(s)$ is the potential along the trajectory.

From Eq. (17) it is evident that temperature plays a fundamental role in determining the static or dynamic features of the system. A classical example is copper fluosilicate hexahydrate $CuSiF_6 \cdot 6\,H_2O$ diluted with the isomorphous zinc salt. An anisotropic ESR spectrum ($g_\parallel$ = 2.46, $g_\perp$ = 2.11) is obsered at 20 K and an isotropic one (g = 2.24) at 90 K[33–35].

For a long time stable distortions were thought to be the only evidence for a JT effect. However the papers by Moffitt et al.[36, 37] Longuett-Higgins et al.[38], Bersuker[39] and O'Brien[40] showed that, even for weak coupling, the vibronic level scheme was much different from the one obtained in an isolated electronic system.

A fundamental contribution was given successively by Ham[16, 41], who demonstrated the possibility of quenching off-diagonal electronic operators by dynamic JT coupling. The Ham effect can be visualized by the following considerations. Let us express the complete wavefunction of the system $\Phi(r, Q)$ as

$$\Phi(r, Q) = \sum_k c_k \Psi_k(r, Q^{(k)}) \tag{19}$$

where $\Psi_k(r, Q^{(k)})$ is the vibronic wavefunction in the k-th minimum:

$$\Psi_k(r, Q^{(k)}) = \phi_k(r, Q^{(k)})\, \Pi\chi_n(Q^{(k)}) \tag{20}$$

$\chi_n(Q^{(k)})$ is the vibrational wavefunction of the n-th state of the harmonic oscillator centered on the k-th minimum. If O is an electronic operator, which is independent of nuclear coordinates, the vibronic matrix element is given by the product of an electronic matrix element and an overlap integral between oscillator wavefunctions:

$$\langle \Psi_k' | O | \Psi_k'' \rangle = \langle \phi_k'(r, Q^{(k)}) | O | \phi_k''(r, Q^{(k)}) \rangle \langle \Pi\chi_n'(Q^{(k)}) | \Pi\chi_n''(Q^{(k)}) \rangle \tag{21}$$

The overlap integral $\langle \Pi\chi_n'(Q^{(k)}) | \Pi\chi_n''(Q^{(k)}) \rangle$ is always less than 1, because the oscillators are centered on different wells, and therefore the original electronic matrix element is reduced.

Spin-orbit coupling, g-factors, nuclear quadrupole splitting, hyperfine interaction are some of the magnitude affected by the Ham effect and one can say that Ham's papers have opened up the wide field of the different spectroscopic techniques to investigate the JT effect.

2.4 Spectroscopic and Structural Manifestations of the Jahn-Teller Effect

In 1939 Van Vleck said[42]: "It is a great merit of the JT effect that it disappears when not needed". Indeed for a long time, even after the first clear experimental evidence[33-35], the JT effect was very elusive and only since the late sixties a great bulk of experimental evidence was attained by structural or spectroscopic investigations.

Structural phase transitions induced by JT interactions (cooperative JT effect) are perhaps the most striking structural manifestations. Owing to the importance of such an effect in solid state physics and chemistry many review articles are available[1, 3, 19, 42-45]. We shall not discuss this subject further, because till now no example of cooperative JT effect has been reported in biomolecules.

Single JT molecules or clusters in a crystalline environment may be found distorted, when they are investigated by X-ray (or other) diffraction methods. In his monumental work Liehr[46] extensively dealt with JT instabilities and static distortions for many types of molecular structures. We have previously seen (Sect. 2.1) that one of the equivalent distorted configurations could be stabilized by any weak low symmetry perturbation[3, 47, 48]. In order to be sure that a given distortion is JT induced and not due to lattice forces a comparison with isomorphous compounds containing non-JT ions or an analysis of analogous compounds are necessary[42, 43, 49].

The spectroscopic behavior of a JT system is to a large extent governed by the quenching of electronic operators (Ham effect), which causes a shift of the absorption (or emission) bands and a modification of their shapes. Moreover the vibronic mixing of different electronic states can strongly affect relaxation processes, which also modify spectral band shapes.

ESR was the experimental technique, which first gave clear evidence for the JT effect and also showed a static or dynamic behavior of the same system at different temperatures[33, 35]. In fact the rather long characteristic time of measurement is of the right order of magnitude to reveal both static and dynamic JT effects in dependence on temperature.

Figure 3 is particularly useful to understand how the seeming "static" or "dynamic" features of a system are governed by temperature and/or experimental techniques: Cu^{2+} in elpasolite-type crystals, $Ba_2Zn_{1-x}Cu_xWO_6$, gives an anisotropic ESR spectrum at 4.2 K and an isotropic one at $T > 77$ K (Fig. 3a), whereas the electronic absorption spectrum (a "fast" technique) shows always anisotropic features[42] (Fig. 3b). If the relaxation time τ between different configurations is long compared with $(2\pi\Delta\nu)^{-1}$, where

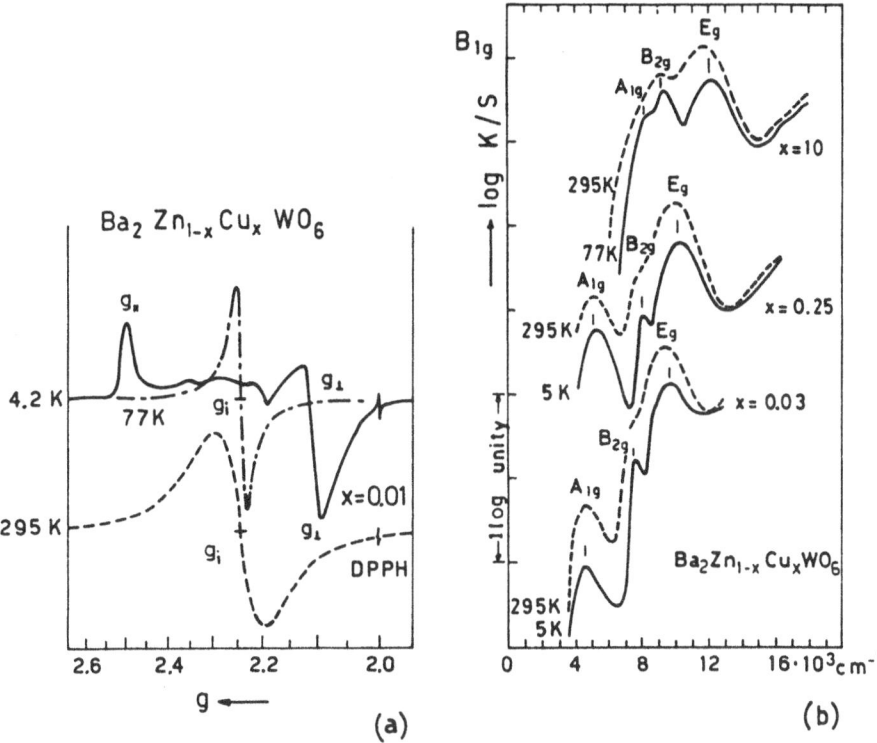

Fig. 3a, b. Static or dynamic JT effect as a function of temperature or spectroscopic technique. ESR (a) and ligand field (b) spectra of elpasolite-type mixed crystals $Ba_2Zn_{1-x}Cu_x WO_6$ at different temperatures (adapted from Ref. 42)

$\Delta \nu$ is the frequency difference between two anisotropic ESR resonance lines, the resulting spectrum is the superposition of the individual configurations. On the contrary, if $\tau \ll (2\pi\Delta\nu)^{-1}$, we have an isotropic spectrum; the resonance frequency is the average of the anisotropic components of the individual configurations[12]. As discussed in detail by Ham[12], "motional" narrowing can be produced by three relaxation mechanisms, which are characterized by a different temperature dependence: an Arrhenius-type dependence $(\tau^{-1} = \nu_0 e^{-E/kT})$ for an Orbach process, and a linear dependence or proportional to T^5 for direct and Raman processes, respectively. Therefore, the temperature dependence of the isotropic spectrum gives information about the relaxation mechanism and consequently on the vibronic level scheme.

Electronic absorption spectroscopy is also able to give useful informations about dynamic JT effects through both the quenching of off-diagonal operators, like spin-orbit coupling, and the relative intensities of vibronic transitions. A nice example of such effects is given by $KMgF_3 : V^{2+}$, which was firstly investigated by Sturge[50]. The zerophonon line, in the transition $^4A_{2g} \rightarrow {}^4T_{2g}$, is split by spin-orbit interaction into four components, whose separation is strongly reduced with respect to a computation based on the free-ion spin-orbit constant[59]. Moreover the relative intensities are not the ones expected following static crystal field theory, since the high-energy components are less

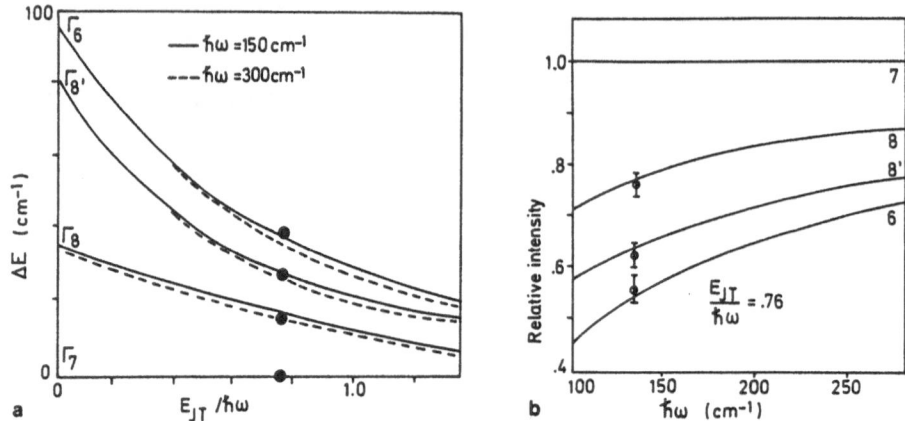

Fig. 4a, b. Dynamic JT effect in KMgF$_3$: V^{2+}. **a** Zero-phonon energy levels ($^4A_{2g} \rightarrow {}^4T_{2g}$) as a function of $E_{JT}/\hbar\omega$ for two values of $\hbar\omega$. Energy zero is referred to the lowest level Γ_7. The *dots* give the experimental splittings. **b** Computed (degeneracy-normalized) relative intensities of the components of the $^4A_{2g} \rightarrow {}^4T_{2g}$ zero-phonon line as a function of $\hbar\omega$ for $E_{JT}/\hbar\omega = 0.76$. *Circles* indicate experimental intensities (from Ref. 51)

intense. Both observations are in accord with a dynamical treatment of the $^4T_2 \otimes \varepsilon$ JT problem including spin-orbit coupling[51]: not only the spin-orbit splitting is quenched (Fig. 4a), but such a quenching is consistent with the intensity borrowing from the high-energy components (Fig. 4b).

Infrared and Raman spectra are also sensitive to the JT effect, which induces a breakdown of the usual selection rules of infrared absorption[52] or modifies the Raman excitation profiles[53-56].

Good information about the JT effect can be obtained by other experimental techniques too, like Zeeman experiments, magnetic circular dichroism (MCD)[57-59], optically detected ESR and luminescence. These two latter are particularly suitable to investigate JT systems in their excited states and it was just by these two methods that coexisting distortions of different symmetry were revealed[60, 61].

Finally, I wish to recall that the vibronic mixing of thermally accessible electronic levels can affect to a large extent the temperature dependence of magnetic susceptibility[62-64] and of Mössbauer spectra[3, 65], as it will be better seen in Sect. 3.

3 Jahn-Teller Effect and Vibronic Coupling in Biomolecules: Experimental Findings and Theoretical Interpretation

The term *biomolecules* is nowadays widely used and misused, particularly since researchers have used the label "bio", "life", "energy" and so on to obtain financial support from government agencies. Quite often any molecule which could be of biological interest is defined as a biomolecule, so that even water or oxygen may be considered as biomolecules in such a meaning! However, the present paper is restricted to molecules, which really exist in living system or which are closely related to these by structural similarities or reaction mechanism.

In his book[1], published in 1972, Englman quoted only one example[6] of a biomolecule, ferrihemoglobine azide, where the JT coupling was introduced to account for the large anisotropy in the ESR spectra. Since then many other experimental results have been interpreted invoking the JT effect (Table 1). It seems worth while to examine, to what extent the JT effect is actually effective in the considered systems or instead, if it represents a plausible explanation of findings, which can hardly be understood otherwise.

Indeed there is some reluctance in applying JT effect in biomolecules, since one thinks that this effect can be safely neglected in those systems, where low-symmetry fields formally remove the electronic degeneracy. However we have seen in Sect. 2.2 that the pseudo-JT effect can couple also non-degenerate electronic levels and therefore symmetry restrictions may not be of essential importance; furthermore the vibronic mixing of close-lying states may induce large changes of physical properties.

It is useful to consider the contributions to the total Hamiltonian:

$$H = H_{EE} + H_{LF} + H_{SO} + H_{LS} + H_{JT} \tag{22}$$

[H_{EE}, H_{LF}, H_{SO}, H_{LS} and H_{JT}: electron-electron, ligand-field, spin-orbit, low-symmetry, and JT interactions, respectively] with respect to their energetic importance (Table 2).

Table 1. Jahn-Teller effect in biomolecules: historical survey

Molecule	Phenomenology	References
Metalloporphyrins	Optically detected ESR,	87, 88
	microwave induced delayed phosphorescence,	87, 89
	Zeeman spectroscopy,	94
	MCD,	95
	resonance Raman spectra,	97, 100, 101
	ESR,	88, 96
	photolysis	142
HbO_2	Deoxygenation process	102, 103, 104, 135–137, 140
Hemoprotein complexes with di-atomics (CN, CO, NO, O_2)	Photolysis	142
Hb, Mb	Magnetic susceptibility,	5, 117, 118
	Mössbauer spectra,	117, 118
	NMR,	122
	heme geometry	136–140
NO-Fe(II) horseradish peroxidase	MCD	125
Ferrihemoglobin azide	ESR	6, 105
Copper-proteins, containing "blue" centers	ESR	148, 153
	Absorption spectra	148
	Chromophore geometry	148, 151, 152
Cubane-like complexes	Chromophore geometry	161–166

Table 2. Approximate magnitude of the Hamiltonian operators in Eq. (22)

Hamiltonian	Interaction	Magnitude (cm^{-1})
H_{EE}	Electron-electron	10^5
H_{LF}	Ligand field	10^4
H_{SO}	Spin-orbit	$10^2 - 10^3$
H_{LS}	Low-symmetry static field	$0 - 10^3$
H_{JT}	Vibronic	$0 - 10^3$

Weaker interactions, like spin-spin, nuclear quadrupole, electron and nuclear Zeeman etc. have been neglected in Eq. (22).

Inspecting Table 2 there is no doubt that electron-electron interactions and high-symmetry contribution of the ligand field have to be taken into account first. The problem arises when one wants to introduce vibronic interactions, which are often greater than or comparable with spin-orbit coupling or low-symmetry static fields. In such cases JT effect cannot be neglected, but has to be treated simultaneously with the other interactions.

3.1 When and Where has the Jahn-Teller Effect to be Considered?

Even an approximate knowledge of the magnitude of the vibronic interactions is most important to interpret experimental results or to make theoretical predictions.

The first attempt of calculating JT coupling constants on the basis of the point charge model was made by Van Vleck[9] for octahedral aquo-complexes. More reliable results can be obtained by ligand field or LCAO MO treatments[66–69]; however, the resulting expressions are rather inconvenient to handle owing to the large number of parameters. Instead the vibronic coupling constants can be expressed as a function of a very limited number of parameters in the framework of the Angular Overlap Model (AOM)[70, 71], which is intercorrelated with the LCAO MO theory in the Wolfsberg-Helmholz approximation[72–75]. The corresponding parameters can be derived from experiment and are closely related to the strength of the σ or π bond. Furthermore they are transferable from one chromophore to another of different symmetry, if the bond lengths are unchanged, and thus allow to make predictions within a series of similar compounds.

The general expression for the linear JT coupling constant, in the AOM, is given by[70, 71]:

$$C_\gamma^\Gamma = \left\langle \phi_i \left| \left(\frac{\partial V}{\partial S_\gamma^\Gamma} \right)_0 \right| \phi_j \right\rangle = \sum_{\lambda\omega} \sum_n^N \left\{ \frac{\partial}{\partial r_n} [e_{\lambda n} F_{\lambda\omega}(\phi_i, X_n) F_{\lambda\omega}(\phi_j, X_n)] \right\} \frac{\partial r_n}{\partial S_\gamma^\Gamma} \quad (23)$$

S_γ^Γ are symmetry coordinates; r_n stands for the generic polar coordinate, R_n, ϕ_n, θ_n, of the n-th X_n ligand; $e_{\lambda n}$ is the energy change of a given metal orbital, due to the interaction with a ligand orbital; $F_{\lambda\omega}$, the angular overlap factor, is a fraction of the maximum overlap integral and is expressed as a function of the angular coordinates (ϕ_n, θ_n) of the ligands; λ indicates the bonding symmetry with respect to the metal-ligand axis (σ, π, δ) and ω specifies the particular orbital (for $\lambda > 0$).

As $\left(\dfrac{\partial V}{\partial S_\gamma^\Gamma}\right)$ is an one electron operator, the JT constants are calculated for any config-

uration of a general p^n, d^n or f^n ion starting from Eq. (23)[70, 71, 76, 77]. In Table 3 the results for transition metal ions commonly involved in metalloproteins are reported[5]. In general larger values of e_λ are found for cations of higher oxidation states, and e_σ is always greater than e_π. Quite often e_π is negligibly small, as in ammonia or aliphatic amines, sometimes quite large, however, as in aromatic or more "soft" ligands[78]. Moreover the coupling constants may change appreciably, for the same point group symmetry, as a function of the angle θ between the main symmetry axis and the equatorial bond M-X. It has been shown[5] for example that for a square-pyramidal (C_{4v} point group) high-spin iron(II) chromophore (see Hb or Mb) the coupling constant between the electronic state ^5E and the bending out-of-plane β_1 reaches its maximum for θ in the range 105°–120°, while it vanishes when the iron lies in the equatorial plane ($\theta = 90°$). Therefore the experimental value of about 104° for Hb and Mb[79–81] allows an appreciable vibronic coupling and leads to an additional energetic stabilization.

By an inspection of Table 3 some qualitative predictions can be made. Strong coupling is expected in copper-proteins, particularly in pseudo-tetrahedral sites (the so-called "blue" centers), in low-spin iron(III) and high-spin iron(II) proteins and molybdo-proteins, where the high strong formal charge provides large e_σ values[5].

3.2 Metallo-Porphyrins: A Clear Experimental Evidence

Porphyrins have been widely studied owing to their importance in many biological processes[82]. The porphin structure consists of four pyrrole rings linked together by four methine bridges (= CH-). The resulting 16-membered ring occurs in Hb, Mb, cytochromes, chlorophyll and so on. At present a renewed interest in the optical properties of porphyrins has arisen, because they act as photosensitizers and also induce photoprocesses, which are used for photochemotherapy and diagnosis of malignant tumors[83–85].

Since 1959[86] Gouterman and co-workers developed theoretical concepts for the interpretation of the spectroscopic behavior of porphyrins. However only at the beginning of the seventies Gouterman[87] and van der Waals et al.[88, 89], recognized the importance of the JT effect with respect to metalloporphyrins (MeP's).

The point group symmetry of MeP's can be safely assumed to be D_{4h}. Accordingly in diamagnetic MeP's such as ZnP or PdP the first and second excited singlet states (S_1 and S_2) and the lowest triplet state (T_0) are orbitally degenerate and bear symmetry labels E_u. The $S_0 \rightarrow S_1$ and $S_0 \rightarrow S_2$ transitions correspond to the so-called Q band and Soret band, respectively, in the absorption spectrum. The latter band is very strong (oscillator strength ~ 1) and appears around $\lambda = 400$ nm, while the former is approximately ten times weaker and occurs between 500 and 560 nm. Such intensity difference was reasonably accounted for in the free electron picture, which was originally put forth by Simpson[90] and then developed by Platt[91] and Gouterman[87, 92].

Table 3. Jahn-Teller coupling constants for transition metal ions commonly involved in metalloproteins[5]

Electronic configuration	Ion	Point group	Electronic level[a]	JT active vibrational mode[b]	JT coupling matrix elements[c]
d^1	Mo^{5+}	C_{4v}	$^2E(\xi, \eta)$	$\nu(\beta_1)$	$\left\langle \xi \left\| \frac{\partial V}{\partial S_{\beta_1}} \right\| \xi \right\rangle = -\frac{3}{4}\sin^2 2\theta \cdot \frac{\partial e_\sigma}{\partial R} + (\cos^2\theta - \cos^2 2\theta)\cdot \frac{\partial e_\pi}{\partial R}$
				$\delta(\beta_1)$	$\left\langle \xi \left\| \frac{\partial V}{\partial S_{\beta_1}} \right\| \xi \right\rangle = -\frac{1}{R}\left[\frac{3}{2}\sin 4\theta \cdot e_\sigma + (\sin 2\theta - 2\sin 4\theta)\cdot e_\pi\right]$
				$\delta(\beta_2)$	$\left\langle \xi \left\| \frac{\partial V}{\partial S_{\beta_2}} \right\| \eta \right\rangle = \frac{2}{R}(\cos^2 2\theta - \cos^2\theta)\cdot e_\pi$
		C_{3v}	$^2E(\xi, \eta)$	$\nu(\varepsilon)$	$\left\langle \xi \left\| \frac{\partial V}{\partial S_\xi^\varepsilon} \right\| \xi \right\rangle = -\frac{1}{2}\sqrt{\frac{3}{2}}\left[\frac{3}{4}\sin^2 2\theta \cdot \frac{\partial e_\sigma}{\partial R} - (\cos^2\theta - \cos^2 2\theta)\cdot \frac{\partial e_\pi}{\partial R}\right]$
				$\delta(\varepsilon)^d$	$\left\langle \xi \left\| \frac{\partial V}{\partial S_\xi^\varepsilon} \right\| \xi \right\rangle = \frac{1}{R}\sqrt{\frac{3}{2}}\left[-\frac{3}{4}\sin 4\theta \cdot e_\sigma + \left(-\frac{1}{2}\sin 2\theta + \sin 4\theta\right)\cdot e_\pi\right]$
				$\delta(\varepsilon)^e$	$\left\langle \xi \left\| \frac{\partial V}{\partial S_\xi^\varepsilon} \right\| \xi \right\rangle = -\frac{1}{R}\sqrt{\frac{3}{2}}\left[\frac{3}{4}\sin^2 2\theta \cdot e_\sigma + (\cos^2 2\theta - \cos^2\theta)\cdot e_\pi\right]$
d^5	Fe^{3+}	C_{4v}	$^2E(\xi^+\eta^2\zeta^2, \xi^2\eta+\zeta^2)$	$\nu(\beta_1)$	$\left\langle \xi \left\| \frac{\partial V}{\partial S_{\beta_1}} \right\| \xi \right\rangle = \frac{3}{4}\sin^2 2\theta \cdot \frac{\partial e_\sigma}{\partial R} - (\cos^2\theta - \cos^2 2\theta)\cdot \frac{\partial e_\pi}{\partial R}$
				$\delta(\beta_1)$	$\left\langle \xi \left\| \frac{\partial V}{\partial S_{\beta_1}} \right\| \xi \right\rangle = \frac{1}{R}\left[\frac{3}{2}\sin 4\theta \cdot e_\sigma + (\sin 2\theta - 2\sin 4\theta)\cdot e_\pi\right]$
				$\delta(\beta_2)$	$\left\langle \xi \left\| \frac{\partial V}{S_{\beta_2}} \right\| \eta \right\rangle = -\frac{2}{R}(\cos^2 2\theta - \cos^2\theta)\, e_\pi$

d^6 Fe^{2+}, Co^{3+} C_{4v} ${}^5E(\xi^2\eta^+\zeta^+\varepsilon^+\theta^+, -\xi^+\eta^2\zeta^+\varepsilon^+\theta^+)$, $\nu(\beta_1)$,

$$\left\langle\xi\left|\frac{\partial V}{\partial S_{\beta_1}}\right|\xi\right\rangle = -\frac{3}{4}\sin^2 2\theta\cdot\frac{\partial e_\sigma}{\partial R} + (\cos^2\theta - \cos^2 2\theta)\cdot\frac{\partial e_\pi}{\partial R}$$

$\delta(\beta_1)$

$$\left\langle\xi\left|\frac{\partial V}{\partial S_{\beta_1}}\right|\xi\right\rangle = -\frac{1}{R}\left[\frac{3}{2}\sin 4\theta\cdot e_\sigma + (\sin 2\theta - 2\sin 4\theta)\cdot e_\pi\right]$$

$\delta(\beta_2)$

$$\left\langle\xi\left|\frac{\partial V}{\partial S_{\beta_2}}\right|\eta\right\rangle = \frac{2}{R}(\cos^2 2\theta - \cos^2\theta)\cdot e_\pi$$

T_d ${}^5E(\varepsilon^2\theta^+\xi^+\eta^+\zeta^+, \varepsilon^+\theta^2\xi^+\eta^+\zeta^+)$ $\delta(\varepsilon)$

$$\left\langle\theta\left|\frac{\partial V}{\partial S_\theta^\varepsilon}\right|\theta\right\rangle = -\frac{4\sqrt 2}{R}\cdot\varepsilon_\pi$$

d^7 Co^{2+} O_h ${}^2E_g(\xi^2\eta^2\zeta^2\theta^+, \xi^2\eta^2\zeta^2\varepsilon^+)$, $\nu(\varepsilon)$

$$\left\langle\theta\left|\frac{\partial V}{\partial S_\theta^\varepsilon}\right|\theta\right\rangle = \frac{\sqrt 3}{2}\frac{\partial e_\sigma}{\partial R}$$

D_{3h} ${}^2E'(\xi^2\eta^2\zeta^2\varepsilon^+, \xi^2\eta^2\zeta^+\varepsilon^2)$, $\nu(\varepsilon')$

$$\left\langle\varepsilon\left|\frac{\partial V}{\partial S_\zeta^{\varepsilon'}}\right|\varepsilon\right\rangle = -\frac{1}{2}\sqrt{\frac{3}{2}}\left(\frac{3}{4}\frac{\partial e_\sigma}{\partial R} - \frac{\partial e_\pi}{\partial R}\right)$$

$\delta(\varepsilon')^f$

$$\left\langle\varepsilon\left|\frac{\partial V}{\partial S_\zeta^{\varepsilon'}}\right|\varepsilon\right\rangle = \frac{1}{R}\sqrt{\frac{3}{2}}\left(\frac{3}{2}e_\sigma - 2e_\pi\right)$$

C_{4v} ${}^4E(\theta^+\varepsilon^+\xi^+\eta^2\zeta^2, \theta^+\varepsilon^+\xi^2\eta^+\zeta^2)$, $\nu(\beta_1)$

$$\left\langle\xi\left|\frac{\partial V}{\partial S_{\beta_1}}\right|\xi\right\rangle = \frac{3}{4}\sin^2 2\theta\cdot\frac{\partial e_\sigma}{\partial R} - (\cos^2\theta - \cos^2 2\theta)\cdot\frac{\partial e_\pi}{\partial R}$$

$\delta(\beta_1)$

$$\left\langle\xi\left|\frac{\partial V}{\partial S_{\beta_1}}\right|\xi\right\rangle = \frac{1}{R}\left[\frac{3}{2}\sin 4\theta\cdot e_\sigma + (\sin 2\theta - 2\sin 4\theta)\cdot e_\pi\right]$$

$\delta(\beta_2)$

$$\left\langle\xi\left|\frac{\partial V}{\partial S_{\beta_2}}\right|\eta\right\rangle = -\frac{2}{R}(\cos^2 2\theta - \cos^2\theta)\,e_\pi$$

d^9 Cu^{2+} T_d ${}^2T_2(\theta^2\varepsilon^2\xi^+\eta^2\zeta^2, \theta^2\varepsilon^2\xi^2\eta^+\zeta^2, \theta^2\varepsilon^2\xi^2\eta^2\zeta^+)$, $\delta(\varepsilon)$

$$\left\langle\xi\left|\frac{\partial V}{\partial S_\varepsilon^\varepsilon}\right|\xi\right\rangle = \frac{2\sqrt 6}{3R}\left(e_\sigma - \frac{e_\pi}{3}\right)$$

Table 3 (continued)

Electronic configuration	Ion	Point group	Electronic level[a]	JT active vibrational mode[b]	JT coupling matrix elements[c]
		D_{2d}[g]	$^2E(\theta^2\varepsilon^2\xi^+\eta^2\zeta^2,\ \theta^2\varepsilon^2\xi\eta^+\zeta^2)$	$\nu(\tau_2)$	$\left\langle \xi \left\| \dfrac{\partial V}{\partial S_\zeta^{\tau_2}} \right\| \eta \right\rangle = -\dfrac{2}{3}\left(\dfrac{\partial e_\sigma}{\partial R} - \dfrac{1}{3}\dfrac{\partial e_\pi}{\partial R}\right)$
				$\delta(\tau_2)$	$\left\langle \xi \left\| \dfrac{\partial V}{\partial S_\zeta^{\tau_2}} \right\| \eta \right\rangle = \dfrac{2\sqrt{2}}{3R}\left(e_\sigma - \dfrac{7}{3}e_\pi\right)$
				$\delta(\beta_1)$	$\left\langle \xi \left\| \dfrac{\partial V}{\partial S_{\beta_1}} \right\| \xi \right\rangle = \dfrac{\sqrt{6}}{R}\left[\dfrac{3}{4}\sin^2 2\theta \cdot e_\sigma + (\cos^2 2\theta - \cos^2\theta)\cdot e_\pi\right]$
				$\delta(\beta_2)$	$\left\langle \xi \left\| \dfrac{\partial V}{\partial S_{\beta_2}} \right\| \eta \right\rangle = -\dfrac{2}{R}\left[\dfrac{3}{4}\sin 4\theta \cdot e_\sigma - \dfrac{1}{2}(2\sin 4\theta - \sin 2\theta)\cdot e_\pi\right]$
				$\nu(\beta_2)$	$\left\langle \xi \left\| \dfrac{\partial V}{\partial S_{\beta_2}} \right\| \eta \right\rangle = -\dfrac{3}{4}\sin^2 2\theta \cdot \dfrac{\partial e_\sigma}{\partial R} - 2(\cos^2 2\theta - \cos^2\theta)\dfrac{\partial e_\pi}{\partial R}$

[a] Strong field configuration in parenthesis; $\xi, \eta, \zeta, \theta, \varepsilon$ stand for the one-electron levels $d_{yz}, d_{xz}, d_{xy}, d_{z^2}$ and $d_{x^2-y^2}$ respectively
[b] The notation ν and δ refer to the stretching and bending vibrational modes, respectively
[c] Only one matrix element is reported for each case, since the other elements are easily obtained by symmetry considerations[70, 71]; the angle θ corresponds to the angle between the main symmetry axis and the equatorial ligands; contributions from δ-bonds are neglected
[d] Out-of-plane bending
[e] In-plane bending
[f] In-plane bending; in D_{3h} symmetry the coupling constants relative to the out-of-plane bending are zero
[g] The angle θ corresponds to the angle between the S_4 symmetry axis and the M-X bond, where X is the ligand involved in the vibrational mode.

Since MeP's nowadays are the only clear examples of a JT effect experimentally verified in biologically important molecules, I want to discuss the first experimental evidence more extensively[88, 89].

The porphyrin ZnP in a crystalline n-octane matrix at 1.2 K was investigated by optically detected magnetic resonance (ODM) and microwave induced delayed phosphorescence (MIDP)[89].

In microwave phosphorescence double resonance experiments a steady-state population of the triplet state is obtained by continuous optical pumping of the sample at zero-field; then microwaves are slowly swept until resonance makes itself felt as a change in phosphorescence intensity owing to an unbalanced population on the three components of the triplet[93] (Fig. 5). The zero-field spectrum of ZnP shows six transitions, labeled A...F, which are grouped into three sets. The strongest pair, A and B, was attributed to single ZnP molecules, while the other two weaker pairs were attributed to dimers or more complicate aggregates (C and D) and to ZnP molecules at slightly different crystal sites (E and F). The two transitions A and B imply a zero-field level scheme for the lowest vibronic triplet as indicated in Fig. 5 and hence the loss of the four-fold symmetry axis. The conclusions drawn from this experiment were that the lowest *vibronic* level E_u, which originates from the coupling between the *electronic* E_u level and β_{1g} and/or β_{2g} vibrations, was split by a low symmetry crystal field into two components Ψ_0 and Ψ_1 with an energy separation δ. At very low temperature (1.2–4.2 K) only Ψ_0 is populated so that the three spin sublevels of Fig. 5 belong only to Ψ_0. An upper limit for δ (50 cm^{-1}) was obtained from ESR experiments at 77 K ($|D| = 0.035 + 0.009$ cm^{-1}, E ~ 0).

At the same time the dynamics of populating and depopulating the phosporescent triplet state of ZnP was investigated by MIDP[89]. In this experiment the pumping light is shut off at t = 0 and then, at time t_1, a microwave pulse at the frequency ν_A or ν_B is applied. In the meantime the phosphorescence intensity is recorded as a function of time. The microwave induced signal is negative for both frequencies ν_A and ν_B, while it is positive for ν_A when $t_1 \geq 0.75$ s. By an analysis of the decay time, the lifetimes for the

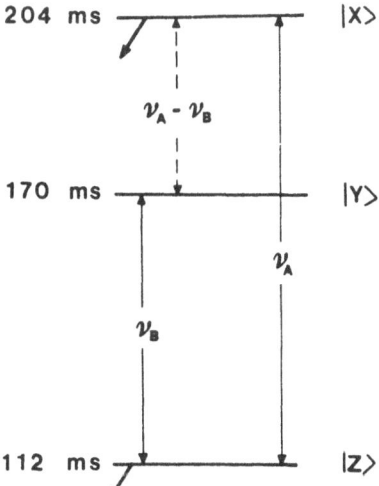

Fig. 5. Zero-field transitions observed by microwave phosphorescence double resonance in the lowest triplet state of ZnP. The lifetimes were obtained from MIDP experiments. *Arrows* indicate two radiative channels for the phosphorescence (adapted from Refs. 88, 89)

three spin sublevels were determined (Fig. 5) and successively the ratio of the popula-
tions of the individual levels under steady-state condition was estimated. The lowest
level, labelled $|z>$, is the most radiative and predominantly populated, while the inter-
mediate level $|y>$ is practically empty. The different behavior of the three levels further
supports the hypothesis of a combined action of the JT effect and crystal field anisotropy.

Successively Van der Waals and his group in Leiden extended their investigations to
other MeP's using also different techniques.

High resolution Zeeman experiments were performed[94] on ZnP, PdP and CuP, for
the first excited singlet (ZnP, PdP) and the lowest phosphorescing states (PdP, S = 1;
CuP, S = 3/2). The above compounds, together with MgP and PtP, were investigated
also by MCD measurements[95], which are particularly suitable to gain information about
the symmetry of the vibrational states involved in the absorption spectrum[58, 59, 92]. For
ZnP an asymmetric quadratic Zeeman effect was observed in the first singlet state and a
definite indication of a moderately strong JT effect ($E_{JT} \sim 30\text{--}40$ cm^{-1}) was obtained.
The frequency of the JT active mode ($\bar{\nu}_{JT} = 170\text{--}180$ cm^{-1}) is of the order of the crystal
field splitting ($\delta = 109$ cm^{-1}). When $\delta \gg \bar{\nu}_{JT}$ or $\delta \ll \bar{\nu}_{JT}$, a symmetric Zeeman effect
occurs, but in the latter case the orbital reduction factor can deviate significantly from
1[94]. The MCD spectra were still consistent with an appreciable JT coupling in ZnP,
whereas this interaction escaped experimental detection in PdP and PtP, because no B
term was seen for the $S_0 \rightarrow S_1$ transition[95]. As for the symmetry of the JT active vibrations, the MCD spectra gave indication that β modes were involved in the vibronic
coupling mechanism and this view was corroborated also by ESR experiments on the
photo-excited triplet, which strongly suggested a JT active β_{1g} vibrational mode[96] [2].

As we have seen in Sect. 2.4 the excitation profiles of the resonance Raman spectra
are very sensitive to the vibronic coupling. Also by this technique a moderate JT effect on
the Q state in MeP's is evident. The first paper in 1977 dealt with etioporphyrin Ni(II)
(NiEtio) and tetraphenylporphyrin Cr(III) chloride (CrTPPCl)[97]. Owing to the non
adiabatic coupling of the Q state to the state, from which the Soret band originates, an
enhanced intensity is expected when the exciting frequency equals the 0–1 electronic
transition, whereas a diminished intensity should occur for 0–0 resonance[98, 99]. On the
contrary the most striking result for NiEtio[97] was the strong 0–0 Raman intensity dis-
played by depolarized (β_{1g}, β_{2g}) and polarized (α_{1g}) vibrational modes, besides the
expected strong 0–1 scattering shown by the inversely polarized modes α_{2g}. Since in the
D_{4h} point group the symmetric product $[E^2]_{sym}$ transforms as a_{1g}, b_{1g} and b_{2g}[1], the different behavior of α_{2g} vibrations compared to the α_{1g}, β_{1g} and β_{2g} modes strongly suggests
the influence of JT effect. Actually a theoretical model well accounts for the observed
excitation profiles in the Q-band, if JT coupling is included; moreover it has to be
remarked that JT effect manifests itself not only in the Q band, but also in the
Soret   band[97, 100].   The   strength   of   the   coupling   is   in   the   order[100]:
NiEtio > CrTPPCl > CuTPP.

---

2 β_{1g} or β_{2g}, depending on the orientation of the reference axes[20]

3.3 Hemoproteins: A Widely and Long Debated Problem

Hemoproteins constitute a large class of biomolecules, which include oxygen carriers, like hemoglobins and myoglobin, and enzymes, like cytochrome c oxidase, peroxidase and catalase. All these molecules contain as prosthetic group an iron porphyrin, which is called heme for iron(II) porphyrin and hemin for iron(III) porphyrin[82]. In the light of the considerations made in Sect. 3.2. vibronic coupling is expected to be effective also in hemoproteins, though the remainder of the protein surrounding the chromophore might strongly complicate all the phenomenology.

The JT effect in this class of molecules was considered to account for two different aspects of the phenomenology: results of physical measurements and conformational changes induced during the oxygenation/deoxygenation process. In the former case the JT treatment has to be considered correct as it can interpret results, which otherwise are hardly understandable. In the latter case the proposal of the JT effect as a triggering mechanism of conformational changes is fascinating, but it is difficult to ascertain and distinguish JT contributions (if any) from other contributions, like purely electronic effects. Indeed it is just on this aspect that the dispute still continues since the first proposal by Weissbluth on the deoxygenation mechanism in HbO_2[102-104].

3.3.1 Magnetic Anisotropy in Ferrihemoglobin: the First Example

To our knowledge ferrihemoglobin azide is the first example of applying the JT effect to a biomolecule[6, 105].

ESR spectra of low spin ferrihemoglobins show a much larger anisotropy within the hemin plane ($g_x = 1.72$; $g_y = 2.22$; $g_z = 2.80$ for ferrihemoglobin azide at 20 K) than that observed for high-spin derivatives. In the high spin case the rhombic crystal field, produced by the axial imidazole and the sixth ligand (Fig. 6), could account for the anisotropy[106], but in the low spin case it was difficult to attribute the increase of the anisotropy to the change of the sixth ligand only. Instead, the experimental results fitted well considering the combined action of the rhombic crystal field and the dynamical JT interaction between the electronic ground state ^2E (C_{4v} point group) and the in-plane

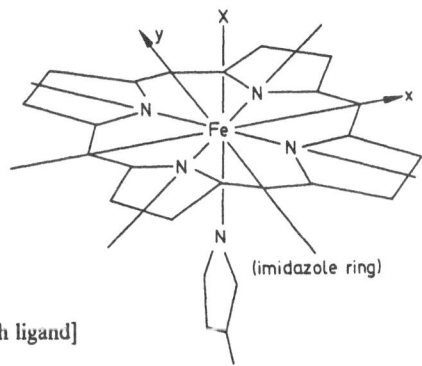

Fig. 6. Molecular geometry of hemoglobin [X: sixth ligand] (from Ref. 105)

stretching mode β_{1g}. The original idea was that static JT distortions could produce the large observed anisotropy; however, a simple static treatment showed comparable energies of the zero-point vibration ($\hbar\omega = 350$ cm^{-1} or 225 cm^{-1}), the JT stabilization ($E_{JT} = 516$ cm^{-1}), the spin-orbit interaction ($\zeta = 435$ cm^{-1}) and the rhombic field due to the azide ion ($H_{Rh} = 60$ cm^{-1}). Therefore all interactions had to be considered simultaneously including dynamic effects. The used Hamiltonian was

$$H = H_{Te} + H_{Rh} + H_{SO} + H_{JT} + V + T \qquad (24)$$

H_{Te} and H_{Rh} are the tetragonal and rhombic crystal fields, respectively; H_{SO} is the spin-orbit interaction; $H_{JT} = \dfrac{\partial V(r, Q_\beta)}{\partial Q_\beta}\,'_\beta$, $V = \dfrac{1}{2} K_\beta Q_\beta^2$ (K_β: force constant for vibration β) and $T = P_\beta^2/2\,M_\beta$ (M_β and P_β: effective mass and momentum conjugate to Q_β, respectively). The eigenfunctions of Eq. (24) can be expressed as linear combination of Born-Oppenheimer functions $\phi_i s_u \chi^{(m)}$

$$\Psi(r, Q) \;=\; \sum_k c_k \phi_{ik} s_{uk} \chi_k^{(m)} \qquad (25)$$

where ϕ_i (i = 1, 0, −1) are the bases of the electronic T_2 state, s_u (u = ± 1/2) the spin functions and $\chi^{(m)}$ the vibrational wavefunctions (m = 0, 1...4), expressed by usual Hermite polynomials centered at the equilibrium position. The fourfold degenerate vibronic ground level is split into two Kramers' doublets by the spin-orbit interaction. By this method, general expressions for the g-values were given; their anisotropy can be strongly enhanced by dynamic coupling even in the presence of weak rhombic fields. The best fit of the experimental g-values gave 220 cm^{-1} and 0.8 for the rhombic field and $[E_{JT}/\hbar\omega]^{1/2}$, respectively. These figures were substantially confirmed by a theoretical estimation of the JT coupling constant and of the rhombic crystal field due to the azide ion in a covalent bond model. On the other hand the calculated value of the rhombic field well explained the g-values of the high-spin catalase azide[107], where the anisotropy is obviously only due to the azide ion.

Very recently[108] it was pointed out that an accurate crystal field treatment for a d^5 ion cannot be limited to the state $^2T_2(t_2^5)$ (O point group), but has to include higher terms such as those arising from (t$_2^4$e) configuration, which are mixed into the ground level through electrostatic and spin-orbit interactions. In this way the ESR spectra of several derivatives of low spin ferrihemoglobin, ferrimyoglobin and ferricytochrome c were explained without including JT effect. However, the authors acknowledge to have neglected such an effect in order to reduce the number of free parameters. They admit its importance, because a crystal field treatment is not equally able to interpret the spectra of ferricytochrome P-450.

3.3.2 Deoxygenated Hemoglobins

Owing to the failure of ESR techniques for iron(II) ions, Mössbauer spectroscopy, together with magnetic susceptibility measurements, has been the most appropriate investigative tool for deoxygenated Hb and Mb. However, a simultaneous fitting of the magnetic susceptibility and Mössbauer data was only possible, if rather large rhombic symmetry components were available[109-113]. Though a not well specified inequivalence of the four nitrogen atoms in the heme plane[112] or the imidazole ring of the axial histidine residue[110] could give rise to a rhombic field, the large value seems unjustified, particularly in the light of ESR measurements on high spin ferric Hb and Mb or cobalt-substituted Hb and Mb, where the rhombic component are absent or negligible[104, 114-116].

Recently it was shown[117, 118] that vibronic interactions might drastically reduce the rhombic components. The procedure was quite analogous to the one described in the previous section, though H_{Rh} in (24) was neglected. Only the low-lying electronic levels $^5E, ^5B_2, ^1A_1, ^3E$ and the first three vibrational quanta of the out-of plane bending mode β_1 were taken into account, to keep the dimensions of the involved matrices reasonably low. In order to compute the magnetic susceptibility and the nuclear quadrupole splitting as function of temperature, the magnetic and electric field gradients were introduced as perturbing terms of the eigenfunctions (25) and a Boltzmann distribution was assumed. The following relationship was obtained for the magnetic susceptibility χ_n:

$$\chi_n^{(T)} = \frac{N \Sigma_m [(E_m^{(0)}/kT) - 2 E_m^{(0)}] \exp(-E_m^{(0)}/kT)}{\Sigma_m \exp(-E_m^{(0)}/kT)} \tag{26}$$

where $E_m^{(1)}$ and $E_m^{(2)}$ are the first and second order Zeeman coefficients and $E_m^{(0)}$ is the eigenvalue of the m-th vibronic level at zero magnetic field.

Neglecting lattice contributions, the temperature dependence of the quadrupole splitting ΔE_Q is given by[119]:

$$\Delta E_Q^{(T)} = \frac{2}{7} e^2 Q (1 - R_0) \langle r^{-3} \rangle_{3d} \alpha^2 F(T) \tag{27}$$

Here α is a reduction factor due to the covalency; the factor F(T) is expressed, in axial symmetry, as follows:

$$F(T) = \frac{\Sigma_m \langle \Psi_m | \ell_{zz} | \Psi_m \rangle \exp(-E_m^{(0)}/kT)}{\Sigma_m \exp(-E_m^{(0)}/kT)} \tag{28}$$

and the other terms have their usual significance[119].

A good fitting was obtained for both Hb and Mb, within the assumed approximations, with $E_{JT} \sim 30 \text{ cm}^{-1}$ and $\hbar\omega \sim 50 \text{ cm}^{-1}$. Apart from the good quality of the fit, which should be partially re-examined because of new experimental findings however[120], the important point is the significant effect of the vibronic coupling mainly at low temperatures (Fig. 7)[118]. Moreover the resulting vibronic level scheme provides as low lying states two non-degenerate and one doubly-degenerate vibronic levels within 4 cm^{-1}.

This result compares well with the far infrared transition observed at 3.5 cm^{-1} in both Hb and Mb, using far infrared magnetic resonance[121]. The presence of at least one excited state within 2 cm^{-1} of the ground state was supposed also by Lumpkin and Dixon[122] to interpret their measurements of proton relaxation rates in Mb at low temperatures. In fact there is an interconnection between the proton spin-lattice relaxation time

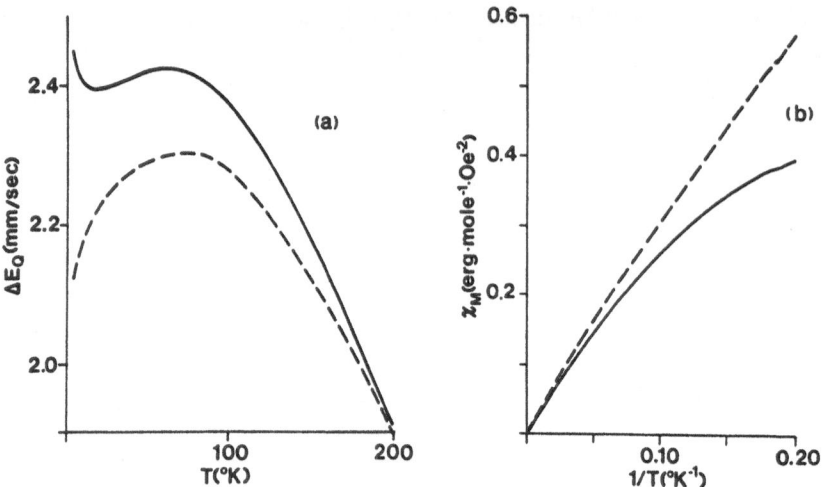

Fig. 7a, b. Vibronic coupling effect [Hb, Mb] on: (a) quadrupole splitting ΔE_Q and (b) magnetic susceptibility χ_M. The *dashed line* is obtained by neglecting vibronic coupling and the *solid line* by including coupling to the β_1 bending with $\hbar\omega = 50 \text{ cm}^{-1}$ and $E_{JT} = 30 \text{ cm}^{-1}$ (from Ref. 118)

T_1 and the electronic relaxation time τ_e such, that the electronic structure of the ferrous ion, like the relative magnitudes of the spacings between electronic levels and kT, are effective for the relaxation time T_1. In the case of Mb, T_1 is proportional to τ_e and from the temperature and magnetic field dependence of T_1 the authors[122] deduce an electron relaxation time, which is longer than the expected one ($10^{-7} < \tau_e < 10^{-4}$ s). The proposed model also requires a large quenching of the spin-orbit interaction due to a dynamic JT coupling with low frequency vibrational modes. Very recent NMR investigations on model compounds of Mb and Hb[123] seem to give further support to the above model.

In 1975 Alpert and Banerjee[124] observed different magnetic moments in phosphate-free ($\mu_{eff} \sim 4.90 \ \mu_B$) and phosphate-bound ($\mu_{eff} \sim 5.30$–$5.45 \ \mu_B$) Hb and suggested that phosphate binding could cancel some vibrational modes and thus induce a level rearrangement. This idea was further developed and put on a quantitative basis[117] by considering the coupling between the electronic states of iron(II) and the out-of-plane bending mode β_1. The calculations showed indeed that an increase of the magnetic moment corresponds to a decrease of the ratio $S = E_{JT}/\hbar\omega$ (Fig. 8). Since the mean-square amplitude δ of a normal coordinate in the harmonic approximation is[5]:

$$\delta = \frac{h}{8\pi^2 c\omega} \coth\left(\frac{hc\omega}{2kT}\right) \tag{29}$$

a constraint on δ, due to the phosphate binding, may lead to an increase of the vibrational quantum ω and, consequently, of the magnetic moment.

Finally the dynamic JT effect has been included to explain the low-temperature MCD of NO-Fe(II) hemoproteins[125]. In 1956 Griffith had already discussed the possible presence of a JT effect in NO-Hb[126], but did not proceed further because the NO molecule

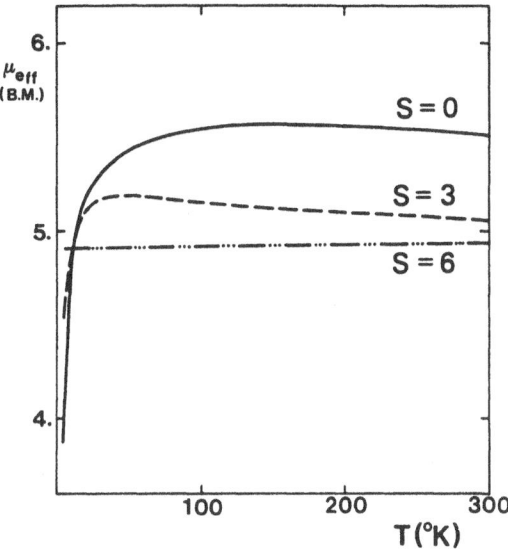

Fig. 8. Variation of effective magnetic moment versus temperature, computed for different values of $S = E_{JT}/\hbar\omega$ (adapted from Ref. 117)

was able by itself to remove the degeneracy. The theoretical analysis developed by Mineyev et al.[125] considers the dynamic coupling of a ^2E state, resulting from iron d and NO π orbitals, with β_1 and β_2 vibrational modes of the chromophore NO-Fe N_4 in C_{4v} symmetry. Under the assumptions: i) rhombic components are neglected, ii) the two equivalent minima are sufficiently deep to exclude thermal transitions even at room-temperature and therefore transitions between the wells occur only by quantum-mechanical tunneling, iii) the lifetime of the excited $\pi - \pi^*$ states is much greater than the tunneling time, an expression for the MCD line shape expected in the Soret band was obtained as a function of temperature and tunneling splitting. Though these assumptions are quite restrictive, preliminary data down to 80 K seem to agree with the computed curve; unfortunately MCD data at very low temperature are lacking, however.

3.3.3 Structural Changes in the Oxy-Deoxy Reaction

Owing to its physiological importance the oxygenation-deoxygenation process in hemoglobin has been widely investigated, but it is beyond the purposes of the present paper to give an account of the studies so far reported (see e.g. Ref. 127 for recent reviews), except for some main aspects.

The hemoglobin molecule consists of four subunits (two α and two β chains), and each chain contains a heme group, which can bind one O_2 molecule. Most interesting is the sigmoidal shape of the oxygenation curve as a function of the partial oxygen pressure. Such a behavior is indicative of a cooperative effect. If oxygen attaches to a particular chain, the successive attachment to another chain is increased; the same behaviour is observed upon deoxygenation. In the deoxygenated molecule, corresponding to the T (or tensed) conformation of the protein, the iron(II) ion is high-spin (S = 2), five-coordinated and placed out of the heme plane[79-81]. In HbO_2, corresponding to the R (or relaxed) conformation, the iron(II) ion is low-spin (S = 0), with a thermally accessible triplet (S = 1)[128-132]; it is hexa-coordinated and lies in the heme plane. Perutz proposed[133-134] that the proximal histidine, i.e. the fifth ligand, following the iron displacement, initiates the conformational change T → R in one of the subunits.

Weissbluth suggested[102-104] that the JT effect could provide a driving force which resulted in a local distortion which in turn acted as a trigger for the following conformational changes. This hypothesis was based on a high-spin state $^5T_2$ (O_h symmetry) in HbO_2, which could be reached by thermal activation from the ground state $^1A_1$. Once this occurs the system is subject to a JT effect and the most likely distortion, owing to local geometrical constraints, is along the Fe-O_2 bond, thus favoring the detachment of the oxygen molecule.

Cianchi et al.[135] criticized the above suggestion on symmetry grounds because the local symmetry of the chromophore is not higher than C_{4v} and therefore the stretching Fe-O_2 vibration is no more JT active. On the other hand recent experimental results[128, 129] and theoretical investigations[130, 132] have shown that there is indeed at least one thermally accessible electronic state above the ground state $^1A_1$ of HbO_2. It has a spin multiplicity of 3, while the levels originated from the $^5T_2$ state are placed at higher energy. However, apart from the above criticism, the interesting idea remains that vibronic coupling might provide a triggering mechanism for the conformational changes occurring at the early stages of enzymatic activity[103].

The pseudo JT effect was then invoked independently by Mizuhashi[136-138] and Bersuker et al.[139, 140] to explain the large out-of-plane displacement of the iron(II) ion in Hb and also the oxy-deoxy reaction mechanism.

If one assumes the symmetry D_{4h} for MeP's (see Sect. 3.2), odd vibrations of a_{2u} type tend to displace the metal ion out of the equatorial plane. In all the MeP's the vibrational modes a_{2u} can couple the ground state with one (or more) excited states of suitable symmetry; however, the pseudo-JT effect does not always lead to distorted configurations. In fact, if only two levels are considered and their energy separation is 2δ, the ground state "softening" induces instability along the coordinate a_{2u} only if $\delta < a^2/K_{a_{2u}}$ [a: pseudo-JT coupling constant]. Using the results of a MO treatment[141] it was possible to explain[140] the instability of MnP and FeP and the stability of CoP, NiP, CuP and ZnP with respect to displacements of the described kind. Furthermore of fifth axial ligand, which can stabilize small out-of-plane displacements even when vibronic softening is negligible, will amplify the displacement.

In Hb the formation of the iron-imidazole bond does not change essentially the energy gap between the two electronic MO states a_{1g} and a_{2u}, the mixing of which may lead to the out-of-plane displacement of the iron atom. Since the pseudo-JT stabilization energy is of the order of 100–1000 cm^{-1}, the minima are expected to be shallow, and the distance between Fe^{2+} and the equatorial plane will be very sensitive to the fifth ligand. On the other hand the energy gap strongly increases in HbO_2 and HbCO so that the inequality $\delta < a^2/K_{a_{2u}}$ is no more verified and the iron atom may move back into the plane position thus triggering the conformational change T → R[136, 137, 140].

On the basis of the above concepts, i.e. the pseudo-JT effect, Bersuker and coworkers have attempted in a recent paper[142] to explain the origin and the mechanism of photolysis of 3 d^n MeP's and hemoprotein complexes with diatomic molecules, like CN,

CO, NO and O_2. It was shown that the most probable coordinate of photo-dissociation is the ε coordinate of the ligand bending mode, which is a superposition of two symmetrized distortions corresponding to the rotation of the ligand and to its displacement parallel to the porphyrin ring. Different APES's are obtained for ligands in linear or bent coordination; the different photolysis quantum yields ($\phi_L \sim 1$ for linear and $\phi_L \sim 10^{-4} \div 10^{-2}$ for bent orientation) were qualitatively accounted for in this model.

3.4 Copper-Proteins

Copper-proteins are wide-spread in both animals and plants and have been related to many metabolic processes, as oxygen transport, electron transfer and hydroxylation[143]. Copper-containing sites are usually classified in three different types[144, 145]: the type 1, or "blue" center is characterized by a combination of properties that has not yet been reproduced in model complexes (an intense absorption band at ~ 600 nm, a very small copper hyperfine coupling constant $A_{\parallel}$ and a high positive redox potential for the Cu(II)/Cu(I) couple); the type 2, or "non-blue" center has properties comparable to those of low molecular weight cupric complexes; the type 3 consists of an antiferromagnetically coupled copper(II) pair.

Recent X-ray structures of plastocyanin[146] and azurin[147] have shown a flattened pseudo-tetrahedral CuN_2S_2 chromophore, probably arising from bonds between the copper atom and a cysteine, a methionine, and two histidine residues, and it is likely that also other blue centers have a similar pseudo-tetrahedral stereochemistry.

Copper (II) is a typical JT ion and a strong coupling is expected in octahedral and in tetrahedral symmetry as well[5, 71]. An AOM calculation has shown[148] that the coupling between the electronic ground state $^2T_2$ and the bending ε, is the most effective in distorting a tetrahedral chromophore. The sign of the coupling is such as to flatten the tetrahedron. Since the ionization potentials of sulfur and nitrogen in thiols and heterocyclic compounds are comparable, to first approximation a mean ligand field produced by four equivalent ligands was assumed for blue copper centers[148]. A simultaneous diagonalization of the ligand field, JT and spin-orbit interactions was performed for different values of the symmetry coordinates S_ε. The AOM approximation allowed to reduce the number of parameters, though the mixing between 3 d and 4 p copper(II) and ligand orbitals was included. The result shows that the JT effect can indeed induce a static flattening of the tetrahedron of the right order of magnitude; moreover the simultaneous action of the JT effect and covalency and/or d–p hybridization can account for both absorption and ESR spectra (Table 4)[148]. The role of the spin-orbit coupling of the ligands in lowering the copper hyperfine constant $A_{\parallel}$ through the covalent bond has been recently stressed[149]: it is likely that this latter effect is not in contradiction to the JT interpretation.

The role of the protein in vibronic activation of redox reactions in blue centers was considered by Brill[151, 152], who put forward the interesting idea that the APES of the system is indeed modulated by the different conformations of the protein. By lowering the temperature a "frozen" distribution of different site geometries is expected, which can be revealed by suitable experiments such as ESR spectra in azurin. In fact this protein contains a single blue center and shows unequal peak separations (55, 56 and

Table 4. Comparison of computed[148] and experimental values for the optical and magnetic parameters of blue copper centers

		Computed	Experimental	Ref.
d-d transitions	v_1	4 800 cm^{-1}	~ 5 000 cm^{-1}	143, 150
	v_2	5 085 cm^{-1}	~ 9 000 cm^{-1}	
	v_3	9 725 cm^{-1}	~ 11 000 cm^{-1}	
	v_4	10 440 cm^{-1}		
	$g_\perp$	2.037	2.025–2.077	145
	$g_\parallel$	2.255	2.226–2.287	
	$|A_\perp|$	0.0053 + k^a cm^{-1}	0–0.0057 cm^{-1}	
	$|A_\parallel|$	0.0035 + k cm^{-1}	0.0035–0.0060 cm^{-1}	
	$|A_\parallel - A_\perp|$	0.0018		
	$\bar{\mu}$(R.T.)	1.83 μ_B		
	$\Delta\theta^b$	+ 12.1°	≈ 25°	146

^{a} k is the isotropic term of the hyperfine interaction
^{b} $\Delta\theta$ is the flattening (for positive values) angle
The calculation was performed assuming e_σ = 9000 cm^{-1}, e_π = 3000 cm^{-1}, K_ϵ = 3500 cm^{-1}/Å^2, no rhombic distortion and a 50% contribution of the copper(II) d orbitals in the $^2T_2$ level

48 G) and different relative heights and shapes of the low field quartet of copper hyperfine lines[153]. A simulation of the experimental spectrum has been obtained by a superposition of spectra generated with spin Hamiltonian parameters calculated over a Gaussian distribution around an out-of-planarity angle η of about 16° (corresponding to $\Delta\theta \sim$ 19°)[151].

3.5 Cubane Like Iron and Cobalt Complexes

In the previous Sections we have seen that it is not a trivial matter to decide whether or to what extent a JT effect is present in complexes containing single metal ions. The situation is even more difficult with polynuclear metal complexes. In fact if single metal ions are in orbitally non-degenerate levels the problem of spin exchange coupling can be satisfactorely resolved using a simple formalism as suggested by Kambe[154, 155] with the Hamiltonian

$$H = -2 \sum_{i,j} J_{ij} S_i \cdot S_j \tag{30}$$

where J_{ij} is the exchange integral and S_i the spin vector of the i-th metal ion. This approach is no more valid, if the interacting ions of the cluster are in degenerate levels, because spin and orbital moment are strongly correlated and vibronic reduction of the Coulomb and exchange interactions may be effective[156–160]. Because complete theory of this kind is lacking as yet, the treatment of complicated systems such as the biological ones is at present purely speculative, even if some suggestions seem interesting and more attention should be paid to deepen them.

In particular we refer to high-potential iron proteins (HP) and to cubane-like model complexes, which are the only examples of JT effect in biomolecules containing metal clusters we have found in literature[161-166].

Polynuclear iron-sulfur clusters of the type Fe_4S_4 have now been recognized in proteins[167]. The cluster can be visualized as two interpenetrating tetrahedra of 4 iron atoms and 4 inorganic sulfur atoms: each iron atom is four-coordinated in a pseudo tetrahedral arrangement, the fourth ligand being an organic sulfur, often a cysteinyl residue. The clusters Fe_4S_4, in model compounds as well as in proteins, are tetragonally compressed with four shorter and equivalent Fe-Fe bonds and two longer ones[167]. Such a lowering of symmetry from T_d to D_{2d}, has been attributed to the JT effect operating on the highest occupied molecular orbitals (HOMO)[161-163]. Yang et al.[164] performed self-consistent field X_α calculations on the simplest known analogue of 4 Fe-4 S proteins, i.e. $[Fe_4S_4(SCH_3)_4]^{2-}$, assuming a perfect T_d symmetry and introducing the tetragonal component as a perturbation. The resulting HOMO in T_d is an orbital triplet t_2 filled with four electrons in the reduced form of HP's. The JT effect splits the t_2 level into e and b_2 levels, so that the four electrons can be accomodated in the lower e orbital and a closed shell electronic configuration is attained. The resulting zero net spin is consistent with the observed low temperature magnetic properties; moreover the paramagnetism observed at higher temperatures was also accounted for, since the small gap between the e and b_2 levels allowed a thermal population of the b_2 state.

This picture was then criticized by Thomson[166], who remarked that the JT effect operates upon the multi-electron state and it is erroneous to apply the JT effect to the one-electron picture. Indeed, owing to the JT interaction the paramagnetic state $^3T_1$ splits into ^3E and $^3B_2$ and therefore the spin multiplicity is unaffected. In order for the spin state to be changed the JT stabilization energy of the corresponding $^1T_1$ state should be large enough as to induce a spin crossover. Thomson invoked the JT effect, however, to explain the ESR spectra of oxidised HP's[166]. In this case a Hückel type molecular orbital calculation provides an electronic ground state ^2E, which is unstable with respect to the ε vibrational modes leading to D_{2d} or D_2 structures. While at low temperature the system is frozen into a JT minimum and an anisotropic ESR spectrum results, an isotropic spectrum might be observed by increasing the temperature [Sects. 2.3 and 2.4]. Unfortunately ESR spectra of HP's at high temperature are not detectable and Thomson's hypothesis cannot be fully verified.

Finally new and more accurate X_α-calculations[168] on $[Fe_4S_4(SH)_4]^{2-}$ and $[Fe_4S_4(SCH_3)_4]^{2-}$ are at variance with the results by Yang et al.[164] and the JT hypothesis about the origin of the chromophore distortion is again criticized.

4 Conclusions

For many years the JT coupling has been either considered as an exotic effect restricted to physics or it was invoked, right or wrong, whenever other reasonable explanations of a given behavior could not be found.

Mean while, as we have seen, both physicists and chemists have become aware of the importance of JT interactions for a better understanding of spectroscopic properties and reaction mechanisms in biomolecules so that new and wider fields of investigation are open.

However, following latin verse[169]

est modus in rebus, sunt certi denique fines,
quos ultra citraque nequit consistere rectum,

I would like to conclude saying that possible JT interactions, not only in biomolecules but in all chemical systems, have to be carefully analysed in comparison with the other terms of the Hamiltonian in order to avoid complicated descriptions by including insignificant terms.

5 References

1. Englman, R.: The Jahn-Teller Effect in Molecules and Crystals, New York, Wiley 1972
2. Azumi, T., Matsuzaki, K.: Photochem. Photobiol. *25*, 315 (1977)
3. Bersuker, I. B.: Coord. Chem. Rev. *14*, 357 (1975)
4. Jahn, H. A., Teller, E.: Proc. Roy. Soc. A*161*, 220 (1937)
5. Bacci, M.: Biophys. Chem. *11*, 39 (1980)
6. Kamimura, H., Mizuhashi, S.: J. Appl. Phys. *39*, 684 (1968)
7. Jahn, H. A.: Proc. Roy. Soc. A*164*, 117 (1938)
8. Van Vleck, J. H.: J. Phys. Chem. *41*, 67 (1937)
9. Van Vleck, J. H.: J. Chem. Phys. *7*, 72 (1939)
10. Ruch, E., Schönhofer, A.: Theor. Chim. Acta *3*, 291 (1965)
11. Blount, E. I.: J. Math. Phys. *12*, 1890 (1971)
12. Ham, F. S.: Jahn-Teller Effects in Electron Paramagnetic Resonance, in: Electron Paramagnetic Resonance (ed. Geschwind, S.) p. 1, New York, Plenum 1972
13. Feynman, R. P.: Phys. Rev. *56*, 340 (1939)
14. Öpik, U., Pryce, M. H. L.: Proc. Roy. Soc. (London) A*238*, 425 (1957)
15. Moffitt, W., Thorson, W.: Calcul des Fonctions d'Onde Moleculaire, (ed. Daudel, R.) p. 141, Rec. Mem. Centre Nat. Rech. Sci., Paris 1958
16. Ham, F. S.: Phys. Rev. *166*, 307 (1968)
17. Hougen, J. T.: J. Mol. Spectrosc. *13*, 149 (1964)
18. Ballhausen, C. J.: Theor. Chim. Acta *3*, 368 (1965)
19. Sturge, M. D.: Solid State Phys. *20*, 91 (1967)
20. Bacci, M.: Phys. Rev. B. *17*, 4495 (1978)
21. Pearson, R. G.: J. Am. Chem. Soc. *91*, 4947 (1969)
22. Mingos, D. M. P.: Nature *229*, 193 (1971)
23. Pearson, R. G.: Proc. Nat. Acad. Sci. USA 72, 2104 (1975)
24. Pearson, R. G.: Symmetry Rules for Chemical Reactions: Orbital Topology and Elementary Processes. New York, Wiley 1976
25. Burdett, J. K.: Molecular Shapes. New York, Wiley 1980
26. Muramatsu, S., Iida, T.: J. Phys. Chem. Solids *31*, 2209 (1970)
27. Bersuker, I. B., Polinger, V. Z.: Phys. Lett A *44*, 495 (1973)
28. Bacci, M. et al.: Phys. Lett A *50*, 405 (1975)
29. Bacci, M. et al.: Phys. Rev. B *11*, 3052 (1975)
30. Ranfagni, A. et al.: Phys. Rev. B *20*, 5358 (1979)
31. Fröman, N., Fröman, P. O.: JWKB Approximation. Amsterdam, North-Holland 1965
32. Bell, R. P.: The Tunnel Effect in Chemistry. London, Chapman and Hall, 1980
33. Bleaney, B., Ingram, D. J. E.: Proc. Phys. Soc. A *63*, 408 (1950)

34. Abragam, A., Pryce, M. H. L.: Proc. Phys. Soc. A 63, 409 (1950)
35. Bleaney, B., Bowers, K. D.: ibid. 65, 667 (1952)
36. Moffitt, W., Liehr, A. D.: Phys. Rev. 106, 1195 (1957)
37. Moffitt, W., Thorson, W.: ibid. 108, 1251 (1957)
38. Longuett-Higgins, H. C. et al.: Proc. Roy. Soc. A (London) 244, 1 (1958)
39. Bersuker, I.B.: Soviet Phys. JETP 16, 933 (1963)
40. O'Brien, M. C. M.: Proc. Roy. Soc. A. (London) 281, 323 (1964)
41. Ham, F. S.: Phys. Rev. A 138, 1727 (1965)
42. Reinen, D., Friebel, C.: Structure and Bonding 37, 1 (1979)
43. Reinen, D.: J. Sol. State Chem. 27, 71 (1979)
44. Goodenough, J. B.: Magnetism and The Chemical Bond. New York, Wiley-Interscience 1963
45. Bersuker, I. B., Vekhter, B. G.: Ferroelectrics 19, 137 (1978)
46. Liehr, A. D.: J. Phys. Chem. 67, 389 (1963)
47. Ammeter, J. H., et al.: Helv. Chim. Acta 64, 1063 (1981)
48. Gazo, J. et al.: Coord. Chem. Rev. 19, 253 (1976)
49. Ammeter, J. H. et al.: Inorg. Chem. 18, 733 (1979)
50. Sturge, M. D.: Phys. Rev. B 1, 1005 (1970)
51. Montagna M., Pilla, O., Viliani, G.: J. Phys. C. 12, L 699 (1979)
52. Thorson, W. R.: J. Chem. Phys. 29, 938 (1958)
53. Tsuboi, M., Hirakawa, A. Y., Muraishi, S.: J. Molec. Spectrosc. 56, 146 (1975)
54. Pawlikowski, M., Zgierski, M. Z.: Chem. Phys. Lett 48, 201 (1977)
55. Pawlikowski, M., Zgierski, M. Z.: J. Raman Spectry. 7, 106 (1978)
56. Pawlikowski, M., Chem. Phys. Lett 80, 168 (1981)
57. Robbins, D. J.: Theoret. Chim. Acta 33, 51 (1974)
58. Stephens, P. J.: Ann. Rev. Phys. Chem. 25, 201 (1974)
59. Stephens, P. J.: Adv. Chem. Phys. 35, 197 (1976)
60. Trinkler, M. F., Zolovkina, I. S.: Phys. Stat. Sol. (b) 79, 49 (1977)
61. Le Si Dang et al.: Phys. Rev. Lett 38, 1539 (1977)
62. Ham, F. S.: J. Physique (Paris) 32, C1-952 (1971)
63. Kahn, O., Kettle, S. F. A.: Theoret. Chim. Acta 27, 187 (1972)
64. Kahn, O., Kettle, S. F. A.: Mol. Phys. 29, 61 (1975)
65. Ham, F. S.: J. Physique (Paris) 35, C6-121 (1974)
66. Lohr, L. L. Jr., Lipscomb, W. N.: Inorg. Chem. 2, 911 (1963)
67. Ballhausen, C. J., de Heer, J.: J. Chem. Phys. 43, 4304 (1965)
68. Nikiforov, A. E., Shashkin, S. Yu., Krotkii, A. I.: Phys. Stat. Sol. (b) 97, 475 (1980)
69. Nikiforov, A. E., Shashkin, S. Yu., Krofkii, A. I.: ibid. 98, 289 (1980)
70. Bacci, M.: Chem. Phys. Lett 58, 537 (1978)
71. Bacci, M.: Chem. Phys. 40, 237 (1979)
72. Jørgensen, C. K., Pappalardo, R., Schmidtke, H. H.: J. Chem. Phys. 39, 1422 (1963)
73. Schäffer, C. E., Jørgensen, C. K.: Mol. Phys. 9, 401 (1965)
74. Schäffer, C. E.: Structure and Bonding 5, 68 (1968)
75. Jørgensen, C. K.: Modern aspects of ligand field theory. North-Holland, Amsterdam 1971
76. Warren, K. D.: Inorg. Chem. 19, 653 (1980)
77. Warren, K. D.: Chem. Phys. Lett 89, 395 (1982)
78. Smith, D. W.: Structure and Bonding 35, 87 (1978)
79. Perutz, M. F.: Nature 228, 726 (1970)
80. Fermi, G.: J. Mol. Biol. 97, 237 (1975)
81. Perutz, M. F. et al.: Nature 295, 535 (1982)
82. Eichhorn, G. L. (ed.): Inorganic Biochemistry, Vol. 2, Part VI, Elsevier, Amsterdam 1973
83. Dougherty, T. J.: J. Natl. Cancer Inst. 52, 1333 (1974)
84. Spikes, J. D.: Ann. N. Y. Acad. Sci. 244, 496 (1975)
85. Jori, G.: The Molecular Biology of Photodynamic Action, in: Lasers in Photomedicine and
    Photobiology (eds. Pratesi, R., Sacchi, C. A.) p. 58, Berlin, Heidelberg, New York, Springer
    1980
86. Gouterman, M.: J. Chem. Phys. 30, 1139 (1959)
87. Gouterman, M.: Ann. N. Y. Acad. Sci. 206, 70 (1973)
88. Chan, I. Y. et al.: Mol. Phys. 22, 741 (1971)

89. Chan, I. Y. et al.: Mol. Phys. *22*, 753 (1971)
90. Simpson, W. T.: J. Chem. Phys. *17*, 1218 (1949)
91. Platt, J. R.: Electronic structure and excitation of polyenes and porphyrins, in: Radiation Biology (ed. Hollaender A.) vol. 3, p. 71, New York, Mc Graw-Hill 1956
92. McHugh, A. J., Gouterman, M., Weiss, C. Jr.: Theoret. Chim. Acta *24*, 346 (1972)
93. Schmidt, J., van der Waals, J. H.: Chem. Phys. Lett *2*, 640 (1968)
94. Canters, G. W. et al.: J. Phys. Chem. *80*, 2253 (1976)
95. Kielman-Van Luijt, E. C. M., Dekkers, H. P. J. M., Canters, G. W.: Mol. Phys. *32*, 899 (1976)
96. Kooter, J. A., van der Waals, J. H.: ibid. *37*, 997 (1979)
97. Shelnutt, J. A. et al.: J. Chem. Phys. *66*, 3387 (1977)
98. Shelnutt, J. A. et al.: ibid. *64*, 1156 (1976)
99. Small, G. J., Yeung, E. S.: Chem. Phys. *9*, 379 (1975)
100. Cheung, L. D., Yu, N., Felton, R. H.: Chem. Phys. Lett *55*, 527 (1978)
101. Shelnutt, J. A., O'Shea, D. C.: J. Chem. Phys. *69*, 5361 (1978)
102. Weissbluth, M.: J. Theor. Biol. *35*, 597 (1972)
103. Weissbluth, M.: Structural characteristics and electronic states of hemoglobin, in: Computational methods for large molecules and localized states in solids (eds. Herman, F., McLean, A. D., Nesbet, R. K.) p. 59, Plenum Press, New York 1973
104. Weissbluth, M.: Hemoglobin, Berlin, Heidelberg, New York, Springer 1974
105. Mizuhashi, S.: J. Phys. Soc. Jpn. *26*, 468 (1969)
106. Kotani, M.: Biopolymers Symposia *1*, 67 (1964)
107. Torii, K., Ogura, Y.: J. Biochem. *64*, 171 (1968)
108. Sato, M., Ohya, T., Morishima, I.: Mol. Phys. *42*, 475 (1981)
109. Eicher, H., Trautwein, A.: J. Chem. Phys. *50*, 2540 (1969)
110. Huynh, B. H., et al.: ibid. *61*, 3750 (1974)
111. Trautwein, A., Zimmermann, R., Harris, F. E.: Theoret. Chim. Acta *37*, 89 (1975)
112. Eicher, H., Bade, D., Parak, F.: J. Chem. Phys. *64*, 1446 (1976)
113. Bade, D., Parak, F.: Biophys. Struct. Mech. *2*, 219 (1976)
114. Chien, J. C. W., Dickinson, L. C.: Proc. Nat. Acad. Sci. USA *69*, 2783 (1972)
115. Dickinson, L. C., Chien, J. C. W.: Biochem. Biophys. Res. Commun. *51*, 587 (1973)
116. Yonetani, T., Yamamoto, H., Iizuka, T.: J. Biol. Chem. *249*, 2168 (1974)
117. Bacci, M.: Chem. Phys. Lett. *48*, 184 (1977)
118. Bacci, M.: J. Chem. Phys. *68*, 4907 (1978)
119. Ingalls, R.: Phys. Rev. A. *133*, 787 (1964)
120. Spartalian, K., Lang, G.: Oxygen transport and storage materials, in: Applications of Mössbauer spectroscopy (ed. Cohen, R. L.) vol. 2, p. 249, New York, Academic Press 1980
121. Champion, P. M., Sievers, A. J.: J. Chem. Phys. *72*, 1569 (1980)
122. Lumpkin, O., Dixon, W. T.: ibid. *68*, 3485 (1978)
123. Mispelter, J., Momenteau, M., Lhoste, J. M.: Biochemie *63*, 911 (1981)
124. Alpert, Y., Banerjee, R.: Biochem. Biophys. Acta *405*, 144 (1975)
125. Mineyev, A. P. et al.: Intern. J. Quant. Chem. *16*, 883 (1979)
126. Griffith, J. S.: Proc. Roy. Soc. A*235*, 23 (1956)
127. Brunori, M., Giardina, B., Bannister, J. V.: Oxygen-transport proteins, in: Inorganic Biochemistry, vol. 1, The Royal Society of Chemistry, London 1979
     Brunori, M., Giardina, B., Kuiper, H. A.: Oxygen-transport proteins, in: Inorganic Biochemistry, vol. 3, The Royal Society of Chemistry, London 1982
128. Cerdonio, M. et al.: Proc. Nat. Acad. Sci. USA *74*, 398 (1977)
129. Cerdonio, M. et al.: ibid. *75*, 4916 (1978)
130. Bacci, M., Cerdonio, M., Vitale, S.: Biophys. Chem. *10*, 113 (1979)
131. Case, D. A., Huynh, B. H., Karplus, M.: J. Am. Chem. Soc. *101*, 4433 (1979)
132. Herman, Z. S., Loew, G. H.: ibid. *102*, 1815 (1980)
133. Perutz, M. F. et al.: Biochemistry *13*, 2174 (1974)
134. Perutz, M. F. et al.: ibid. *13*, 2187 (1974)
135. Cianchi, L. et al.: Phys. Lett. A*59*, 247 (1976)
136. Mizuhashi, S.: J. Theor. Biol. *66*, 13 (1977)
137. Mizuhashi, S.: Rep. Univ. Electro-Comm. *29-1*, 73 (1978)

138. Mizuhashi, S.: J. Phys. Soc. Japn. *45*, 612 (1978)
139. Bersuker, I. B., Stavrov, S. S., Vekhter, B. G.: Biofizika *24*, 413 (1979)
140. Bersuker, I. B., Stavrov, S. S.: Chem. Phys. *54*, 331 (1981)
141. Schaffer, A. M., Gouterman, M., Davidson, E. R., Theor. Chim. Acta *30*, 9 (1973)
142. Bersuker, I. B., Stavrov, S. S., Vekhter, B. G.: Chem. Phys. *69*, 165 (1982)
143. Spiro, T. G. (ed.): Copper-proteins, Wiley-Interscience, New York 1981
144. Malkin, R., Malmstrom, B. G.: Adv. Enzymol. *33*, 177 (1970)
145. Fee, J. A.: Structure and Bonding *23*, 1 (1975)
146. Colman, P. M. et al.: Nature *272*, 319 (1978)
147. Adman, E. T. et al.: J. Mol. Biol. *123*, 35 (1978)
148. Bacci, M.: J. Inorg. Biochem. *13*, 49 (1980)
149. Bencini, A., Gatteschi, D., Zanchini, C.: J. Am. Chem. Soc. *102*, 5234 (1980)
150. Solomon, E. I. et al.: ibid. *102*, 168 (1980)
151. Brill, A. S.: Biophys. J. *22*, 139 (1978)
152. Brill, A. S.: Conformational distribution and vibronic coupling in the blue copper-containing protein azurin, in: Tunneling in biological systems (eds. Chance, B. et al.) p. 561, New York, Academic Press 1979
153. Brill, A. S.: Transition Metals in Biochemistry, p. 63, Berlin, Heidelberg, New York, Springer 1977
154. Kambe, K.: J. Phys. Soc. Jpn. *5*, 48 (1950)
155. Hatfield, W. E.: Properties of Magnetically Condensed Compounds, in: Theory and Applications of Molecular Paramagnetism (eds. Boudreaux, E. A., Mulay, L. N.) New York, Wiley 1976
156. Tsukerblat, B. S., Vekhter, B. G.: Soviet Phys. Sol. State *14*, 2204 (1973)
157. Passeggi, M. C. G., Stevens, K. W. H.: J. Phys. C *6*, 98 (1973)
158. Passeggi, M. C. G., Stevens, K. W. H.: Physica *71*, 141 (1974)
159. Khomskii, D. I., Kugel, K. I.: Solid State Commun. *35*, 409 (1980)
160. Drillon, M., Georges, R.: Phys. Rev. B *26*, 3882 (1982) and references therein
161. Carter, C. W. et al.: J. Biol. Chem. *249*, 6339 (1974)
162. Gall, R. S., Connelly, N. G., Dahl, L. F.: J. Am. Chem. Soc. *96*, 4017 (1974)
163. Gall, R. S., Chu, C. T., Dahl, L. F.: ibid. *96*, 4019 (1974)
164. Yang, C. Y. et al.: ibid. *97*, 6596 (1975)
165. Laskowski, E. J. et al.: ibid. *100*, 5322 (1978)
166. Thomson, A. J.: J. C. S. Dalton 1180 (1981)
167. Stout, C. D.: Iron-Sulfur Protein Crystallography in: Iron-Sulfur Proteins, vol. 4 (ed. Spiro, T. G.) New York, Wiley 1982
168. Aizman, A., Case, D. A.: J. Am. Chem. Soc. *104*, 3269 (1982)
169. Horatius: Satirae *1*, 1, v. 106–107

Crystal Structure Non-Rigidity of Central Atoms for Mn(II), Fe(II), Fe(III), Co(II), Co(III), Ni(II), Cu(II) and Zn(II) Complexes

Fedor Valach[1], Branislav Koreň[1], Peter Sivý[1], Milan Melník[2]

[1] Department of Chemical and Technical Physics and Nuclear Technique, Slovak Technical University, 812 37 Bratislava, Czechoslovakia
[2] Department of Inorganic Chemistry, Slovak Technical University, 812 37 Bratislava, Czechoslovakia

The ability of coordination polyhedra of different central atoms in the solid state to undergo distortions has often been discussed. These distortions are also correlated with the nature of the central atom. This paper deals with the conception of crystal structure non-rigidity for central atoms of some selected complexes by the using of statistical distributions of interatomic distances central atom-nearest ligand atom (M-L; M = Mn(II), Fe(II), Fe(III), Co(II), Co(III), Ni(II), Cu(II) and Zn(II)). Based on the vector equilibrium principle the dispersion of these empirical distributions has been introduced as a measure of non-rigidity of central atoms, decreasing in the order: Cu(II) > Fe(II)$_{HS}$ > Mn(II)$_{HS}$ > Ni(II)$_{LS}$ > Zn(II) > Fe(III)$_{HS}$ ≈ Co(II)$_{HS}$ > Ni(II)$_{HS}$ > Co(III)$_{LS}$, where LS and HS denote low-spin and high-spin states, respectively. Starting from the distribution of M-L distances for certain types of central and ligand atoms, the contribution of atoms of the inner coordination sphere to the vector equilibrium of crystal structures is discussed.

The structures of complexes with all M-L bond lengths of at least one symmetrically independent central atom in the interval of maximum frequency exhibit in the appropriate set the most stable inner coordination sphere. These compounds are classified both with respect to the geometry of their inner coordination sphere fulfilling the above condition and to their crystal structure aspect. The square-planar geometry is found to be the most stable for Cu(II) and Ni(II)$_{LS}$ complexes, tetragonal-pyramidal for Zn(II) complexes, while for Fe(II)$_{HS}$, Mn(II)$_{HS}$, Fe(III)$_{HS}$, Co(II)$_{HS}$, Ni(II)$_{HS}$ and Co(III)$_{LS}$ complexes it is the regular octahedral geometry. Except for one Mn(II)$_{HS}$ complex, in the discussed structures chromophores with the most stable geometry of the coordination polyhedron form part of the final relatively isolated structural unit (island) with expressively chemical bonds between atoms. These facts are illustrated by the stabilization effected by the crystal field and by the properties of the ligands.

Structure and Bonding 55
© Springer-Verlag Berlin Heidelberg 1983

1 Introduction

The metal atom in crystal structures of coordination compounds is surrounded by adjacent atoms located in the apices of a convex polyhedron. In Ref. 1 such polyhedra with a coordination number of four, five, six and eight are cataloged. Usually, polyhedra of coordination compounds are deformed by some holosymmetric form. Such a non-rigidity of complexes[2] in the solid state (if temperature vibrations are not considered) mainly depends on the properties of the central atom, ligands and of the crystal as a whole. Concerning hexacoordinated complexes with tetragonal deformations arising from O_h symmetry, this aspect has been discussed on the basis of the first two points studied by Gažo et al.[3]

The factor refered to the crystal is usually mentioned in the literature as "crystal packing", "close packing" or the "solid-state effect".

The aim of this work is to consider the distortions of coordination polyhedra from the standpoint of common principles in the geometry of infinite, regular packing of ions and atoms, i.e. in the crystal. Let us confine ourselves to the study of local trends of the central atom and its nearest ligand atoms, i.e. to its nearest surroundings. In the series of crystal structures of a certain transition metal the term "trend" means postulated requirements made on the surroundings of central atoms, leading to their most stable geometry. This trend is greater, the smaller the dispersion of geometries. Such a statistical approach with a sufficiently great number of crystal structure data of different nearest surroundings of central atoms of the same type allows us to suppose some averaging of contributions of more distant atoms. The subject of this study is to determine the contribution of the central atom to the crystal structures.

2 Geometrical Principles of Crystal Structures

The model of close packing of energetically most stable structural units[4, 5] described in a number of papers[6-9] has allowed a successful interpretation and classification especially of organic crystals. In these cases, rather universal radii may be assigned to the atoms so that the structural units consist of spheres having a maximum number of contacts with those of the neighbouring structural units. This approach may be applied to crystals consisting of discrete structural units, between which only interactions of the van der Waals type (possibly hydrogen bonds) exist. In addition to the above limitation there must be considered

(a) perfect rigidity of structural units and
(b) interactions only between neighbouring structural units.

In the reviews of Loeb[10] and Hellner[11] the following classification of structural types on the basis of the structure geometry of intermetallic compounds has been made[12]:

(a) *Space principle;* trend to the most effective packing of atoms in space in the form of contiguous rigid spheres.
(b) *Symmetry principle;* trend of the surroundings of every atom to display the highest symmetry.

(c) *Connection principle;* trend of every atom to undergo the shortest interatomic contacts.

These principles are formulated in[12] on the basis of the relative occurrence of groups of structures of metallic elements which comply with these trends.

The connection principle leads to the occurrence of atom groups with the shortest interatomic contacts referred to as connections according to Laves[12]. Owing to the ability of chemical bonds to be coordinated in certain directions, these connections may be finite formations which are called groups and islands, respectively (see Sect. 3). One-dimensional, two-dimensional and three-dimensional infinite connections are called chains, layers and skeletons, respectively.

The structures of the crystals of each compound is the result of a compromise of an infinite number of interatomic attractive and repulsive actions. Moreover, there are some limitations of the structure following from stoichiometry[13, 14]. On the basis of an equilibrium of attractive and repulsive forces, Loeb[13] postulated the Vector Equilibrium Principle (**VEP**) which, for the crystal structures consisting of atoms and ions of different types, may be formulated as follows:

Each atom or ion in crystals tends to be equidistant to as many as possible atoms or ions of each type.

In addition to the needs following from the stoichiometry the **VEP** also takes those requirements of structural geometry into consideration which coincide with the principles of symmetry and connection. The space principle is also discussed here, however, with the difference that we do not start from a model of rigid mutually contacting spheres but also plastical or soft contacts between atoms and ions are admitted. Let us try and delimit the physical meaning of the **VEP** by the following consideration:

We consider two atoms with the interatomic potential as shown in Fig. 1. Regardless of the type of interatomic interaction, there exists only one equilibrium distance, r_0. It is the distance at which the attractive and repulsive forces are compensated. Three atoms then occupy the most stable positions in the apices of an equilateral triangle with the shoulders of r_0. The most stable arrangement for four atoms is in the apices of a regular tetrahedron. In such arrangement each atom has three equidistant neighbours. A square arrangement is less stable, since each atom here has only two equidistant neighbours. In this respect, for the case of five atoms the most stable arrangement is a regular trigonal-bipyramidal configuration with nine equal interatomic distances. In this configuration apex atom also has three equidistant neighbours. Somewhat less stable is e.g. the tetragonal pyramidal arrangement with eight equal interatomic distances. In this arrangement, however, the apex atom has four equidistant neighbours (the highest number) while the other atoms have only three. Thus, the trigonal-bipyramidal geometry complies better with the **VEP**. Similarly, for the case of a six-atom group an octahedral arrangement appears to be the most stable and most suitable one with respect to the **VEP**. If, however, the apices of these geometrically regular coordinations are occupied by ligand atoms with ions of transition metals in their centres, the stability of such coordinations is not only influenced by interactions between the donor atoms, but also by the ligand field stabilization which depends on its strength and the number of d^n electrons of the central atom[15].

3 Contribution of the Nearest Surroundings of the Central Atom to the Vector Equilibrium of Crystal Structures

According to the **VEP** each atom in the unit cell tends to occupy such positions in which it would be surrounded by the greatest possible number of equidistant atoms, which are nearer to it than to every other translationally joint atoms or they are placed in the midpoint of the connecting line between the initial and the translationally joint atoms. Thus, they are placed inside a convex parallelepiped or on its surface, called the Dirichlet domain[16, 17], and the initial atom is in its geometric mean. Such a trend to vector equilibrium may be postulated for atom arrangements in which each interatomic interaction leading to the formation of a bond shows a potential with an absolute minimum (Fig. 1). In other words, for every such interatomic interaction there must exist an equilibrium distance between atoms with a minimum potential energy. This condition is fulfilled for the arrangements of interatomic interactions of predominantly ionic and covalent type. Similarly, also for interactions of hydrogen and van der Waals type bonds an equilibrium interatomic distance exists. It is, however, questionable whether there exists such a distance between the central atom and the donor atom in structures of transition metal complexes. In the positive case, the central and donor atoms with interatomic vectors of M-L also contribute to the vector equilibrium of the crystal structure. Quantum chemical calculations of linear MX_2 and square planar $[MX_4]^{2-}$ systems (M = Mn, Fe, Co, Ni, Cu; X = F, Cl, Br) using the **MO** method revealed that these systems exhibit equilibrium geometry with a minimum of the adiabatic potential[18]. Similarly, halo systems of the $[MX_6]^{n-}$ type in the electronic states A_{1g}[14] and ^2E_g[19, 20], and trans-$[MCl_2(NH_3)_4]$ (M = Ni(II), Co(III), Zn(II))[21] display equilibrium geometry with a minimum adiabatic potential.

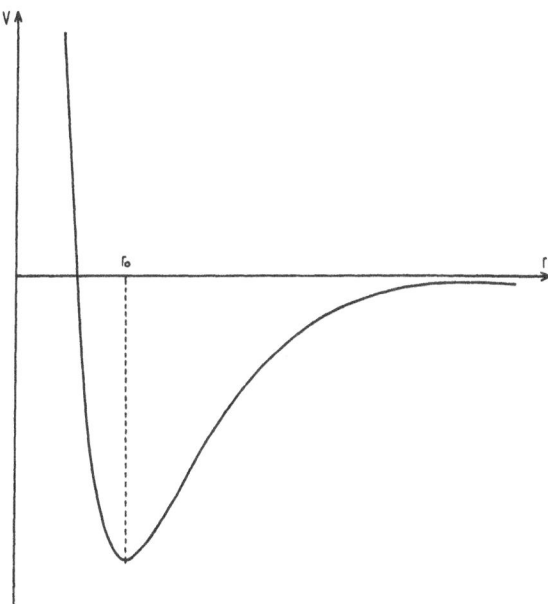

Fig. 1. Interaction potential V as a function of the interatomic distance r at the equilibrium distance r_0

To estimate the equilibrium bond distances M-L of complexes of the central atoms, histograms of M-L distances with the length d = 0.05 Å (1 Å = 100 pm = 0.1 nm) were computed (Fig. 2 a). For the computations data of transition metal compounds with d^n electron configurations of the central atoms as listed in Table 1 were taken. Only data obtained by the diffractometric technique with the values of $R = \Sigma\|F_o\| - |F_c\|/\Sigma|F_o| \leq$ 0.090 were used. Figure 2 shows histograms with significant numbers of data n for the dispersion σ^2 on the significance level of 0.05. Relation (1)[18] was used for testing purposes:

$$n \geq n_{crit} = (z_a)^2 \sigma^2 / d^2 \tag{1}$$

For the selected level $z_a = 1.96$ and n_{crit} is the critical number of data. For the dispersion, the estimation according to Eq. (2) was used.

$$\sigma^2 \approx \left[\sum_{i=1}^{n} (M\text{-}L)_i^2 - \left(\sum_{i=1}^{n} (M\text{-}L)_i \right)^2 / n \right] / (n-1) \tag{2}$$

As Fig. 2 a shows, every histogram exhibits one class with only the greatest relative frequency, limiting at the same time the equilibrium bond lengths. These ranges of M-L bond lengths for individual bond types, closed on the left and open on the right, are compiled in Table 2 with the reliability intervals of their probability. For the computation of the lower and the upper limits of probability relation (3) was applied[23]:

$$g_{l,u} = \frac{1}{n + z_a} \left\{ f_{max} + \frac{z_a^2}{2} \pm z_a \left[\frac{f_{max}(n - f_{max})}{n} \right]^{1/2} + \frac{z_a^2}{4} \right\} \tag{3}$$

where f_{max} is the corresponding maximum frequency.

Thus, atoms contained in chromophores with bonds as given in Table 2 also contribute to the vector equilibrium in the crystal structure which manifests itself by their trend to obtain equilibrium lengths. It is, however, apparent that in the crystal structure these bonds may deviate from equilibrium distances, due to external factors.

Table 1. Central atoms and their d^n electron configurations

d^n	Low-spin			High-spin		
	Central atom	Configuration	Ground state	Central atom	Configuration	Ground state
d^5				Mn(II)	$t_{2g}^3 e_g^2$	$^6A_{1g}$
				Fe(III)	$t_{2g}^3 e_g^2$	$^6A_{1g}$
d^6				Fe(II)	$t_{2g}^4 e_g^2$	$^5T_{2g}$
	Co(III)	$t_{2g}^6 e_g^0$	$^1A_{1g}$			
d^7				Co(II)	$t_{2g}^5 e_g^2$	$^4T_{2g}$
d^8	Ni(II)	$t_{2g}^6 e_g^2$	$^1A_{1g}[D_{4h}]$	Ni(II)	$t_{2g}^6 e_g^2$	$^3A_{1g}$
d^9	Cu(II)	$t_{2g}^6 e_g^3$	^2E_g			
d^{10}	Zn(II)	$t_{2g}^6 e_g^4$	$^1A_{1g}$			

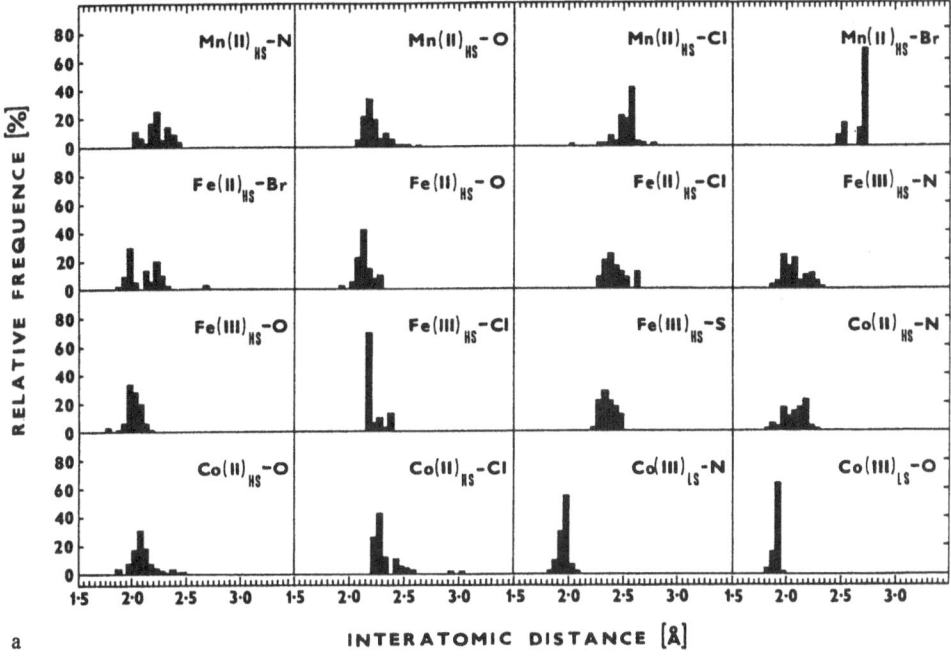

a

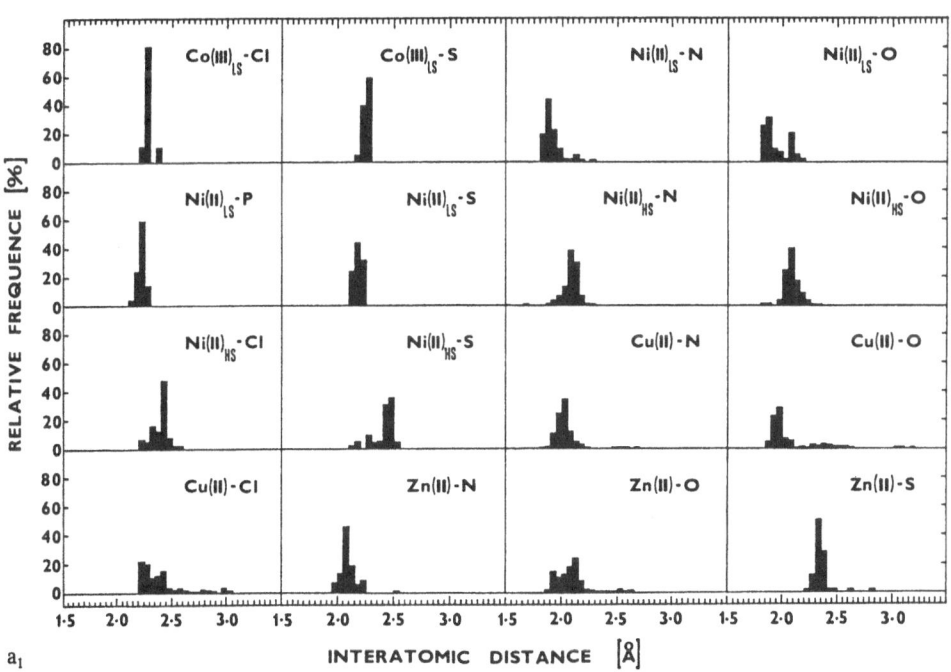

a_1

Fig. 2a. Histograms of interatomic distances M-L with certain central and donor atoms

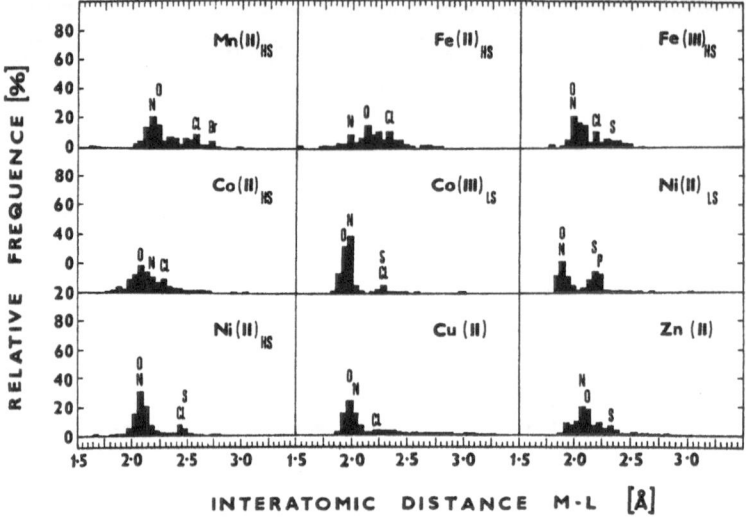

RELATIVE FREQUENCE [%]

INTERATOMIC DISTANCE M-L [Å]

Fig. 2b. Histograms of interatomic distances M-L with certain central atoms and nearest ligand atoms

Figure 2b shows histograms of lengths of interatomic vectors coordinating the central atom to the nearest ligand atom. Classes limiting equilibrium bond lengths in the histograms in Fig. 2a are marked by symbols of the respective donor atoms. As it is seen from Fig. 2b, the maximum frequencies, for the most part, correspond to these intervals which is a consequence of the trend to maximum equidistance also for donor atoms, in the sense of the **VEP**. The deviation of the M-L vector lengths from the maximum equidistance, for complexes of a certain central atom is a measure of the dispersion of their empirical distribution[23]:

$$\sigma^2 = \sum_{i=1}^{n} f_i[(M\text{-}L)_i - \langle M\text{-}L \rangle]^2 \tag{4}$$

where $(M\text{-}L)_i$ is the middle of the i-th class with the relative frequency of f_i and $\langle M\text{-}L \rangle$ the mean value according to Eq. (5):

$$\langle M\text{-}L \rangle = \sum_{i=1}^{n} f_i(M\text{-}L)_i \tag{5}$$

σ^2 dispersions and classes of M-L vector lengths with the greatest relative frequencies are listed in Table 3. The tendency of M-L vectors to attain the maximum equidistance for the complexes under investigation increases in the contrary order to the dispersion values:

$$\text{Cu(II)} < \text{Fe(II)}_{HS} < \text{Mn(II)}_{HS} < \text{Ni(II)}_{LS} < \text{Zn(II)} < \text{Fe(III)}_{HS} \approx \text{Co(II)}_{HS} < \text{Ni(II)}_{HS} <$$
$$< \text{Co(III)}_{LS} \tag{6}$$

Table 2. Classes of bond distances, M-L, with maximum frequencies

Bond	n	n_{crit}	Class [Å]	Relative frequency [%]	g_l [%]	g_u [%]
Mn(II)$_{HS}$-N	74	56	2.20–2.25	25.7	17.1	36.7
Mn(II)$_{HS}$-O	281	26	2.15–2.20	33.5	28.2	39.2
Mn(II)$_{HS}$-Cl	116	21	2.55–2.60	40.5	32.0	49.6
Mn(II)$_{HS}$-Br	27	20	2.70–2.75	66.7	47.8	81.4
Fe(II)$_{HS}$-N	66	69	1.95–2.00	28.8	19.3	40.6
Fe(II)$_{HS}$-O	46	13	2.10–2.15	41.3	28.3	55.7
Fe(II)$_{HS}$-Cl	25	24	2.35–2.40	24.0	11.5	43.4
Fe(III)$_{HS}$-N	57	31	1.95–2.00	22.8	13.8	35.2
Fe(III)$_{HS}$-O	124	12	1.95–2.00	33.9	26.1	42.6
Fe(III)$_{HS}$-Cl	33	13	2.15–2.20	69.7	52.7	82.6
Fe(III)$_{HS}$-S	43	12	2.30–2.35	27.9	16.7	42.7
Co(II)$_{HS}$-N	198	49	2.15–2.20	22.2	17.0	28.5
Co(II)$_{HS}$-O	253	38	2.05–2.10	31.2	25.8	37.2
Co(II)$_{HS}$-Cl	82	51	2.25–2.30	41.5	31.4	52.3
Co(III)$_{LS}$-N	345	19	1.95–2.00	53.9	48.6	59.1
Co(III)$_{LS}$-O	91	2	1.90–1.95	60.4	50.2	69.9
Co(III)$_{LS}$-Cl	10	5	2.25–2.30	80.0	49.0	94.3
Co(III)$_{LS}$-S	26	2	2.25–2.30	57.7	38.9	74.5
Ni(II)$_{LS}$-N	149	19	1.85–1.90	40.3	32.7	48.3
Ni(II)$_{LS}$-O	45	30	1.85–1.90	31.1	19.5	45.8
Ni(II)$_{LS}$-P	29	3	2.20–2.25	58.6	40.7	74.5
Ni(II)$_{LS}$-S	87	3	2.15–2.20	43.7	33.7	54.1
Ni(II)$_{HS}$-N	342	12	2.05–2.10	37.4	32.5	42.7
Ni(II)$_{HS}$-O	307	12	2.05–2.10	39.1	33.8	44.6
Ni(II)$_{HS}$-Cl	62	13	2.40–2.45	48.4	36.4	60.6
Ni(II)$_{HS}$-S	81	25	2.45–2.50	35.8	26.2	46.7
Cu(II)-N	719	40	2.00–2.05	35.2	31.8	38.7
Cu(II)-O	1159	207	1.95–2.00	29.3	26.7	31.9
Cu(II)-Cl	206	58	2.20–2.25	22.3	17.2	28.5
Zn(II)-N	94	15	2.05–2.10	45.7	36.0	55.8
Zn(II)-O	279	53	2.10–2.15	24.0	19.4	29.4
Zn(II)-S	51	19	2.30–2.35	51.0	37.7	64.1

Table 3. Classes of interatomic lengths of vectors, central atom-nearest ligand atom with maximum frequencies and dispersions of empirical distributions

Vector	n	n_{crit}	Class [Å]	Relative frequency [%]	g_l [%]	g_u [%]	σ^2 [Å²]
Mn(II)$_{HS}$-L	515	124	2.15–2.20	20.8	17.5	24.5	0.036
Fe(II)$_{HS}$-L	220	65	2.10–2.15	14.6	10.5	19.8	0.043
Fe(III)$_{HS}$-L	259	37	1.95–2.00	21.2	16.7	26.6	0.024
Co(II)$_{HS}$-L	581	45	2.05–2.10	18.6	15.6	21.9	0.024
Co(III)$_{LS}$-L	487	25	1.95–2.00	38.6	34.4	43.0	0.012
Ni(II)$_{LS}$-L	346	55	1.85–1.90	22.0	17.9	26.6	0.033
Ni(II)$_{HS}$-L	818	34	2.05–2.10	30.3	27.3	33.6	0.022
Cu(II)-L	2180	106	1.95–2.00	23.8	22.1	25.6	0.069
Zn(II)-L	480	44	2.05 2.10	19.6	16.3	23.4	0.028

Robinson et al.[20] found a linear correlation between the mean elongation of M-L bonds and the L-M-L angle variance for a great number of minerals. These parameters characterizing the distortion of the bond lengths and angles of distorted octahedral coordination polyhedra are

$$\langle \lambda_{oct} \rangle = \sum_{i=1}^{6} (l_i/l_o)^2/6 \tag{7}$$

$$\sigma^2_{\theta(oct)} = \sum_{i=1}^{12} (\theta_i - 90°)^2/11$$

while for tetrahedral polyhedra they are

$$\langle \lambda_{tet} \rangle = \sum_{i=1}^{4} (l_i/l_o)^2/4 \tag{8}$$

$$\sigma^2_{\theta(tet)} = \sum_{i=1}^{6} (\theta_i - 109.47°)^2/5$$

where θ_i are angles of L-M-L and l_i M-L bond lengths of the coordination polyhedron; l_o is the M-L length of an undistorted polyhedron of the same volume as the distorted one. Even though the linear correlation $\langle \lambda \rangle$ vs σ^2_θ needs not be generally valid, as e.g. for the greater distortions of Cu(II) complexes, a positive correlation between the distortions of lengths and those of angles may be assumed, according to the VEP. Then, for the studied central atoms the tendency to such distortions increases in the contrary order relative to Eq. (6).

External factors can, however, also lead to stabilization of the M-L bonds and thus also to the geometry stabilization of the coordination polyhedron. In order to estimate such an influence of the central atom surroundings let us further consider the crystal

Table 4. Mean bond lengths, M-L, and cis-angles, L-M-L, with their standard deviations[a] for the chromophores with bond lengths in the maximum frequency range

Central atom	$\langle M\text{-}L \rangle$ [Å]	$\sigma(M\text{-}L)$ [Å]	$\langle L\text{-}M\text{-}L \rangle$ [°]	$\sigma(L\text{-}M\text{-}L)$ [°]
Mn(II)$_{HS}$	2.176	0.006	90.0	1.4
Fe(II)$_{HS}$	2.128	0.017	90.0	0.9
Fe(III)$_{HS}$	1.972	0.016	90.0	3.9
Co(II)$_{HS}$	2.078	0.015	90.0	1.5
Co(III)$_{LS}$	1.963	0.018	90.1	3.2
Ni(II)$_{LS}$	1.869	0.013	89.4	5.4
Ni(II)$_{HS}$	2.068	0.013	90.0	8.0
Cu(II)	1.964	0.009	90.0	4.0
Zn(II)	2.077	0.013	94.0	6.1

[a] computed by use of Eq. (2)

structures of those compounds which have all the M-L vector lengths of at least one symmetrically independent central atom in the range of bond lengths with a maximum frequency (Fig. 2b, Table 3). The mean bond distances and the *cis*-bonding angles of such chromophores are collected in Table 4; and the crystal data of compounds which fulfill the above condition are listed in Table 5. As we can see in Table 4, with the exception of Zn(II) complex, the mean bond angles are practically equal to 90°. This fact shows that the trend of the central atom to attain the maximum equidistance with atoms of the nearest surroundings leads to the tetragonal angles of L-M-L. The finite structure unit, being with the other units in the crystal, bonded only by hydrogen and/or van der Waals forces and containing a chromophore which fulfills the above condition, is called "island". The chemical formula of this structure is given in angular brackets. The other finite structural units are termed as "groups".

3.1 High-Spin Mn(II) Complexes

An X-ray structure analysis of *catena*-bis(μ-glycine-O,O') manganese(II) bromide dihydrate,

$$Mn(^+H_3NCH_2COO^-)_2 \cdot (H_2O)_2Br_2$$

shows that this complex exists in two forms, the monoclinic and the orthorhombic form[25]. Both of these forms exhibit a *chain-like* crystal structure. The arrangement of these chains which are connected by hydrogen bonds via Br^- ions in the monoclinic form is shown in Fig. 3. The distance of Mn . . . Mn = 5.42 Å does not indicate any intermetallic interaction. The Mn(II) atoms occupy centrosymmetric positions with an octahedral coordination, the distances of Mn-O(1)(2x), Mn-O(2^{vi})(2x) and Mn-O(w)(2x) being 2.169(5), 2.177(5) and 2.183(6) Å. The Mn(II) atoms are bridged by a "syn-syn" coordination of the glycine ions with the formation of chains containing MnO_6 chromophores whereby the bond angle O(1)-Mn-O(2^{vi}) is obtuse to the value of 92.0(2)° and the remaining angles O(1)-Mn-O(w) and O(w)-Mn-O(2^{vi}) are somewhat smaller, namely 90.9(2)° and 90.8(2)°. As Fig. 4 shows the orthorhombic form consists of more complicated chains than its monoclinic isomer. For both the monoclinic and the orthorhombic forms the chains are mutually linked by hydrogen bonds. The Mn(II) atoms occupy asymmetric positions and only half the bond lengths of Mn-O are in the range of 2.15–2.20 Å with a maximum relative frequency (Table 3), while the bond angles of O–Mn–O exhibiting values of 81.9(3)–100.2(3)° appear in a much broader range than in the monoclinic form.

3.2 High-Spin Fe(II) Complexes

From the Fe(II) complexes only two fulfill the conditions given in the introduction of Sect. 3. These are hexakis(pyridine-1-oxide)iron(II)perchlorate, [Fe(pyNO)$_6$](ClO$_4$)$_2$[26], and hexadeuteriumcyanideiron(II) ditetrachloroiron(III), (FeCl$_4$)$_2$[Fe(NCD)$_6$][27].

   [Fe(PyNO)$_6$](ClO$_4$)$_2$ consists of [Fe(pyNO)$_6$]$^{2+}$ ions of the symmetry $\bar{3}$(S$_6$) and distorted tetrahedral ClO$_4^-$ groups. Figure 5 shows the structure of this complex projected

Table 5. Crystal data of the compounds with bond lengths, M-L, in the maximum frequency range

Compound	Crystal class[a]	Space group	Z	a [Å] b [Å] c [Å]	α [°] β [°] γ [°]	R[b]	Ref.
$Mn(^+H_3NCH_2COO^-)_2(H_2O)_2Br_2$	m	$P2_1/c(C_{2h}^5)$	2	11.943(3) 6.060(2) 8.979(2)	111.65(3)	0.040	25
$[Fe(pyNO)_6](ClO_4)_2$	tg	$R\bar{3}(C_{3i}^2)$	1	9.640(1) 10.29(1)	81.06(1)	0.032	26
$(FeCl_4)_2[Fe(NCD)_6]$	tg	$P\bar{3}(C_{3i}^1)$	1	6.283(6)		0.086	27
$[Fe(H_2O)_6](NO_3)_3 \cdot 3H_2O$	m	$P2_1/c(C_{2h}^5)$	4	13.989(1) 9.701(1) 11.029(1)	95.52(1)	0.042	29
$(C_{10}H_9N_2)[Fe(C_{10}H_8N_2)_3](ClO_4)_4$	tr	$P\bar{1}(C_i^1)$	2	16.607(3) 13.496(3) 10.254(2)	101.13(2) 92.08(2) 102.77(1)	0.056	30
	tg	$R\bar{3}(C_{3i}^2)$	3	12.512(3) 19.044(3)		0.037	31
$[Co(pyNO)_6](ClO_4)_2$	tg	$R\bar{3}(C_{3i}^2)$	1	9.619(1)	81.19(1)	0.030	26
$[Co(TTA)_2(methanol)_2]$	tr	$P\bar{1}3(C_i^1)$	2	12.016(3) 10.405(2) 10.329(3)	97.76(20) 113.31(22) 98.79(19)	0.057	32
$[Co(NH_3)_6]Cl_3$	m	$C2/m(C_{2h}^3)$	12	12.46(1) 21.30(2) 12.74(1)	112.96(8)	0.038	34
$(+)_D$-$[Co(penten)]Co(CN)_6 \cdot 2H_2O$	or	$P2_12_12_1(D_2^4)$	4	15.471(4) 16.036(5) 9.253(3)		0.083	35
$[Co(H_2O)(NH_3)_5]_2(S_2O_6)_3 \cdot 2H_2O$	m	$P2_1/a$	2	13.538(4) 15.820(9) 6.593(2)	95.99(3)	0.039	36

Compound		Space group	Z	Cell dimensions	Angles	R	Ref.
$[Co(NH_3)_3dien](S_2O_6)Cl$	m	$P2_1/n(C_{2h}^5)$	4	8.345(2) 14.229(4) 12.459(5)		0.057	37
$(-)_{589}$-(lel_2ob)-$[Co(chxn)_3]Cl_3 \cdot 5\,H_2O$	hx	$P6_1(C_6^2)$	6	12.472(1) 32.594(2)	93.93	0.040	38
$(+)_{589}$-(ob_3)-$[Co(chxn)_3]Cl_3 \cdot H_2O$	m	$C2(C_2^3)$	2	13.922(4) 10.720(4) 8.777(4)	108.83(4)	0.034	42
$(-)_{589}$-$[Co(R,R\text{-}chxn)_2(3,3'\text{-}dmbpy)]Br(ClO_4)_2 \cdot H_2O$	or	$P2_12_12_1(D_2^4)$	4	12.298(1) 30.394(4) 8.599(1)		0.052	39
$(+)_{589}$-$[Co(3,3'\text{-}dmbpy)(en)_2]Cl(ClO_4)_2 \cdot H_2O$	or	$P2_12_12_1(D_2^4)$	4	16.754(2) 18.311(2) 8.129(1)		0.036	44
$[Co(en)_3](tart)Cl \cdot 5\,H_2O$	tr	$P1(C_1^1)$	1	8.261(3) 8.507(3) 8.149(3)	101.20(1) 95.27(1) 102.31(1)	0.022	45
$(-)_{589}$-$[Co((R)\text{-}MeTACN)_2]I_3 \cdot 5\,H_2O$	tg	$R32(D_3^7)$	3	8.799(1) 32.180(4)		0.039	46
$[Ni(OAOH)_2] \cdot 2\,H_2O$	m	$P2_1/n$	2	3.969(1) 13.006(4) 11.512(3)	99.45(2)	0.037	47
$[Ni(OAOH)_2]H_2O \cdot DMF$	tr	$P\bar{1}(C_1^1)$	2	8.116(2) 9.902(2) 9.856(1)	94.87(2) 99.98(1) 100.98(2)	0.051	48
$[Ni(C_{18}H_{14}N_4)]$	m	$P2_1/c(C_{2h}^5)$	4	19.456(4) 5.228(1) 14.868(3)	112.28(1)	0.041	49
$[Ni\{Me_2Ga(N_2C_3H_3)_2\}_2]$	m	$P2_1/c(C_{2h}^5)$	2	8.530(6) 17.939(10) 7.415(6)	106.88(7)	0.049	50

Table 5 (continued)

Compound	Crystal class[a]	Space group	Z	a [Å] b [Å] c [Å]	α [°] β [°] γ [°]	R[b]	Ref.
[Ni(EMG)₂]	m	P2₁/c(C₂ₕ⁵)	2	4.7471(5) 11.7409(30) 11.9895(20)	91.611(16)	0.048	51
[Ni(C₃H₈N₄)₂]Cl₂	m	B2/b(C₂ₕ⁶)	4	16.37(1) 16.37(1) 5.121(3)	109.93(4)	0.081	52
K₂Pb[Ni(NO₂)₆]	c	Fm3(Tₕ³)	4	10.5775(8)		0.016	53
β-[Ni(NO₃)₂ · 4 H₂O]	tr	P1̄(Cᵢ¹)	2	7.484(2) 10.277(3) 5.476(1)	101.06(3) 101.53(3) 68.77(3)	0.049	54
[Ni(pyNO)₆](BF₄)₂	tg	R3(C₃⁴) R3̄(C₃ᵢ²)	3	12.487(5) 18.92(1)		0.0419 0.0485	55
Ca[Ni(C₂H₄N₂(CH₂COO)₄] · 4 H₂O	m	P2₁/b(C₂ₕ⁵)	4	6.683(5) 24.84(1) 10.243(7)	95.6(1)	0.068	56
[Ni(quin)₂(C₆H₅COO)₂]	m	P2₁/a	2	14.06(2) 9.75(2) 10.36(2)	113.32(2)	0.045	58
[Cu(C₃H₈N₄)₂](ClO₄)₂	m	I2/m	2	10.727(4) 9.719(3) 7.739(2)	92.12(2)	0.036	52
[Cu(C₃N₃O₂H₂)(H₂O)₂]	tr	P1̄(Cᵢ¹)	1	6.774(2) 8.348(2) 4.878(2)	92.75(2) 102.16(3) 74.94(2)	0.031	59

[Cu$_2$(C$_{12}$H$_{13}$N$_2$O$_2$)$_2$]Cl(CH$_3$OH)$_{0.5}$	tr	P$\overline{1}$(C$_i^1$)	2	11.861(2) 8.265(2) 14.076(3)	104.27(2) 104.68(2) 100.88(2)	0.065	60
[Zn(TPP)(ClO$_4$)]	m	P2$_1$/c(C$_{2h}^5$)	4	13.638(4) 13.017(5) 20.452(8)	107.63(2)	0.073	62

a c, cubic; hx, hexagonal; m, monoclinic; or, orthorhombic; tg, trigonal; tr, triclinic

b R = $\Sigma||F_o| - |F_c||/\Sigma|F_o|$

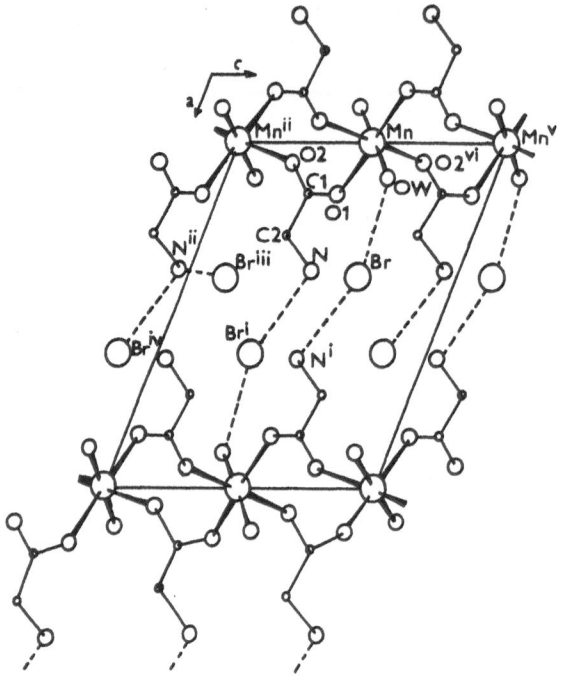

Fig. 3. Crystal structure of the monoclinic $Mn(^-OOCCH_2NH_3^+)_2(H_2O)_2Br_2$ viewed along the b axis. The hydrogen bonds are denoted by the *broken lines* (Ref. 25)

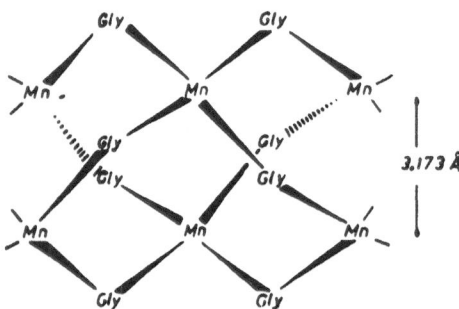

Fig. 4. Schematic representation of the infinite chain in the crystal structure of orthorhombic $Mn(^-OOCCH_2NH_3^+)_2(H_2O)_2 \cdot Br_2$. Water molecules are omitted (Ref. 25)

into the plane of the metal atoms, the chlorine atoms of the perchlorate anions deviating from this plane by 0.34 Å. The structure of this complex may be understood as an approximately cubic packing of such "layers". The angle between the plane of the least squares of the nearly planar pyridine-1-oxide ligand and the plane of Fe-O-N is 71.8°, the bond angles of Fe-O-N being 119.7(1)° and the bond length of Fe-O = 2.112(2)(6x) Å. The octahedron of FeO_6 is very slightly trigonally distorted with an angle O-Fe-O of 89.41(6)°.

An X-ray analysis[27] of the iron(II)-iron(III) complex $(FeCl_4)_2[Fe(NCD)_6]$ reveals that in this complex the Fe(III) atom has a tetrahedral and the Fe(II) atom an octahedral configuration (Fig. 6). In the crystal structure the tetrahedral groups of $(FeCl_4)^-$ and the

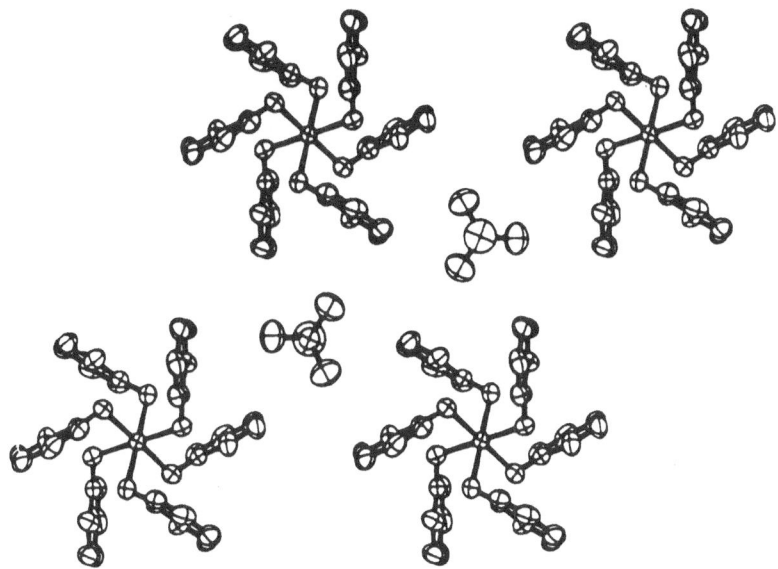

Fig. 5. Packing of $[Fe(pyNO)_6]^{2+}$ and ClO_4^- (Ref. 26)

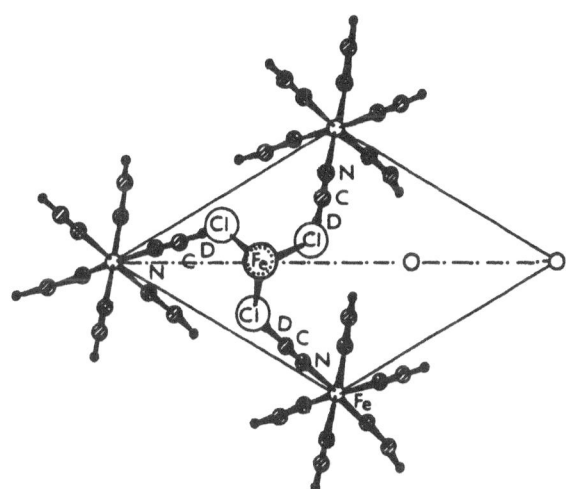

Fig. 6. Projection of the crystal structure $(FeCl_4)_2[Fe(CND)_6]$ along the C_3 axis (Ref. 27)

islands of $[Fe(NCD)_6]^{2+}$ are mutually joined together by $-Fe(II)-N-C-D\ldots Cl-Fe(III)-$ bonds as illustrated in Fig. 7. While the Fe(III) atoms occupy positions of the $3(C_3)$ point symmetry those of Fe(II) belong to the symmetry $\bar{3}(S_6)$. The bond lengths of Fe(III)-Cl are 2.194(6) and 2.190(4) Å and the Cl-Fe(III)-Cl angles 108.7(1) and 110.2(1)°. Apparently, as a consequence of the rigidity of the practically linear N-C-D group, the bond between the tetrahedral groups and the octahedral islands (Fig. 7) causes a deviation of the bond angle Fe(II)-N-C from 180°. The resulting tension in the octahedral chromophore causes a small tetragonal distortion of the FeN_6 octahedron and an obtuseness of the respective angle of N-Fe(II)-N of 91.1(1)°.

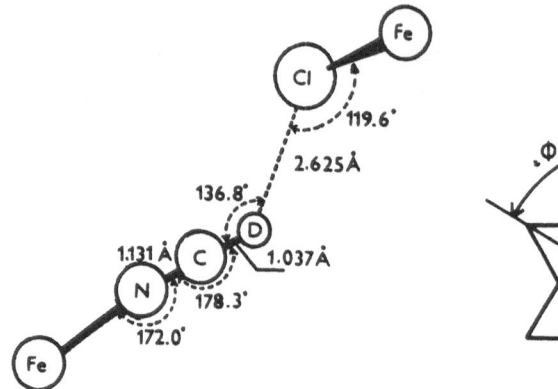

Fig. 7. Schematic representation of the Fe-N-C-D . . . Cl-Fe bond in the crystal structure of (FeCl$_4$)$_2$[Fe(NCD)$_6$]. Atoms Fe-N-C-D . . . Cl are in the same plane. (Ref. 27)

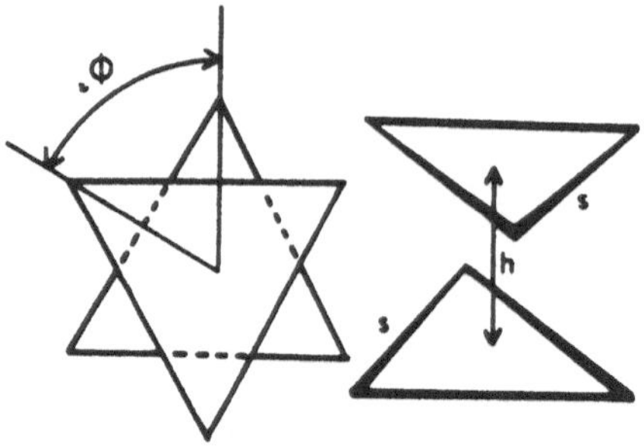

Fig. 8. Representation of the twist angle ϕ triangular faces of octahedron, intertriangular distance h and length of the triangles leg s

It should be noted that in the islands of both of the complexes, [Fe(pyNO)$_6$]$^{2+}$ and [Fe(NCD)$_6$]$^{2+}$, the octahedral inner coordination sphere of Fe(II) shows a very slight trigonal distortion. The twisted angle of the two triangular planes of the octahedron (Fig. 8) is in both cases fixed by the S$_6$ symmetry. Then, the ratio of s/v[28], being 1.197 for [Fe(pyNO)$_6$]$^{2+}$ and 1.190 for [Fe(NCD)$_6$]$^{2+}$, may be taken as a measure of the trigonal distortion of the octahedron. For an ideal octahedron the ratio of s/v = $\sqrt{3/2}$ = 1.225. As these facts show, for both complexes the geometries of the slightly trigonally distorted inner coordination spheres around Fe(II) are compressed.

3.3 High-Spin Fe(III) Complexes

High-spin Fe(III) complexes include hexaaquairon(III) nitrate trihydrate, [Fe(H$_2$O)$_6$](NO$_3$)$_3 \cdot$ 3 H$_2$O[29], and 2,2'-bipyridilium-tris(2,2'-bipyridyl)iron(III) tetraperchlorate, (C$_{10}$H$_9$N$_2$) Fe(C$_{10}$H$_8$N$_2$)$_3$(ClO$_4$)$_4$[30], which, according to the results of an X-ray analysis comply with the conditions described on p. 10. In the structure of [Fe(H$_2$O)$_6$](NO$_3$)$_3 \cdot$ 3 H$_2$O[29] (Fig. 9) each [Fe(H$_2$O)$_6$]$^{3+}$ complex ion is coordinated with NO$_3^-$ ions and water molecules by means of hydrogen bonds. Two types of centrosymmetric, crystallographically non-equivalent [Fe(H$_2$O)$_6$]$^{3+}$ ions differing by the geometry of their coordination polyhedra have been detected (Fig. 10). While in the Fe(1)O$_6$ chromophore the bond lengths of Fe(1)-O are 1.974(3), 1.986(3) and 1.985(3) Å the bond angles of O-Fe(1)-O are in the range of 89.1(1)–90.0(1)°; the Fe(2)O$_6$ chromophore exhibits bond lengths of 1.966(3), 1.992(3) and 2.014(3) Å with bond angles between 87.8(1)–90.1(1)°. The above values show that the [Fe(1)(H$_2$O)$_6$]$^{3+}$ islands have an almost octahedral coordination around Fe(1) while the [Fe(2)(H$_2$O)$_6$]$^{3+}$ groups exhibit a greater deviation from a regular octahedral coordination. The elongation of the Fe(2)-O(6) bond from the interval of 1.95–2.00 Å (Table 3) is probably caused by hydrogen bond interactions of the O(6) atom with two acceptor atoms of the uncoordinated water molecules. This is also confirmed by the fact that the twist of the plane O(6) of the water molecule

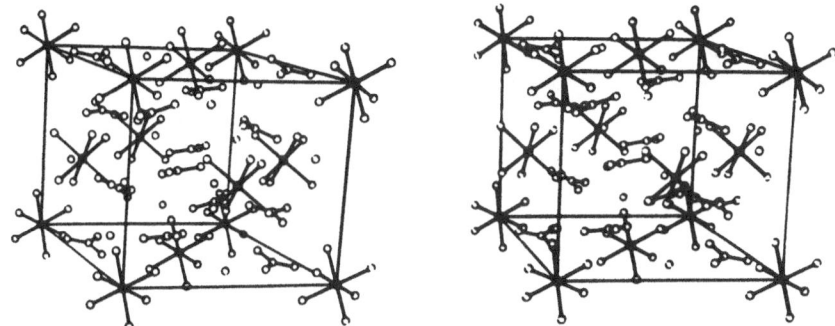

Fig. 9. Stereoscopic view of the unit cell of $[Fe(H_2O)_6](NO_3)_3 \cdot 3\,H_2O$. The Fe atoms are darkened and the hydrogen atoms are omitted (Ref. 29)

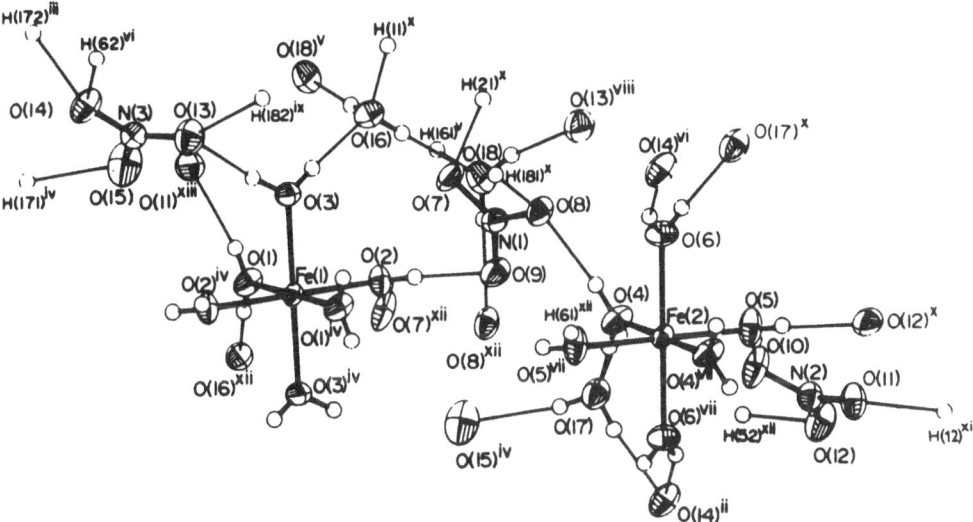

Fig. 10. Perspective view of the asymmetric unit in the crystal structure of $[Fe(H_2O)_6](NO_3)_3 \cdot 3\,H_2O$. Covalent bonds and hydrogen bonds are represented by thickened and thin lines, respectively. The bond distances Fe(1)-O(1), Fe(1)-O(2), Fe(1)-O(3) are 1.974(3), 1.986(3) and 1.985(3) Å, and Fe(2)-O(4), Fe(2)-O(5), Fe(2)-O(6) are 1.966(3), 1.992(3) and 2.014(3) Å (Ref. 29)

around the axis of Fe(2)-O(6) with respect to the O(5)-Fe(2)-O(6) plane is 37°, while the mean angle of remaining water molecules in the inner coordination sphere of Fe(1) and Fe(2) is 13°.

In the structure of $(C_{10}H_9N_2)\,[Fe(C_{10}H_8N_2)_3] \cdot (ClO_4)_4{}^{30)}$, the packing of the structural units of $[Fe(C_{10}H_8N_2)_3]^{3+}$ islands, ClO_4^- anions and $C_{10}H_9N_2^+$ cations in the basic cell is illustrated in Fig. 11. The 2,2'-bipyridyl molecules in the islands are essentially planar. The angle between the planes of two rings of the same ligand is at most 6°. The inner coordination sphere around Fe(III) is almost octahedral and exhibits a pseudoaxis C_3. The mean value of the bond lengths of Fe-N is 1.961(1) Å and that of the N-Fe-N bond angles 82.3(1)°.

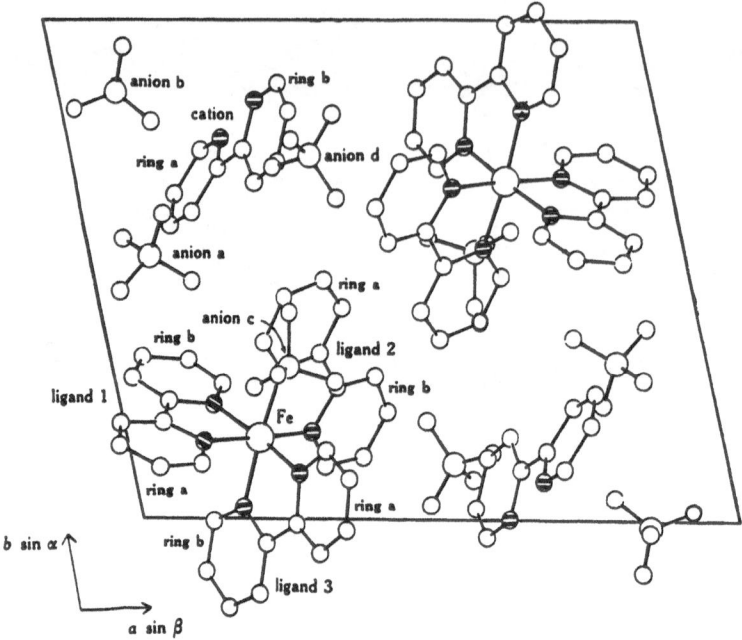

Fig. 11. Unit cell of $(C_{10}H_9N_2)[Fe(C_{10}H_8N_2)_3](ClO_4)_4$ (Ref. 30)

It should be noted that from the high-spin Fe(III) and Fe(II) complexes only those with a purely homogeneous coordination sphere comply with the conditions presented in the introduction of this chapter while, for example, for the high-spin Mn(II) complexes (see p. 10), also complex compounds with heterogeneous coordination spheres fulfil the respective conditions.

3.4 High-Spin Co(II) Complexes

Two types of high-spin Co(II) complexes show Co-L bond lengths in the interval of maximum relative frequencies (Table 3).

One of them is hexakis(pyridine-1-oxide) cobalt(II) perchlorate, $[Co(pyNO)_6]$-$(ClO_4)_2$, the structure of which was already detected[31] and[26] as one of the isostructural complexes $[M(pyNO)_6](ClO_4)_2$ (M = Cu, Co and Fe). This cobalt(II) complex also has islands of the symmetry $\overline{3}(S_6)$ (Fig. 5). In this case the bond length of Co-O is 2.088(2)[26] and 2.090(1) Å[31], respectively the O-Co-O bond angle being 89.97(4)°[26] and 89.96(6)°[31]. The trigonal distortion[28] (Fig. 8) is then s/h = 1.224.

The second complex fulfilling this condition is the adduct cis-cis[1-(2-thienyl)-4,4,4-trifluoro-1,3-butanedionato]-bis(methanol)cobalt(II), $[Co(TTA)_2(methanol)_2]$[32]. The structure of the island displays cis-octahedral geometry (Fig. 12). The thienyl rings are essentially flat. The chelate rings, however, exhibit deformations which are probably caused by hydrogen bonding interactions in the island (Fig. 12). Ring A has a twist-boat conformation while B exhibits an envelope form[33]. The dihedral angle between the least-

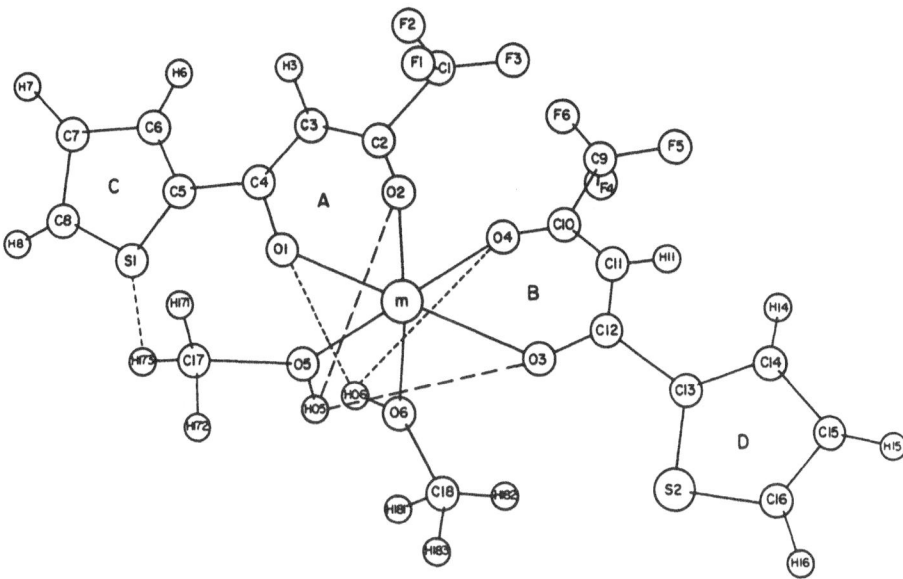

Fig. 12. Schematic representation of the island [Co(TTA)₂(methanol)₂]. The hydrogen bonds and steric influence between S(1) and H(173) are denoted by *broken lines* (Ref. 32)

squares planes of chelate A and the thienyl ring C is 12.0° and between the planes B and D this angle is 6.8°. These structural units are linked by hydrogen bonds. The major interaction between the two methanolic hydrogen atoms H(O 5) and H(O 6) of two adjacent islands is illustrated by stereodiagram in Fig. 13. The mean value of the Co-O bond lengths is 2.066(2) Å. The bond angles of O-Co-O in the rings A and B are slightly acuted to the values of 88.8(2) and 87.9(2)°. The remaining O-Co-O angles are in the range of 86.8–94.4°.

3.5 Low-Spin Co(III) Complexes

The bond distances of low-spin Co(III) complexes, with a homogeneous as well as with a heterogeneous coordination sphere fall into the range of maximum frequencies compiled in Table 3. The [Co(NH₃)₆]³⁺ islands of the crystal structure of hexaamminecobalt(III) chloride, [Co(NH₃)₆]Cl₃[34], exhibit an essentially ideal octahedral inner coordination sphere of Co(III). The arrangement of these islands and Cl⁻ ions in the unit cell is seen in Fig. 14. In the asymmetrical unit there are four crystallographically independent islands, each of which being surrounded by ions located in the corners of a distorted rhombic dodecahedron. Two symmetrically independent Co(II) atoms occupy positions with the symmetry 2(C₂) and the other two cobalt(II) atoms occupy places in the symmetry 2/m(C₂ₕ). The mean values of the Co-N bond lengths are: 1.969(3), 1.959(4), 1.969(3) and 1.966(4) Å. The maximum deviation of N-Co-N angles from their ideal values is 2.1(7)°.

The crystal structure of (+)ᴅ-N,N,N′,N′-tetrakis(2-aminoethyl)-1,2-diaminoethane)-cobalt(III)hexacyanocobaltate(III) dihydrate, (+)ᴅ-[Co(penten)]Co(CN)₆ · 2 H₂O³⁵⁾,

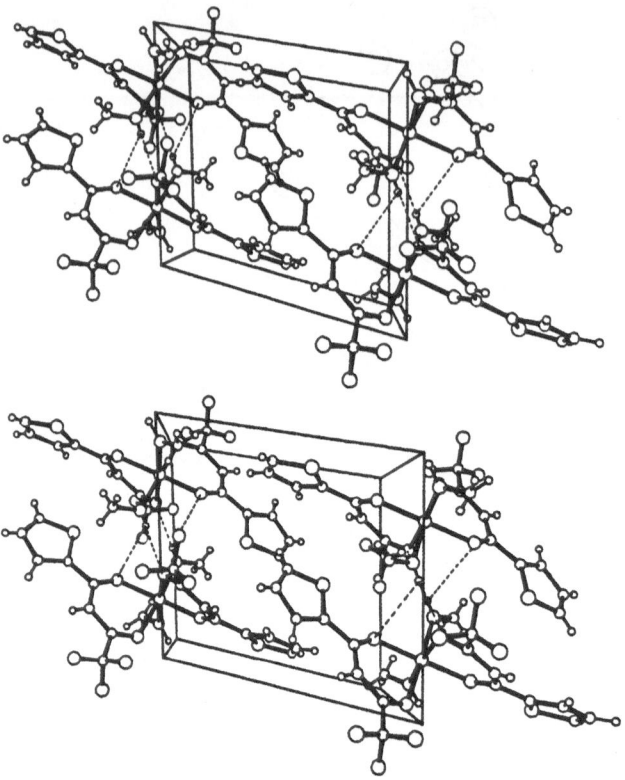

Fig. 13. Stereoscopic view along the [001] direction of unit cell of [Co(TTA)$_2$(methanol)$_2$]. Hydrogen bonds connected to the islands are denoted by *dotted lines* (Ref. 32)

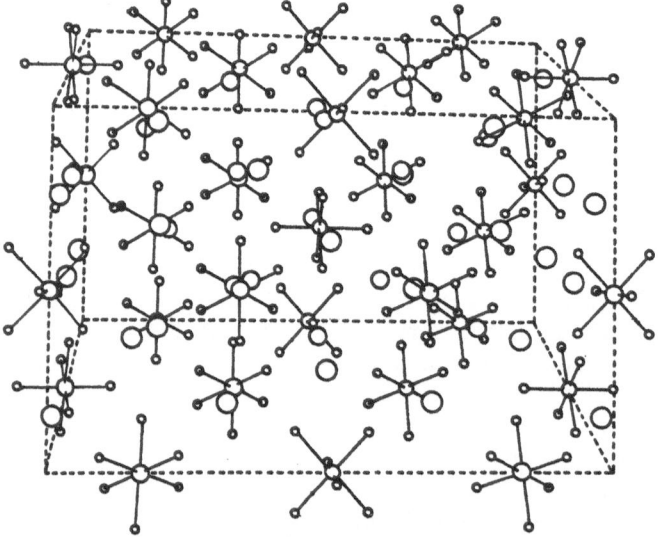

Fig. 14. View of the unit cell of [Cu(NH$_3$)$_6$]Cl$_3$. Cl, Co and N are represented by spheres of decreasing radii (Ref. 34)

consists of $[Co(penten)]^{3+}$ islands, $(Co(CN_6))^{3-}$ groups and water molecules linked with one another by hydrogen bonds (Fig. 15). As the structure of the $[Co(penten)]^{3+}$ island shows (Fig. 16), the chelate rings A, B and C form a girdle around the central atom. The mean bond length of Co-N is 1.982(6) Å while the angles of N-Co-N in the chelate pentacycles A, B, C, D, and E are 89.5(6), 83.2(6), 86.3(6), 85.9(6) and 85.3(6)°. The Co(III) central atom in the $(Co(CN_6))^{3-}$ group shows an approximate octahedral coordi-

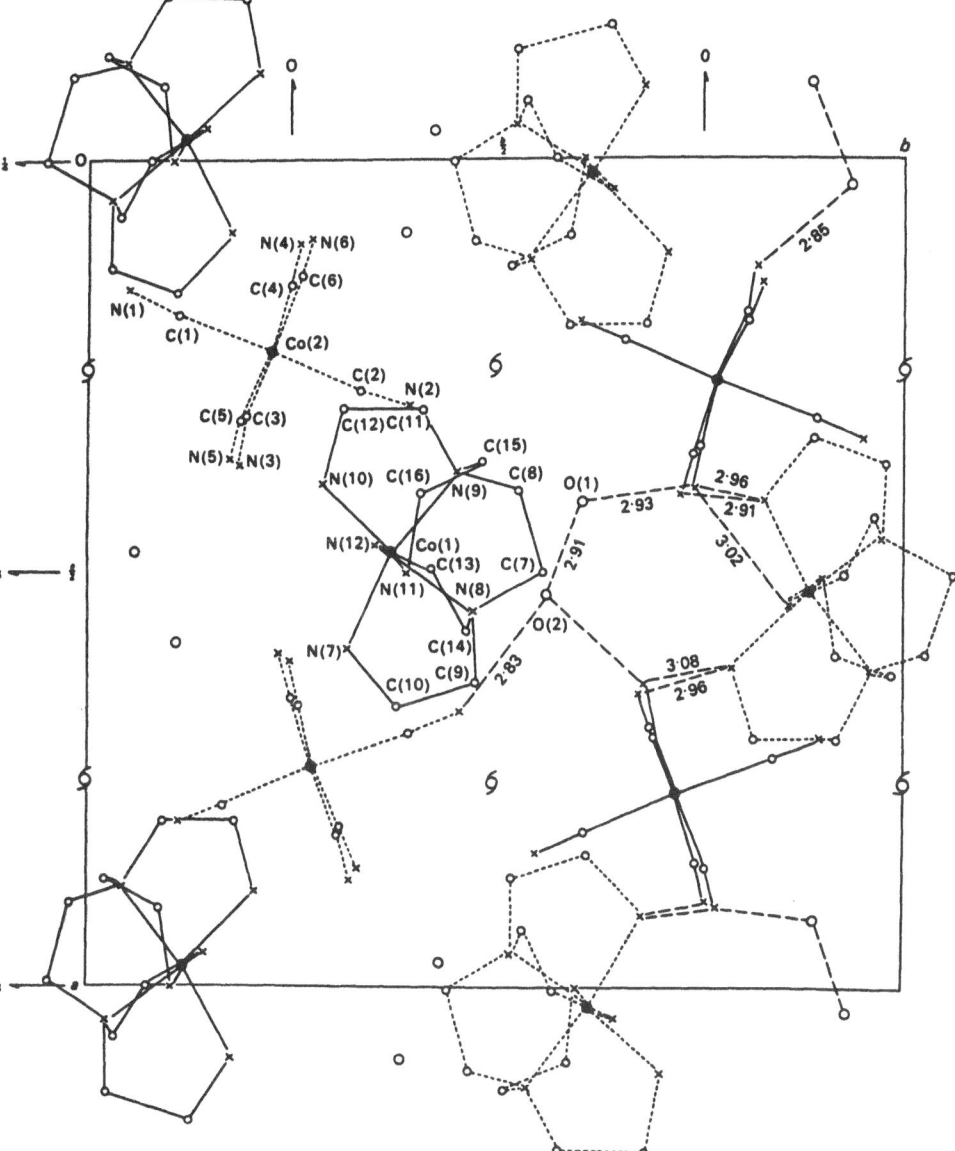

Fig. 15. Projection of the structure of $(+)_D$-[Co(penten)]Co(CN)$_6 \cdot 2\,H_2O$ along the C axis. *Broken lines* indicate close contacts. *Full* and *broken lines* denote complex ions in which the central cobalt atoms are at the approximate heights of 1/2 and 0, respectively (Ref. 35)

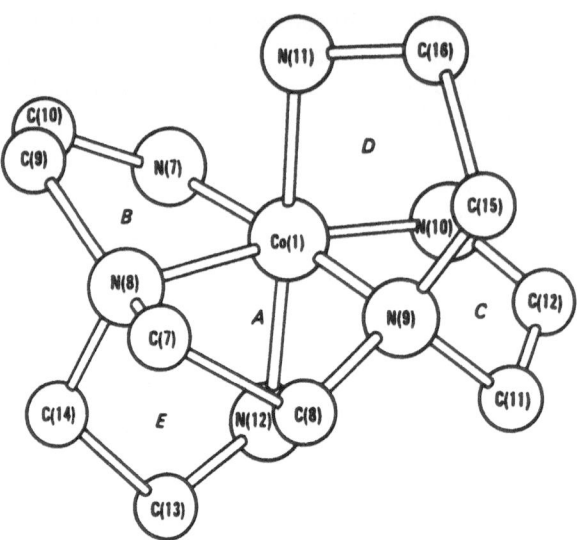

Fig. 16. Perspective drawing of the island $(+)_D$-$[Co(penten)]^{3+}$ (Ref. 35)

nation with a mean Co-C bond length of 1.915(8) Å; the bond angles of C-Co-C are in the range of 85.3(8)–95.9(9)°.

The crystal structure of pentaammineaquacobalt(III) dithionate dihydrate, $[Co(H_2O)(NH_3)_5](S_2O_6)_3 \cdot 2H_2O^{36)}$, consists of $[Co(H_2O)(NH_3)_5]^{3+}$ islands and two types of symmetrically independent $(S_2O_6)^{2-}$ groups, half of which being centrosymmetrical and water molecules. The linkage of these structural units via hydrogen bonds is shown in Fig. 17. The mean bond length of Co-N in the essentially octahedral inner coordination sphere around Co(III) is 1.9600(4) Å; the N-Co-N bond angles are in the range of 87.1(1)–91.6(1)°.

A slightly distorted octahedral coordination of Co(III) was found in the crystal structure of triamminediethylenetriaminecobalt(III) dithionate chloride, $[Co(NH_3)_3dien]$-$(S_2O_6)Cl^{37)}$. This complex consists of asymmetrical $[Co(NH_3)_3dien]^{3+}$ islands, Cl^- anions and two types of symmetrically independent $(S_2O_6)^{2-}$ groups. These units are linked via hydrogen bonds as shown in Fig. 18. The mean Co-N bond length is 1.989(2) Å. The N-Co-N bond angles in both hexametallocycles have become more acute to the values of 86.6(3) and 81.6(3)°. The remaining N-Co-N angles are in the range of 87.5(3)–94.1(3)°.

The X-ray analysis of the complexes $(-)_{589}$-bis[(s,s-trans-1,2-diaminocyclohexane)]-(R,R)-trans-1,2-diaminocyclohexanecobalt(III) chloride pentahydrate, $(-)_{589}$-(lel_2, ob)-$[Co(chxn)_3]Cl_3 \cdot 5H_2O^{38)}$ and $(-)_{589}$-bis[(1R,2R)-1,2-diaminocyclohexane](3,3'-dimethyl-2,2'-bipyridyl)cobalt(III) bromide diperchlorate monohydrate, $(-)_{589}$-$[Co(R,R-chxn)_2(3,3'-dmbpy)]Br(ClO_4)_2 \cdot H_2O^{39)}$, shows that in both cases the molecules are coordinated with Co(III) atoms as chelates. While in $(-)_{589}$-(lel_2, ob)-$[Co(chxn)_3]Cl_3 \cdot 5H_2O$, the $[Co(chxn)_3]^{3+}$ islands are formed by three chxn molecules (Fig. 19), in $(-)_{589}$-$[Co(R,R-chxn)_2(3,3'-dmbpy)]Br(ClO_4)_2 \cdot H_2O$, besides two chxn molecules, also the 3,3'-dmbpy molecule participates in the formation of $(-)_{589}$-$[Co(R,R-chxn)_2(3,3'-dmbpy)]^{3+}$ islands (Fig. 20). The islands of these complexes are connected with Cl^- ions in $(-)_{589}$-(lel_2, ob)-$[Co(chxn)_3]Cl_3 \cdot 5H_2O$, by tetrahedral ClO_4^- groups and Br^- ions in

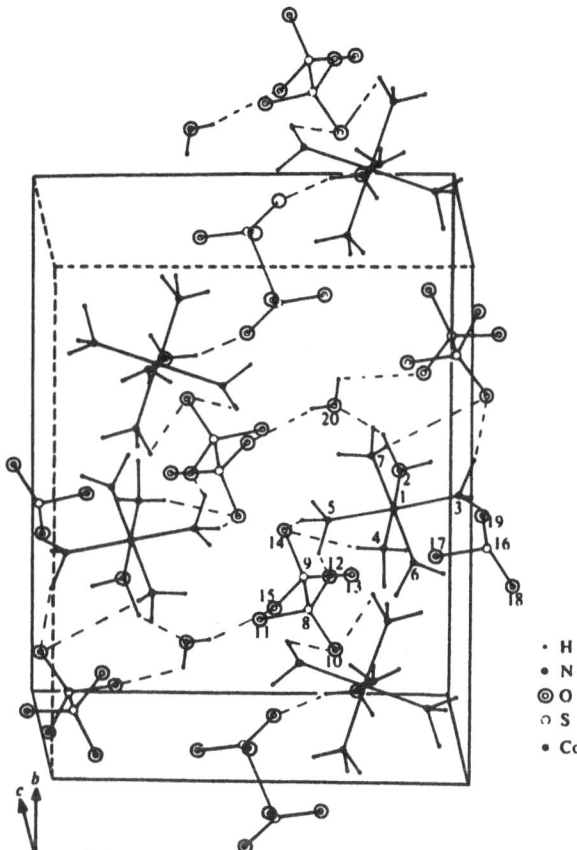

Fig. 17. View of the unit cell of [Co(H₂O)(NH₃)₅]₂(S₂O₆)₃ · 2 H₂O (Ref. 36)

$(-)_{589}$-[Co(R,R-chxn)$_2$(3,3'-dmbpy)]Br(ClO$_4$)$_2$ · H$_2$O and to water molecules via hydrogen bonds (Figs. 19 and 21). From three chxn molecules in [Co(chxn)$_3$]$^{3+}$ two exhibit an *lel* and one exhibits an *ob* conformation (Fig. 22)[40]. The mean Co-N bond lengths in the chelate rings of the chxn *lel* conformation are 1.973(2) Å and of *ob* conformation 1.969(3) Å. The corresponding mean bond angles N-Co-N are 84.6(4) and 84.3(2)°. Thus, the different ligand conformation (*lel* and *ob*) does not cause any significant change of the Co-N bond lengths nor of the bond angles N-Co-N. The bond angles in the chelate rings N(1)-Co-N(4), N(2)-Co-N(5) and N(3)-Co-N(6) are 85.0(2), 84.3(2) and 84.1(2)°, respectively. The octahedral inner coordination sphere around Co(III) is approximately trigonally distorted at an angle of $\phi \doteq 6.4°$ (Fig. 8). The upper triangle formed by three nitrogen atoms (Fig. 19) creates with a lower triangle an angle of about 1.9°. From the Λ-*lel*$_3$[41] and Λ-*ob*$_3$[42] isomers only the second one exhibits all Co-N bond lengths in the range of the maximum relative frequency (Table 3). In $(-)_{589}$-(*ob*)$_3$-[Co(chxn)$_3$]Cl$_3$ · H$_2$O[42] islands display 2(C$_2$) symmetry, and the octahedral coordination is approximately trigonally distorted with a twist angle of $\phi \doteq 9°$. In the island of $(-)_{589}$-[Co(R,R-chxn)$_2$(3,3'-dmbpy)]$^{3+}$ the chelate rings exhibit the conformation of δ^{43}[43]. The mean Co-N bond length and mean angle in the chelate ring of Co-dmbpy are 1.959(5) Å and 82.7(3)°,

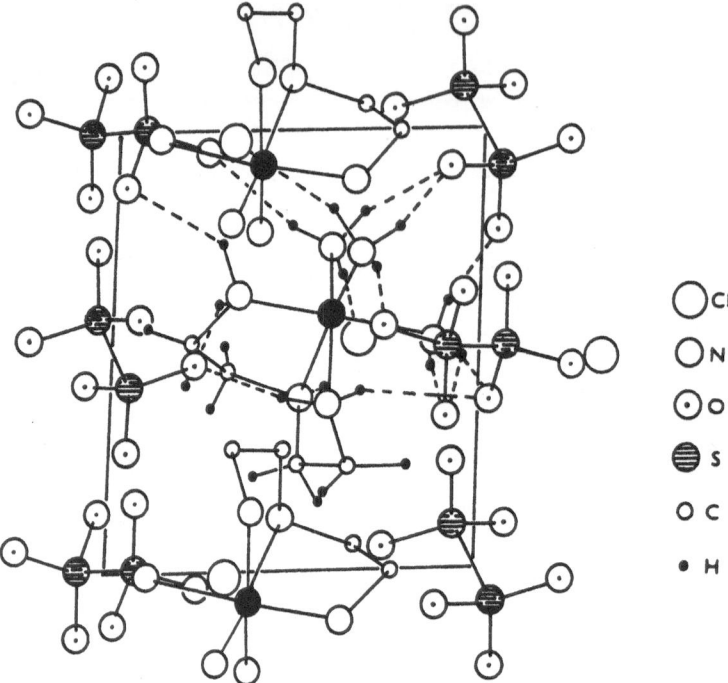

Fig. 18. Projection of the structure of [Co(NH$_3$)$_3$dien](S$_2$O$_6$)Cl on to the plane (001) (Ref. 37)

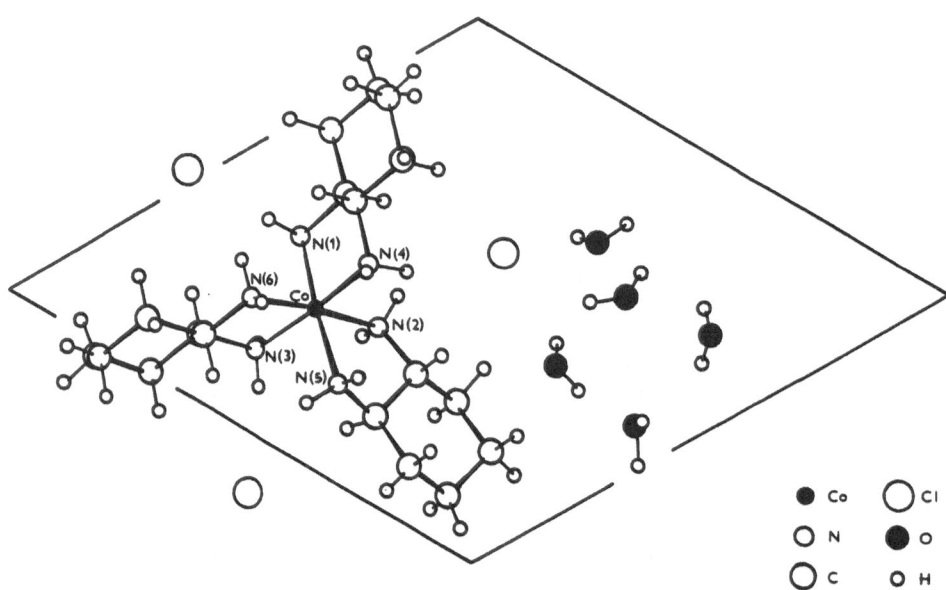

Fig. 19. Partial projection of the structure of $(-)_{589}$-(lel_2ob)-[Co(chxn)$_3$]Cl$_3$ · 5 H$_2$O along the c axis (Ref. 38)

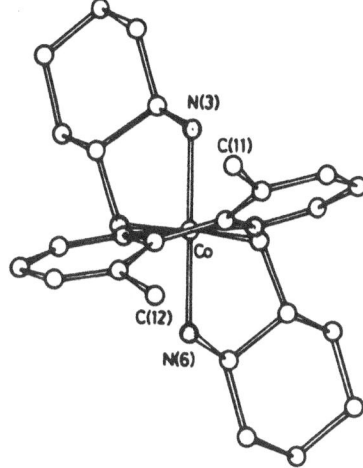

Fig. 20. Projection of an island in the crystal structure of $(-)_{589}$-[Co(R,R-chxn)$_2$(3,3'-dmbpy)]Br(ClO$_4$)$_2$ · H$_2$O in the direction of the pseudotwofold axis (Ref. 39)

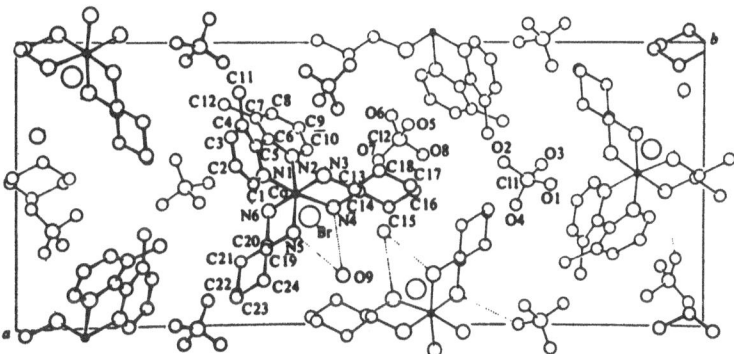

Fig. 21. Projection of the crystal structure of $(-)_{589}$-[Co(R,R-chxn)$_2$(3,3'-dmbpy)]Br(ClO$_4$)$_2$ · H$_2$O along the c axis. Hydrogen bonds are indicated by *dotted lines* (Ref. 39)

respectively, and the corresponding mean values of the chelate ring Co-chxn are 1.967(8) Å and 85.3(3)°, respectively. In principle, these values are the same as the analogous values for $(-)_{589}$-(lel_2ob)-[Co(chxn)$_3$] · Cl$_3$ · 5 H$_2$O. The twisted angle of the two planar pyridine rings around the C(5)-C(6) bond is 29.4°. The slight obtuseness of the tetragonal angles of N-Co-N of the octahedral chromophore CoN$_6$ thus lies in the chelate rings. The twist of the two pyridine rings in the dmbpy molecules is apparently caused by a mutual repulsion of the methyl groups (the interatomic distance of C(11)...C(12) is 2.95(3) Å and this hinders the repulsion of the chxn molecules, thus stabilizing the almost regular octahedral inner coordination sphere of Co(III). 3,3'-Dimethyl-2,2'-bipyridyl as a bidentate ligand also occurs in $(+)_{589}$-(3,3'-dimethyl-bipyridyl)bis(ethylenediamine)-cobalt(III) chloride diperchlorate monohydrate, $(+)_{589}$-[Co(3,3'-dmbpy)(en)$_2$]Cl(ClO$_4$)$_2$ · H$_2$O[44]. The structure consists of [Co(3,3'-dmbpy)(en)$_2$]$^{3+}$ cation islands, of essentially tetrahedral (ClO$_4$)$^-$ groups, Cl$^-$ ions and water molecules which are mutually linked by hydrogen bonds (Fig. 23). One of the en-ligands exhibits some conformational disorder, viz. 90% $\lambda \rightleftarrows$ 10% δ, similarly as in [Co(R,R-chxn)$_2$(3,3'-dmbpy)]$^{3+}$[39]. The dmbpy

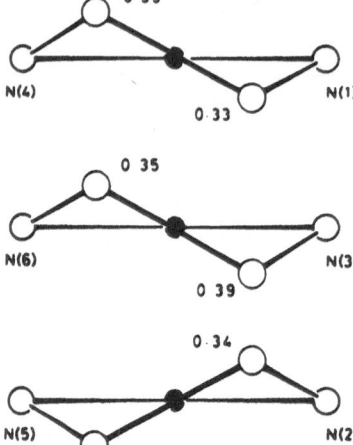

Fig. 22. Conformation of the chelate rings in the crystal structure of $(-)_{589}$-(lel_2ob)-[Co(chxn)$_3$]Cl$_3 \cdot$ 5 H$_2$O. Numbers attached to each carbon atom indicate the distances from the coordination plane (Ref. 38)

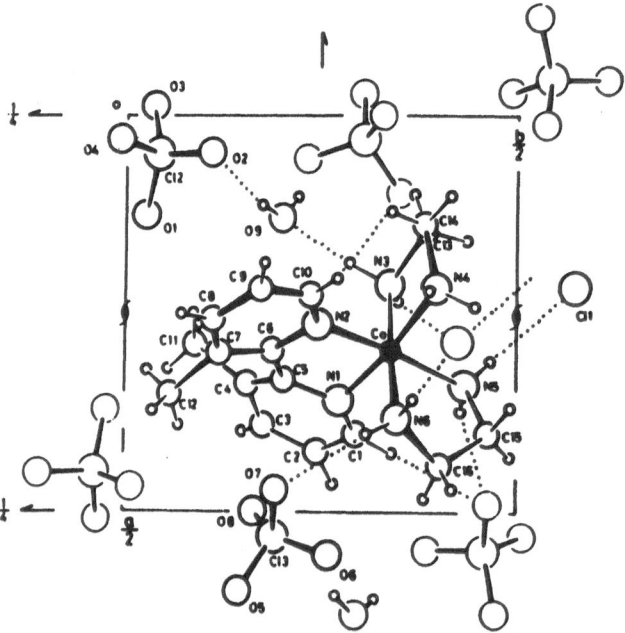

Fig. 23. Partial projection of the structure $(+)_{589}$-[Co(3,3'-dmbpy)(en)$_2$]Cl(ClO$_4$)$_2 \cdot$ H$_2$O along the c axis. The *dotted lines* indicate hydrogen bonds (Ref. 44)

ligand also contains two planar pyridine rings twisted around the C(5)-C(6) bond due to mutual repulsion of the methyl groups. The dihedral angle formed by the planes of two pyridine rings is 32.4°. The inner coordination sphere of Co(III) is essentially octahedral. The N-Co-N bond angles in Co-dmbpy, Co-en(δ) and Co-en(λ) rings are 82.5(1), 84.4(1) and 85.3(1)°. The mean Co-N bond lengths in these rings are 1.964(1), 1.963(2) and 1.965(2) Å.

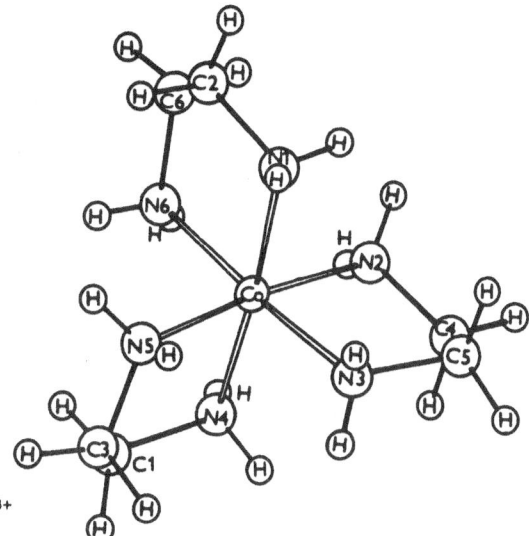

Fig. 24. Structure of the island $[Co(en)_3]^{3+}$ in $[Co(en)_3](tart)Cl \cdot 5 H_2O$ (Ref. 45)

(+)-Tris(ethylenediamine)cobalt(III) chloride (+)-tartrate pentahydrate, $[Co(en)_3]$-(tar)Cl $\cdot$ 5 H$_2$O[45], consists of $[Co(en)_3]^{3+}$ islands, as shown in Fig. 24. The chloride ions are near the threefold pseudoaxis of the almost octahedral configuration of Co(III) formed by nitrogen atoms from ethylenediamine molecules in the conformation of metalocycles (Fig. 24). A slight distortion of the tetragonal angles of the regular octahedral coordination sphere of Co(III) is manifested in more acute N-Co-N angles in the chelate rings up to the mean value of 85.50(6)°. The remaining bond angles are in the range of 89.9(1)–93.2(1)°.

Cyclic tridentate ligands with nitrogen donor atoms are coordinated with the central atoms in the structure of bis[(R)-2-methyl-1,4,7-triazacyclononane]cobalt(III) iodide pentahydrate, $(-)_{589}$-[Co{(R)-MeTACN}$_2$]I$_3$ $\cdot$ 5 H$_2$O[46]. The structure consists of [Co{(R)-MeTACN}$_2$]$^{3+}$ islands, I$^-$ ions and water molecules. The packing of these structural units in the basic cell is shown in the projection in Fig. 25. All the six five-membered chelate rings exhibit the conformation λ. The methyl group is coordinated to one of the three chelate rings in such a way that the island exhibits a positional disorder. The I$^-$ ions and the water molecules near the island are also positionally disordered. Each of the Co atoms is localized in positions of the symmetry 32(D$_3$) and is almost regular octahedrally coordinated by six nitrogen atoms from the two tridentate ligands exhibiting Co-N bond lengths of 1.974(5)(6x) Å. The N-Co-N bond angle in the chelate pentacycle is 85.0°. The mutual clockwise twisting of the two cyclic ligands from the same island can be caused by repulsion of non-bonding hydrogen atoms. Such twisting of ligands leads to a trigonal distortion of the coordination polyhedron with the angle $\phi = 7.6°$ (Fig. 8).

3.6 Low-Spin Ni(II) Complexes

The X-ray structural analyses of bis(oxamide oximato) nickel(II) dihydrate, Ni(OAOH)$_2$ $\cdot$ 2H$_2$O[47], and bis(oxamide oximato)nickel(II) monohydrate dimethylformamide,

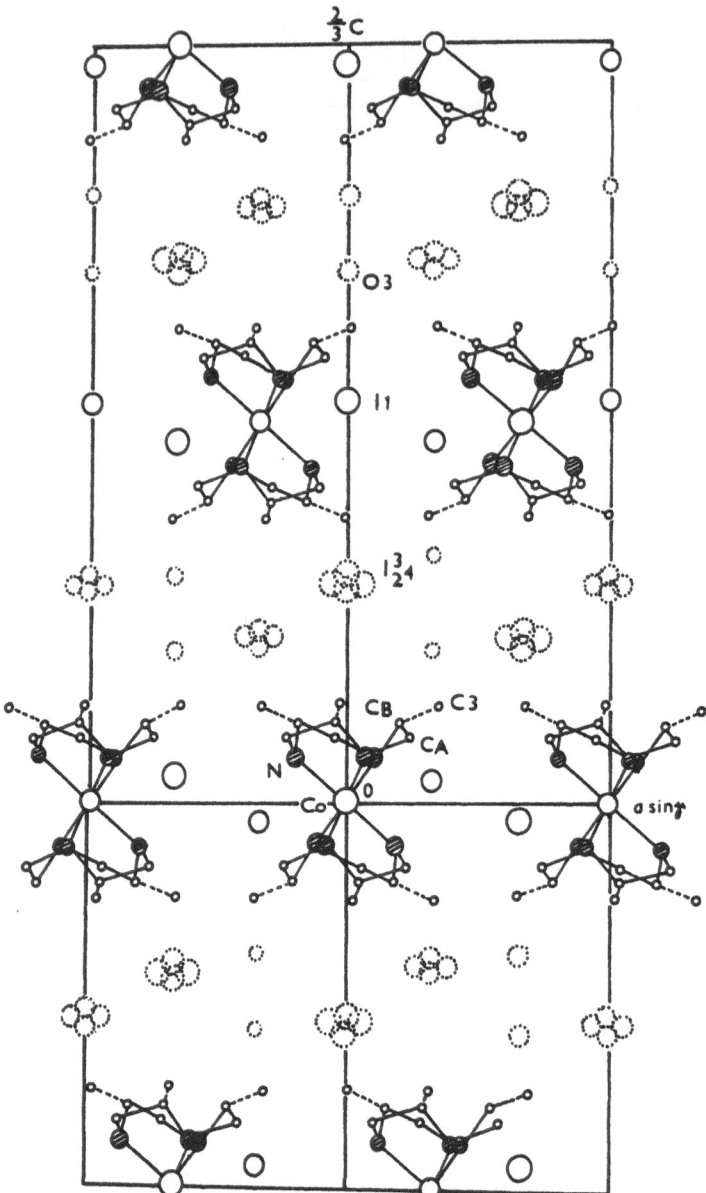

Fig. 25. Structure of $(-)_{589}$-[Co{(R)-MeTACN}$_2$]I$_3$ · 5 H$_2$O viewed along the b axis. Atoms exhibiting a positional disorder are indicated by *broken circles*. Disordered methyl groups are connected by *broken lines*. For clarity, two oxygen atoms are not illustrated (Ref. 46)

Ni(OAOH)$_2$H$_2$O · DMF[48]) complexes, show that the [Ni(OAOH)$_2$] islands of these complexes are isostructural. While in the dihydrate complex these islands are planar and centrosymmetrical (Fig. 26) and connected to water molecules via hydrogen bonds (Fig. 27), the monohydrate complex contains two types of crystallographically independent centrosymmetrical islands; their packing with DMF and H$_2$O is shown in Fig. 28. As

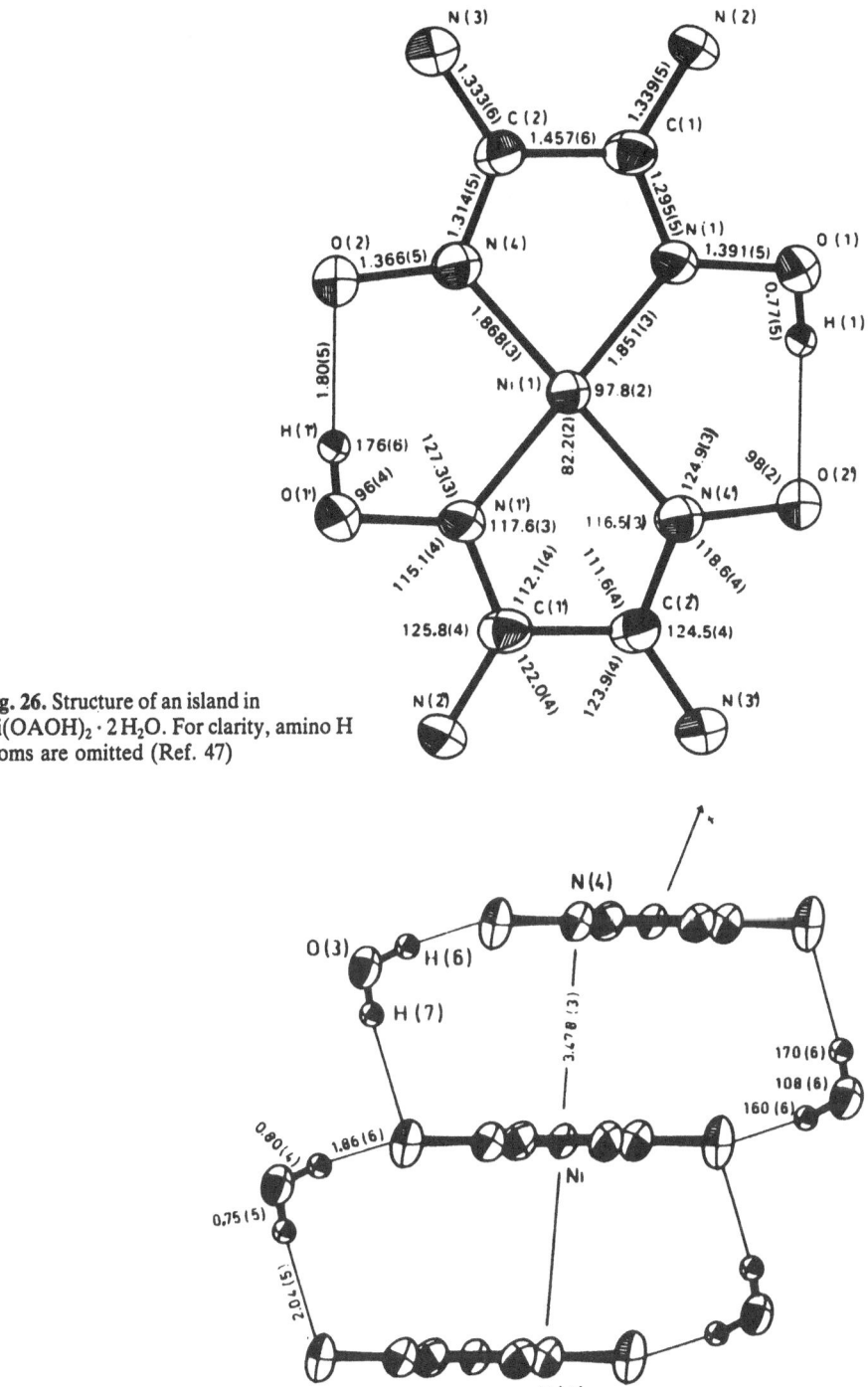

Fig. 26. Structure of an island in
Ni(OAOH)$_2$ · 2 H$_2$O. For clarity, amino H
atoms are omitted (Ref. 47)

Fig. 27. Mutual bonding of [Ni(OAOH)$_2$] islands by hydrogen bonds through water molecules
(Ref. 47)

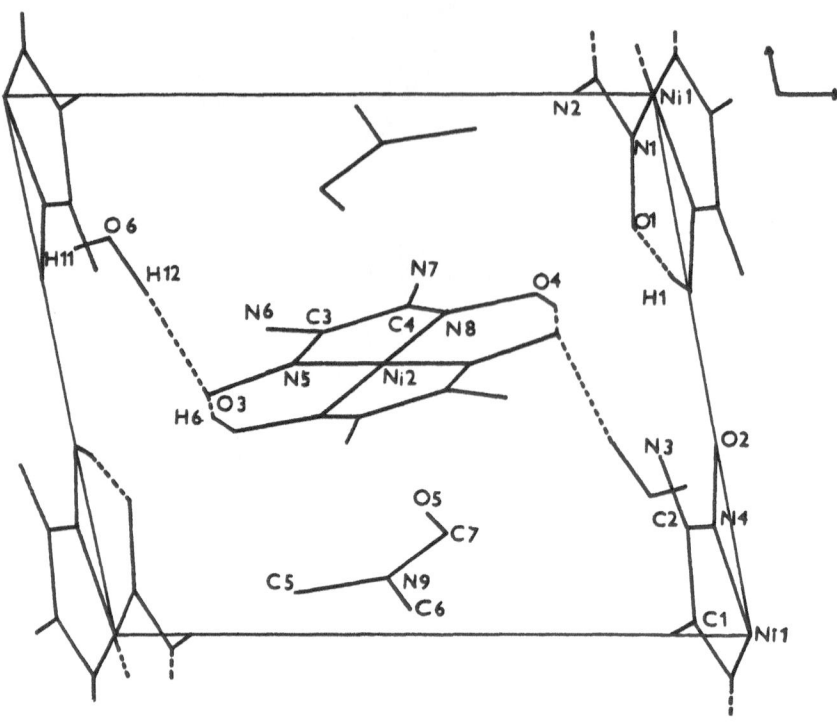

Fig. 28. Projection of the structure of Ni(OAOH)$_2$ · H$_2$O · DMF onto the ab plane. Hydrogen bonds are represented by *broken lines* (Ref. 48)

it can be seen from the island packing in the basic cell of Ni(OAOH)$_2$ · 2 H$_2$O complex (Fig. 27), the O(2) atom exhibits a deviation of 0.205 Å from the plane of the least squares of the non-hydrogen atoms, apparently due to the hydrogen bonding interactions. The nitrogen atoms from ligands of the two islands are normal to the island plane below and above the Ni(II) atom (Fig. 27). According to the histogram of relative frequencies (Fig. 2), this interatomic contact cannot be taken as a bonding and thus the inner coordination sphere around Ni(II) is essentially square-planar. The mean Ni-N bond length and the respective N-Ni-N bond angle in the chelate pentacycles for the dihydrate complex are 1.860(2) Å and 82.20(7) while for the monohydrate these values are 1.858(2) Å and 82.8(1)° for one island and 1.854(3) Å and 83.1(1) for the other one.

Two crystallographically independent island types can be found in the structure of dibenzo(b,i)(1,4,8,11)tetraaza(14) annulenenickel(II), Ni(C$_{18}$H$_{14}$N$_4$)[49]. Both island types (A, B) (Figs. 29, 30) are centrosymmetrical and planar. The maximum deviation from the least squares planes of atoms of both island types is 0.04 Å. As Fig. 29 shows, with respect to their bond lengths and angles both island types are essentially equivalent. While the bond angles of N-Ni-N in pentacycles are more acute, for the hexacycles they appear to be more obtuse.

Figure 31 describes the structure of the [Ni{Me$_2$Ga(N$_2$C$_3$H$_3$)$_2$}$_2$]$^+$ island of the bis-[dimethyl bis(pyrazol-1-yl)gallato]nickel(II) complex[50] in which the islands are only mutually held together by means of van der Waals forces. The interatomic distance of

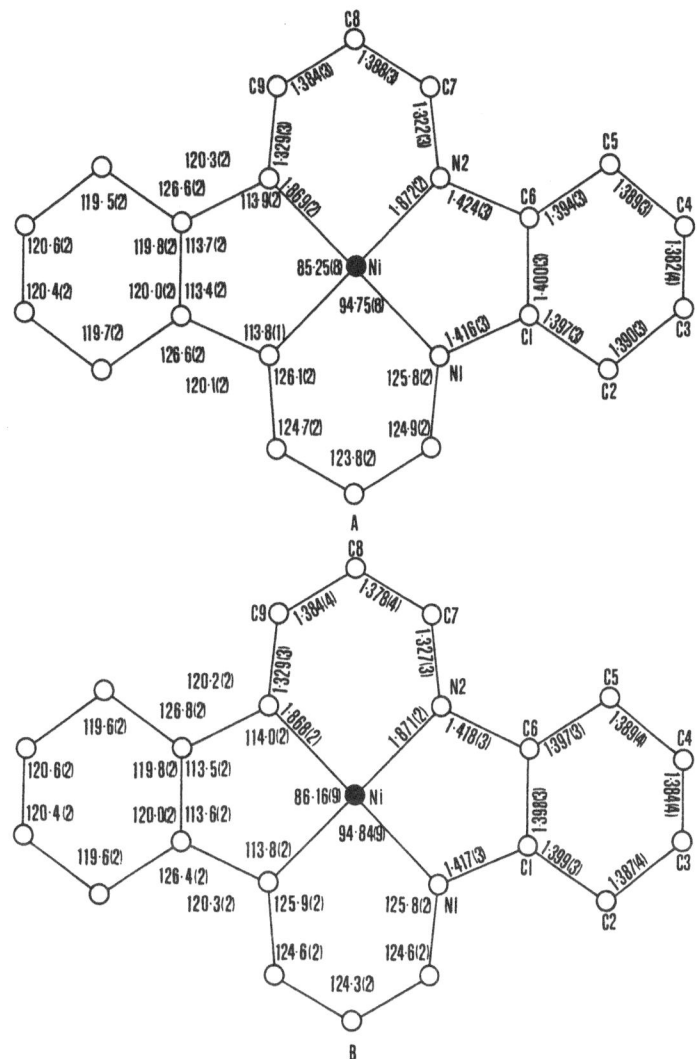

Fig. 29. Structure of the islands for type A and B in Ni(C₁₈H₁₄N₄) (Ref. 49)

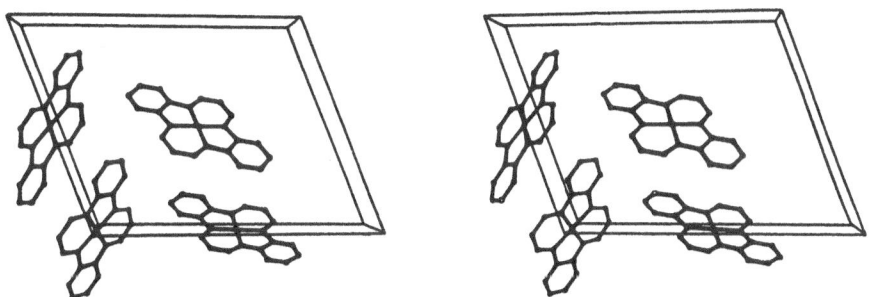

Fig. 30. Stereogram of the unit cell of Ni(C₁₈H₁₄N₄) viewed along the b axis (Ref. 49)

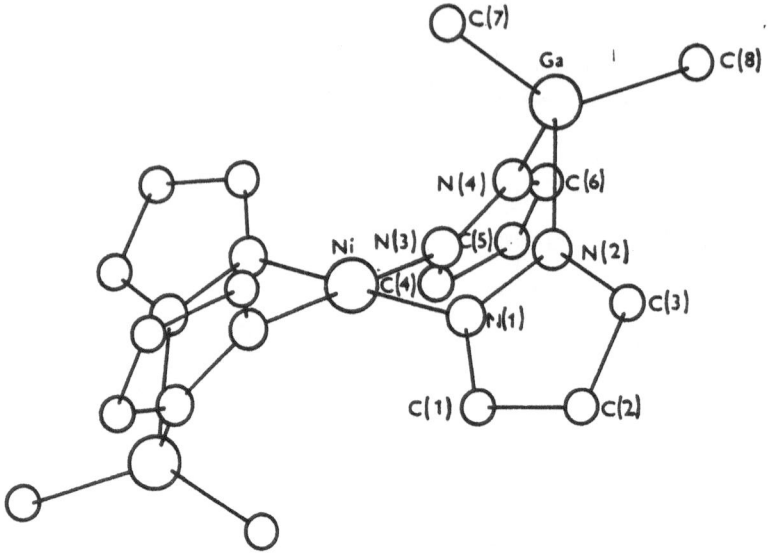

Fig. 31. Structure of the island [Ni{Me$_2$Ga(C$_3$H$_3$N$_2$)$_2$}$_2$] (Ref. 50)

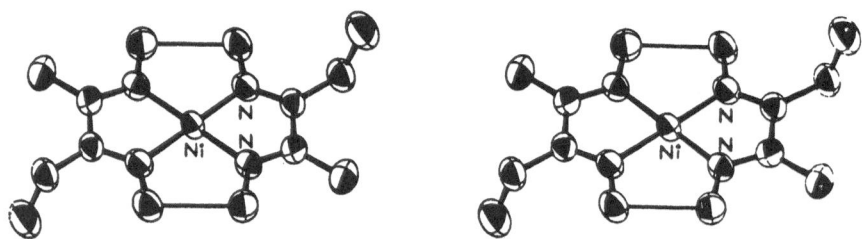

Fig. 32. Stereo illustration of the island [Ni(EMG)$_2$] (Ref. 51)

Ni...Ga in the island is 3.432 Å. The Ni(II) atoms occupy centrosymmetrical positions; the Ni-N bond lengths are 1.899(5) and 1.891(5) Å, the bond angle of N(1)-Ni-N(3) in the chelate hexacycle is 92.4(2)°.

Ethylmethylglyoximenickel(II), [Ni(EMG)$_2$][51], consists of approximately planar centrosymmetrical [Ni(EMG)$_2$] islands (Fig. 32); their packing in the basic cell is illustrated in Fig. 33. In this packing the Ni(II) atom appears approximately above the oxygen atom of the adjacent island. The distance between these islands is 3.286 Å. The Ni-N bond lengths and N-Ni-N angles in chelate pentacycles are 1.861(4) and 1.862(4) Å, and 82.5(2)°, respectively.

The X-ray analysis of bis(malondiaminidine)nickel(II) dichloride, [Ni(C$_3$H$_8$N$_4$)$_2$]Cl$_2$, reveals that this complex exhibits an OD structure with staking faults[52]. The [Ni(C$_3$H$_8$N$_4$)$_2$]$^{2+}$ islands of 2/m(C$_{2i}$) symmetry are mutually held together by Cl$^-$ ions (Fig. 34). The bond lengths of Ni-N and the N-Ni-N angles in both hexacycles are 1.87(1) Å and 91.3(4)°, respectively.

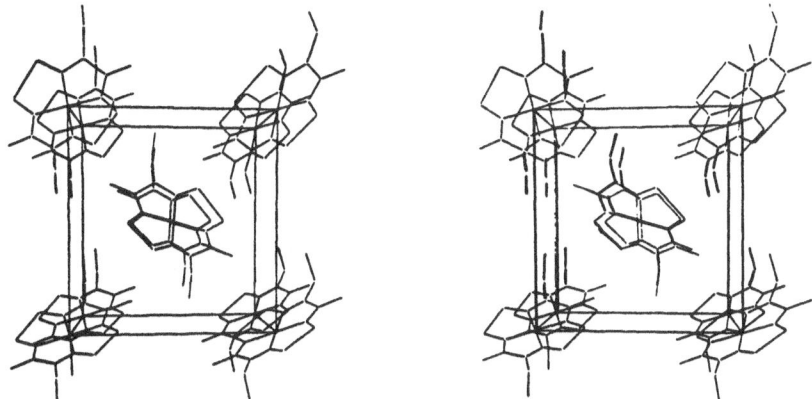

Fig. 33. Stereo view along the a axis of the island [Ni(EMG)$_2$] (Ref. 51)

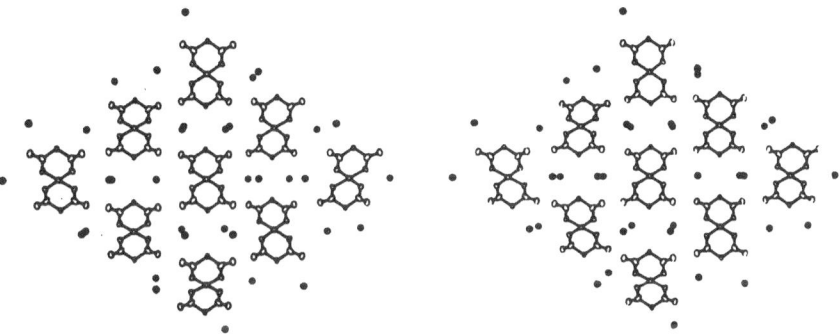

Fig. 34. View of the (001) plane of the OD structure of [Ni(C$_3$H$_8$N$_4$)$_2$]Cl$_2$ (Ref. 52)

3.7 High-Spin Ni(II) Complexes

It is interesting that while for low-spin Ni(II) complexes only those with a homogeneous inner coordination sphere comply with the conditions given in the introduction of chap. 3, for high-spin Ni(II) complexes, both types, i.e. those with a homogeneous and with a heterogeneous coordination sphere, fulfill the respective conditions. Potassium lead hexanitronickel(II), K$_2$PbNi(NO$_2$)$_6$[53], consists of [Ni(NO$_2$)$_6$]$^{4-}$ islands of m3(C$_{3v}$) symmetry (Fig. 35), K$^+$ and Pb^{2+} ions. The packing of the structural units in the basic cell is shown in Fig. 36.

The structure consisting of two crystallographically non-equivalent, centrosymmetrical islands of [Ni(NO$_3$)$_2$(H$_2$O)$_4$] which are linked with one another via hydrogen bonds belongs to β-nickel(II) nitrate tetrahydrate, β-Ni(NO$_3$)$_2 \cdot$ 4H$_2$O[54]. The nitrate groups are in *trans* positions (Fig. 37) with the mean bond lengths of 2.064(2) and 2.059(2) Å for Ni(1)-O and Ni(2)-O; the angles of O-Ni(1)-O and O-Ni(2)-O are in the ranges of 84.1(1)° 95.0(2)° and 85.1(2)–97.3(2)°, respectively. The slight distortion of the octahedral coordination around Ni(II) results from the suppression of both octahedra in the

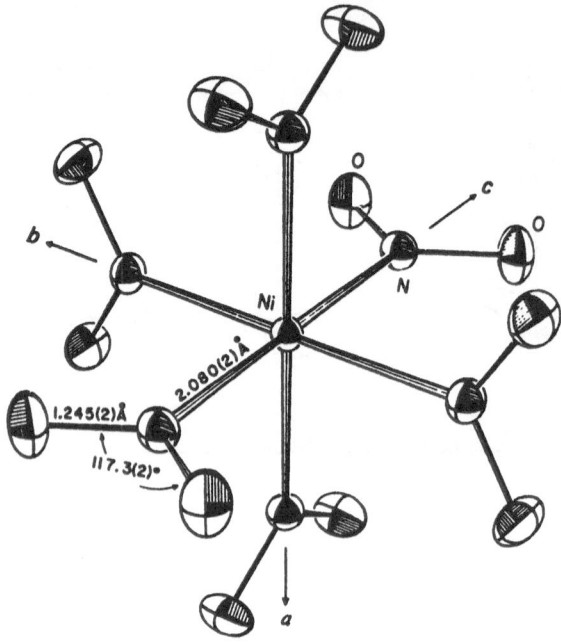

Fig. 35. Structure of the island [Ni(NO₂)₆] in the crystal structure of K₂PbNi(NO₂)₆ (Ref. 53)

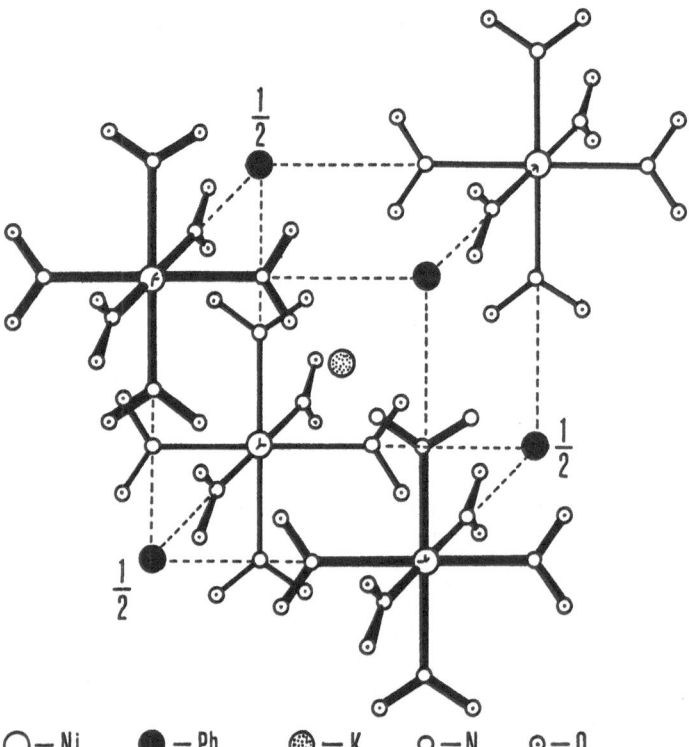

○ − Ni   ● − Pb   ⊕ − K   ○ − N   ⊙ − O

Fig. 36. Packing of the structural units in the unit cell of K₂PbNi(NO₂)₆ (Ref. 53)

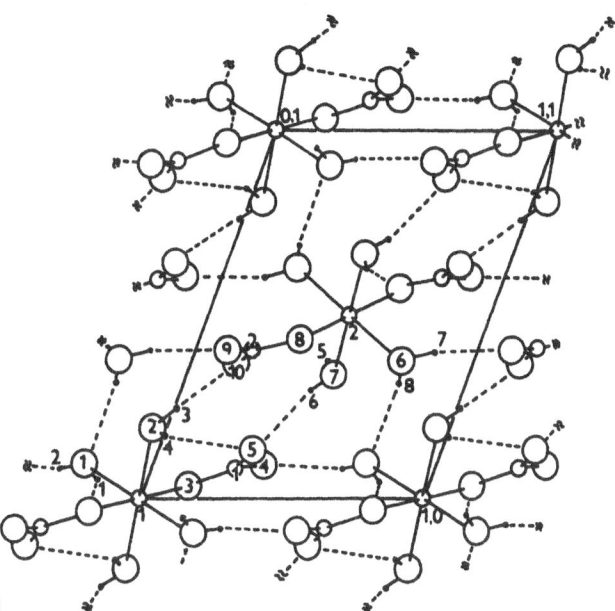

Fig. 37. Crystal structure of β-[Ni(NO$_3$)$_2 \cdot$ 4 H$_2$O] viewed along the [001] direction. The *dashed lines* indicate hydrogen bonding interactions (Ref. 54)

direction of the long diagonal (direction [111]) (Fig. 37). This direction is probably caused by the different character of the hydrogen bond systems of two symmetrically independent islands.

The structure of hexakis(pyridine-1-oxide)nickel(II) tetrafluoroborate, Ni(py-NO)$_6$(BF$_4$)$_2$[55], is isomorphous with the structures of Fe(pyNO)$_6$(ClO$_4$)$_2$[26] and Co(py-NO)$_6$(ClO$_4$)$_2$[26, 34].

The almost tetrahedral BF$_4^-$ ions are disordered in two positions with slight differences in their geometry. The inner coordination sphere around Ni(II) is essentially octahedral with a Ni-O bond length of 2.060(1) Å and O-Ni-O angle of 89.7(1)°. The compression ratio of s/h is 1.234.

Calcium ethylenediaminetetraacetatonickel(II) tetrahydrate, Ca[Ni(C$_2$H$_4$N$_2$)-(CH$_2$COO)$_4$] $\cdot$ 4 H$_2$O[56], exhibits a crystal structure composed of asymmetrical islands (Fig. 38), Ca^{2+} ions and water molecules. The anion islands and the crystal water molecules are held together by hydrogen bonds. The mean Ni-O and Ni-N bond lengths in an approximately octahedral coordination of the central atom formed by donor atoms are 2.063(7) and 2.059(7) Å. The O-Ni-N bond angles in glycine metallocycles lie in the interval 80.8(2)–83.8(2)°. The bond angle of N-Ni-N in the ethylenediamine metallocycle is 87.6(2)°. The remaining O-Ni-O and O-Ni-N angles are in the range 87.4(2)–94.4(2)° (some distances and angles were computed by means of the BONDLA program[57]).

An X-ray structural analysis of bis(benzoato)bis(quinoline)-nickel(II), Ni(quin)$_2$-(C$_6$H$_5$COO)$_2$[58], shows that the centrosymmetrical islands of [Ni(quin)$_2$(C$_6$H$_5$COO)$_2$] (Fig. 39) are held together only by van der Waals forces (Fig. 40). The dihedral angle between the plane of O$_2$C and that of the respective phenyl ring is 11.0°. The coordinated polyhedron of Ni(II) is a distorted octahedron. The bond lengths of Ni-O(1), Ni-O(2)

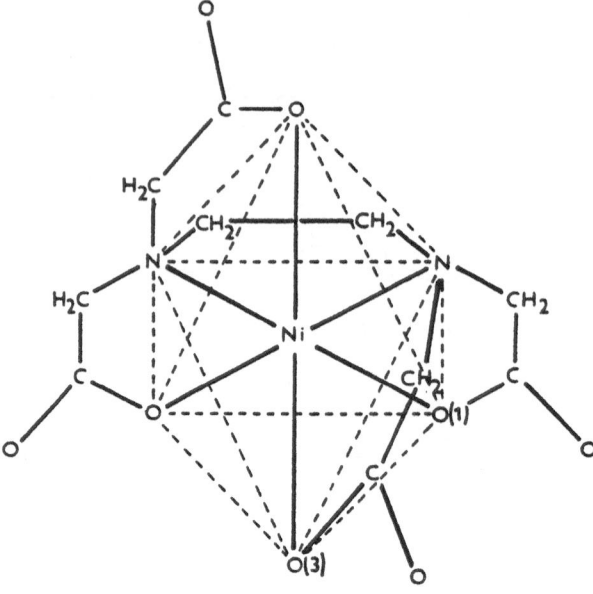

Fig. 38. Schematic representation of the island [Ni(C$_2$H$_4$N$_2$(CH$_2$COO)$_4$)] in the crystal structure of Ca[Ni(C$_2$H$_4$N$_2$(CH$_2$COO)$_4$)] · 4 H$_2$O (Ref. 56)

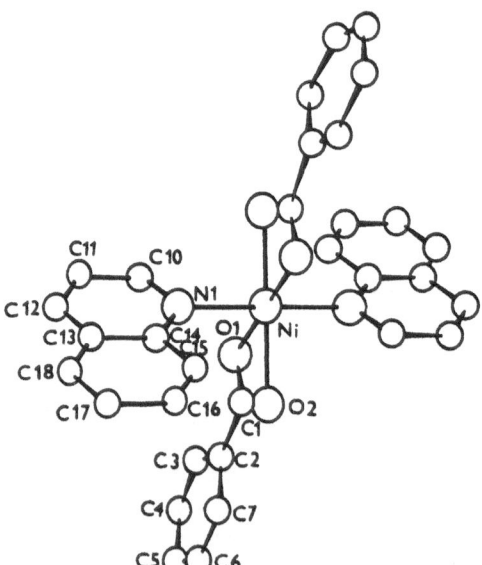

Fig. 39. Structure of the island [Ni(quin)$_2$(C$_6$H$_5$COO)$_2$] (Ref. 58)

and Ni-N(1) are 2.096(3), 2.083(8) and 2.080(4) Å, and the bond angle of O(1)-Ni-O(2) in the chelate tetracycle is 62.4(1)° while the remaining angles of O(2)-Ni-N(1) and O(1)-Ni-N(1) are much larger, namely 87.0(1) and 90.3(1)°.

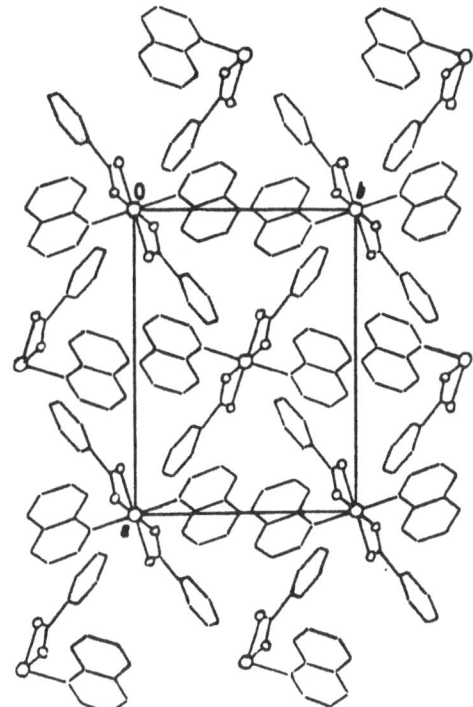

Fig. 40. Packing of the islands [Ni(quin)$_2$-(C$_6$H$_5$COO)$_2$] in the projection along the c axis (Ref. 58)

3.8 Cu(II) Complexes

In investigating the crystal structure of the bis(malodiamine)copper(II) diperchlorate complex, [Cu(C$_3$H$_8$N$_4$)$_2$](ClO$_4$)$_2$, at 200 K it was found[52], similarly as for its low-spin chloride Ni(II) analogue, that it contains [Cu(C$_3$H$_8$N$_4$)$_2$]$^{2+}$ islands which display 2/m(C$_{2i}$) symmetry. This structure, however, appears to be ordered. The Cu-N bond lengths are 1.956(1) Å while the N-Cu-N angles in the chelate hexacycles are 89.38(6)°. The islands are linked with tetrahedral perchlorate anions via hydrogen bonds (Fig. 41).

Diaquabis(6-azauracilato)copper(II), [Cu(C$_3$N$_3$O$_2$H$_2$)$_2$(H$_2$O)$_2$][59], also contains centrosymmetrical islands held together by hydrogen bonds (Fig. 42). Each of the Cu(II) atoms is square-planar and trans-coordinated by two oxygen atoms from the water molecules, and by two nitrogen atoms from approximately planar 6-azauracilate molecules. The bond lengths of Cu-O and Cu-N are 1.964(4) and 1.972(2) Å, the O-Cu-N(3) bond angle is 90.4(1)°. It is possible that in this case the hydrogen bonds between the islands (Fig. 42) do not cause a deformation of the square-planar arrangement of Cu(II).

The macrocyclic Cu(II)-Cu(I) complex with two Schiff's bases as ligands[60] consists of [Cu$_2$(C$_{12}$H$_{13}$N$_2$O)$_2$]$^+$ islands (Fig. 43), approximately tetrahedral ClO$_4^-$ groups and disordered CH$_3$OH molecules. The nearly planar islands in the crystal are arranged parallelly (Fig. 44). The distance between the least-squares planes of the nitrogen atoms from the two neighbouring islands is 3.37 Å. The Cu(2) atoms are disordered in two different

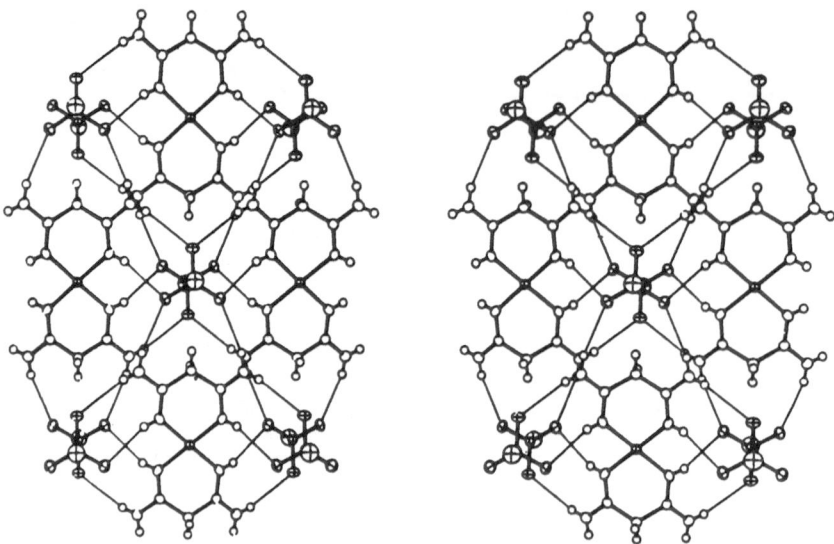

Fig. 41. Part of the structure $[Cu(C_3H_8N_4)_2](ClO_4)_2$ viewed along the [101] direction. Hydrogen bonds are represented by *thin lines*. The O atom of the perchlorate ion sticking out of the hydrogen-bonded layer is in close contact with the methylene H atom of the next layer (Ref. 52)

positions at a mutual distance of 0.51 Å. The donor atoms coordinated with Cu(2), however, do not exhibit any disorder. There are two types of islands with bond lengths and angles as shown in Fig. 45. Thus, only the inner coordination sphere of Cu(1) has Cu-L distances in the range 1.95–2.00 Å with a maximum relative frequency (Table 3). Cu(2 a) shows a population of 35% while for Cu(2 b) it is 65%. The atoms of Cu(1), O(1), O(21), N(1) and N(21) do not show great deviations from their least-squares planes ($\chi^2 = 428.4$).

$$\chi^2 = \sum_i p_i^2/\sigma^2(p_i)$$

where p_i denotes the distance between an atom and its plane and $\sigma(p_i)$ is the standard deviation. These values were computed by means of the NRC 22 program[61]. For the least-squares planes of Cu(2 A) and Cu(2 B), O(1), O(21), N(2), N(22) the χ^2 parameter is 29 048.9 and 3961.4, respectively.

3.9 Zn(II) Complexes

From the Zn(II) compounds only perchloratotetraphenylporphinatozinc(II), $Zn(TPP)(ClO_4)$[62], contains ZnN_4O_2 chromophores with M-L bond lengths in the range 2.05–2.10 Å with a maximum relative frequency (Table 3). This structure consists of $[Zn(TPP)(ClO_4)]$ islands (Fig. 46). In the tetraphenylporphyrinate ligand only the pyrrole rings I–IV are nearly planar. The nitrogen atoms of this ligand are planar with a maximum deviation of 0.003 Å from the least-squares plane. These donor atoms from

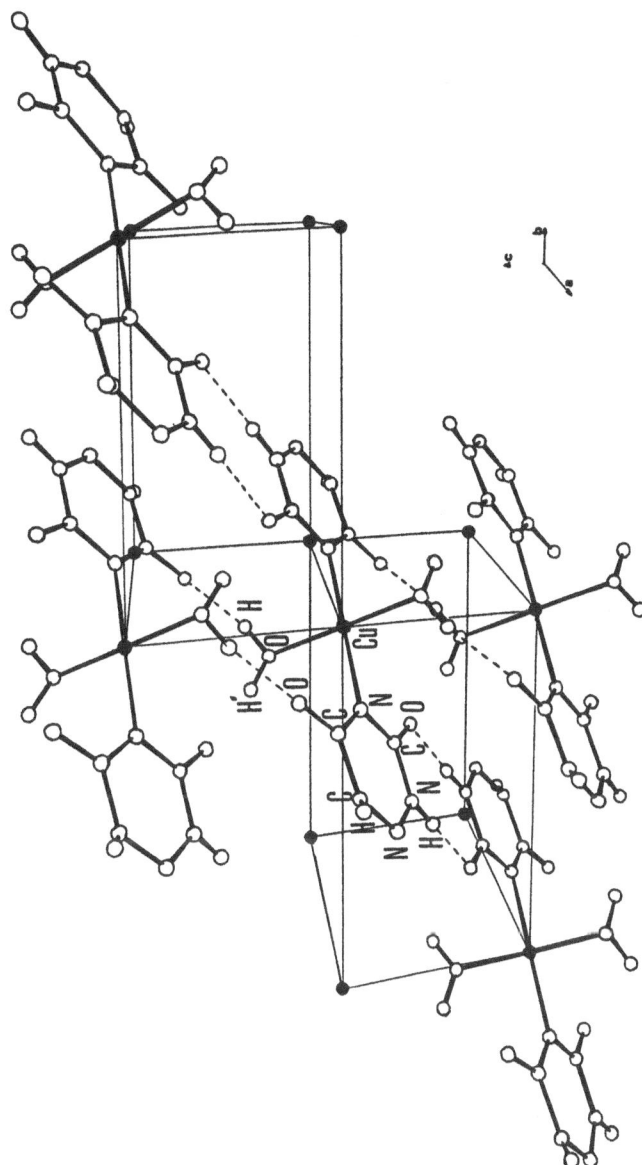

Fig. 42. System of the hydrogen bonds in the structure of $Cu(C_3N_3O_2H_2)_2(H_2O)_2$ (Ref. 59)

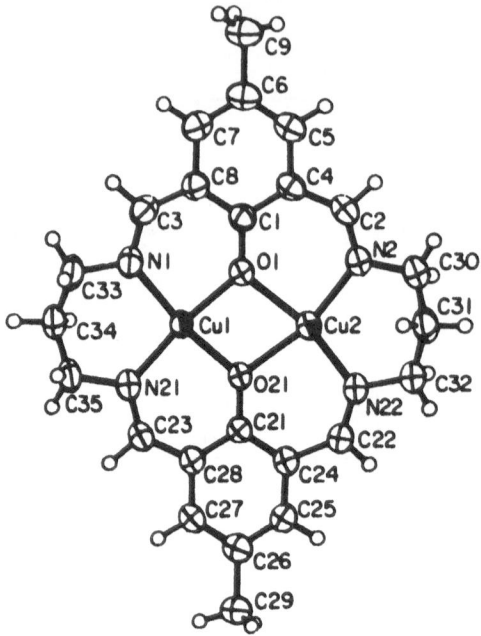

Fig. 43. Structure of the island
$[Cu_2(C_{12}H_{13}N_2O)_2]^+$ in the structure of
$[Cu_2(H_{12}H_{13}N_2O)_2](ClO_4)(CH_3OH)$
(Ref. 60)

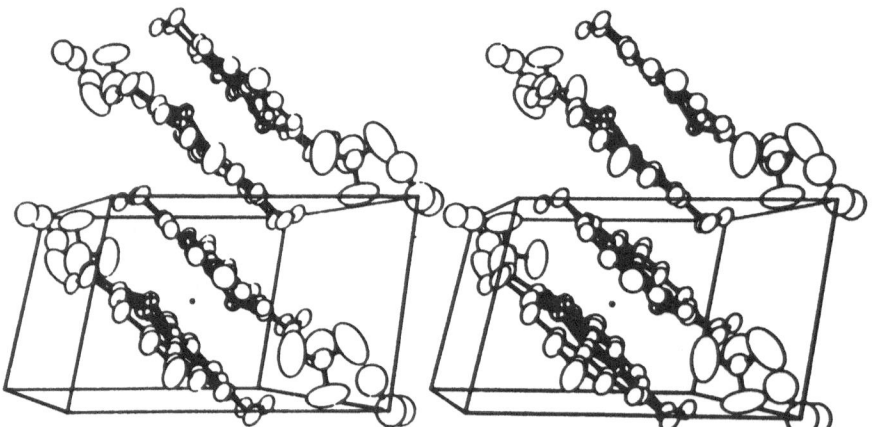

Fig. 44. Packing of the islands $[Cu_2(C_{12}H_{13}N_2O)_2]^+$ in the unit cell of $[Cu_2(C_{12}H_{13}N_2O)_2]$ $(ClO_4)(CH_3OH)$. Two of the methanol oxygens, related by a center of symmetry to the two shown, are omitted (Ref. 60)

the inner coordination sphere of Zn(II) form the basic plane of the tetragonal-pyramidal coordination around the central atom. The geometry of the central atom coordination is shown in Fig. 47. Probably for steric and crystal-packing reasons, the quadridentate tetraphenylporphyrinate ligand cannot coordinate the central atom with Zn-N bond lengths in the basic plane of the coordination polyhedron (Fig. 47). Therefore, the Zn(II) atom occupies a more stable position, and this circumstance leads to its displacement

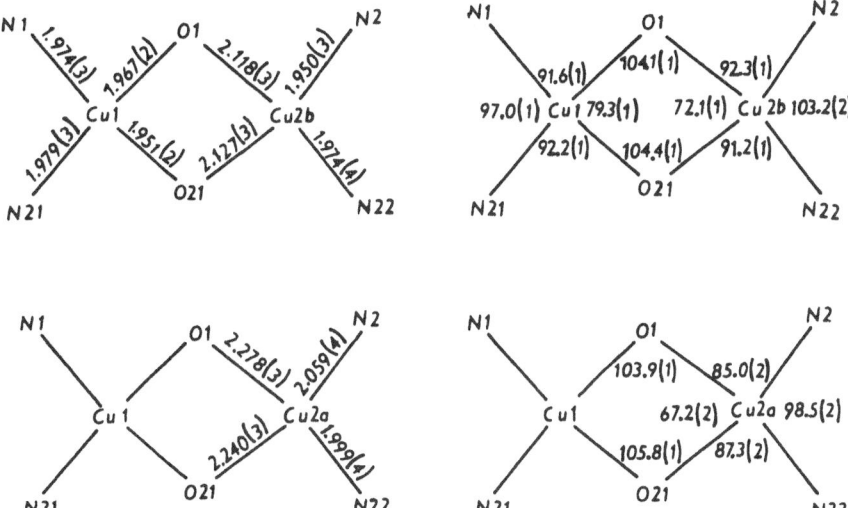

Fig. 45. Bond distances and angles of coordination spheres of three copper atom types in [Cu$_2$(C$_{12}$H$_{13}$N$_2$O)$_2$](ClO$_4$)(CH$_3$OH) (Ref. 60)

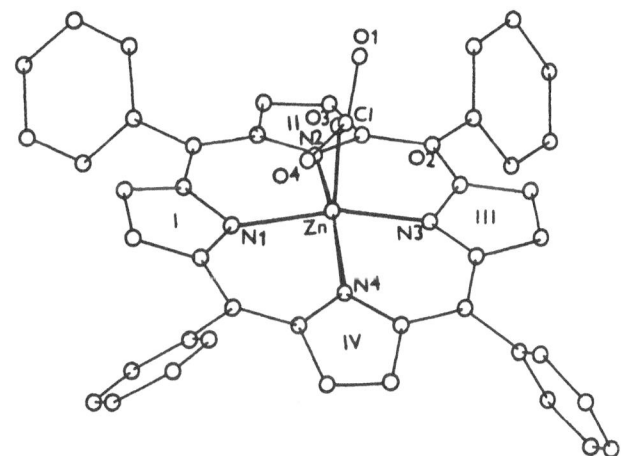

Fig. 46. Structure of the island in Zn(TPP)(ClO$_4$) (Ref. 62)

from the basic plane by 0.347 Å and at the same time to a dilatation of the Zn-N bonds. The symmetrical position of the perchlorate group with respect to the O(3)-Zn-O(4) plane is the result of repulsive interactions between the uncoordinated oxygen atoms of the perchlorate group and the nitrogen atoms of the tetraphenylporphyrinate ligand. According to Spaulding et al.[62], if the perchlorate group would be twisted around Zn-O(3) bond by 45° against this position in such a way that the pseudomirror planes of the perchlorate group and of the six-membered chelate ring would coincide, then the interatomic distances of O(2)-N(4) and O(4)-N(4) would have the minimum value of 2.1 Å.

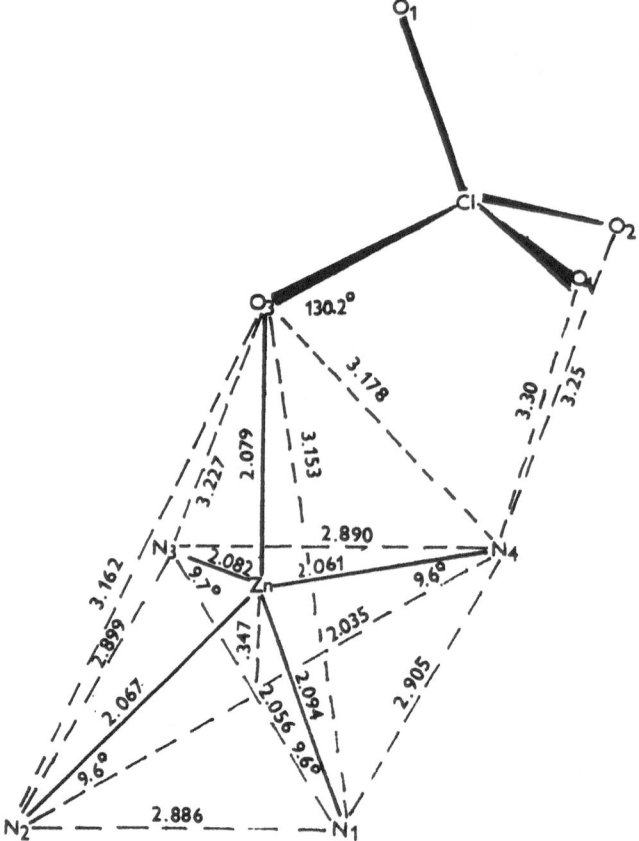

Fig. 47. Coordination of Zn(II) in Zn(TPP)(ClO₄) (Ref. 62)

4 Conclusions

The investigated crystal structures of complex compounds exhibit in the complex groups discussed equilibrium M-L bond lengths at least in one symmetrically independent chromophore. If the hydrogen bonds in each of these structures are not taken into consideration, except for the high-spin Mn(II) complex, the nearest surroundings of the central atom is part of the finite group of atoms held together by chemical bonds, i.e. it is contained in the island. From the point of view of the crystal field theory[15] and with respect to the octahedral and tetrahedral high-spin complexes having the electron config-urations of d^5 and d^{10}, the stabilization by the ligand field has a minimum. In the structure of $Mn(^+H_3NCH_2COO^-)_2(H_2O)_2 \cdot Br_2$ the essentially octahedral coordination of Mn(II) is mainly stabilized by the effects of "syn-syn" coordinated glycine ligands in an infinite chain. In [Zn(TPP)(ClO₄)] the almost regular tetragonal pyramidal coordination with a deviation of the central atom from the basic plane is determined by steric effects of ligands. In the low-spin [Co(NH₃)₆]Cl₃ complex the almost regular coordination is probably stabilized by the electrostatic influence of Cl^- ions which surround $[Co(NH_3)_6]^{3+}$ ions. In $[Co(H_2O)(NH_3)_5](S_2O_6)_3 \cdot 2H_2O$ the coordination around Co(III) is probably stabilized by the closepacking of ions. In the remaining low-spin Co(III) compounds, the chelated ligands are involved in the stabilization of the inner coordination spheres. Concerning the geometry of the coordination polyhedra of chromophores with bond lengths which are in the range of maximum frequencies (Table 3) it follows that

(a) The central atoms of $Mn(II)_{HS}$, $Fe(II)_{HS}$, $Fe(III)_{HS}$, $Co(II)_{HS}$, $Co(III)_{LS}$, and $Ni(II)_{HS}$ exhibit an almost regular octahedral coordination.

(b) Central atoms of $Ni(II)_{LS}$ and of Cu(II) exhibit an almost regular square-planar coordination.

Compounds in which the central atoms are coordinated by chelate ligands are listed in Table 6. The coordination types of the central atoms of these compounds are shown in Fig. 48. The mean L-M-L bond angles in hexa- and pentametallocycles are $92 \pm 3°$ and $84 \pm 2°$. In the tetrametallocycles of $[Ni(quin)_2(C_6H_5COO)_2]$ (type F) this angle becomes acute down to the value of $62.4(1)°$. Such angular deformations of the octahedral and square-planar coordination are thus caused by the coordination of the ligands in the chelate with the central atom, mainly in the penta- and tetrametallocycles. However, it appears that in addition to this chelate effect also some other factors influence the distorsion of the regular coordination around the central atom. The small angular deviations from the regular octahedral coordination of the central atoms contained in the islands of $[Fe(H_2O)_6] (NO_3)_3 \cdot H_2O$, $(FeCl_4)[Fe(NCD)_6]$, $[Co(H_2O)(NH_3)_5](S_2O_6)_3 \cdot$

Table 6. Type of coordination around central atoms of chelate ligands

Compound	Type of coordination (Fig. 48)
$(C_{10}H_9N_2)[Fe(C_{10}H_8N_2)_3]$	C
$[Co(TTA)_2(methanol)_2]$	D
$(+)_D$-$[Co(penten)]Co(CN)_6 \cdot 2H_2O$	A
$[Co(NH_3)_3dien]Cl(S_2O_6)$	E
$(-)_{589}$-(lel_2ob)-$[Co(chxn)_3]Cl_3 \cdot 5H_2O$	C
$(+)_{589}$-(ob_3)-$[Co(chxn)_3]Cl_3 \cdot H_2O$	C
$(-)_{589}$-$[Co(R,R-chxn)_2(3,3'-dmbpy)]Br(ClO_4)_2 \cdot H_2O$	C
$(+)_{589}$-$[Co(3,3'-dmbpy)(en)_2]Cl(ClO_4)_2 \cdot H_2O$	C
$[Co(en)_3](tart)Cl \cdot 5H_2O$	C
$(-)_{589}$-$[Co\{(R)-MeTACN\}_2]I_3 \cdot 5H_2O$	B
$[Ni(OAOH)_2] \cdot 2H_2O$	I
$[Ni(OAOH)_2] \cdot H_2O \cdot DMF$	I
$[Ni(C_{18}H_{14}N_4)]$	G
$[Ni\{Me_2Ga(N_2C_3H_3)_2\}_2]$	H
$[Ni(EMG)_2]$	I
$[Ni(C_3H_8N_4)_2]Cl_2$	H
$Ca[Ni(C_2H_4N_2(CH_2COO)_4)] \cdot 4H_2O$	A
$[Ni(quin)_2(C_6H_5COO)_2]$	F
$[Cu(C_3H_8N_4)_2](ClO_4)_2$	H

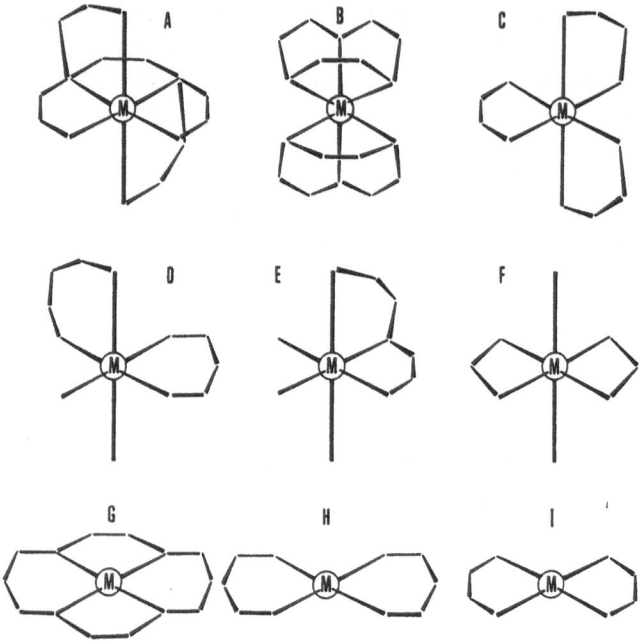

Fig. 48. Schematic representation of the coordination types around the central atoms of chelate ligands

$2 H_2O$ and $[Ni(NO_3)_2(H_2O)_4]$ are predominantly caused by hydrogen or deuterium bond interactions between islands and other groups. On the other hand, the $[Co(py-NO)_6](ClO_4)_2$ complex shows a regular octahedral central atom coordination. The slight trigonal distortions of the octahedral coordinations of $[Fe(pyNO)_6](ClO_4)_2$ and $[Ni(py-NO)_6](BF_4)_2$ are most probably caused by the the close packing of cations and anions in the crystal. A regular octahedral geometry with angular deviations which are within experimental errors has been found for four symmetrically independent islands of the structure $[Co(NH_3)_6]Cl_3$. An ideal octahedral coordination around the central atom of the symmetry O_h has been observed for $K_2Pb[Ni(NO_2)_6]$. In the structure of this compound influence of the external factors is apparently so symmetrical, that no distortion of the octahedral coordination, stabilized by the ligand field, takes place.

In $Ni(II)_{LS}$ and $Cu(II)$ complexes the fifth and sixth coordination sites in the axial direction are not occupied owing to the close packing of structural units in the crystals. In the macrocyclic $Cu(II)$-$Cu(I)$ complex with Schiff's bases, $Cu(II)$ atoms occupy a more stable position than $Cu(I)$. The coordination of $Cu(II)$ is planar with the slight deviations of L-M-L angles of metallocycles from 90°, being in the range 91.6(1)–97.0(1)°. The lowering of the O-Cu-O angle from 90° to 79.3(1)° is apparently caused by the bridging of the $Cu(II)$ and $Cu(I)$ atoms through oxygen atoms.

Chromophores with equilibrium vector lengths of M-L, limited by the ranges given in Table 3 show the most stable geometry of the coordination polyhedra. External factors can destabilize chromophores by deviations of the M-L vectors from the equilibrium distances. However, the magnitude of these deviations depends on external factors, as

well as on the proton number of the central atom and its oxidation and spin states[3, 63].
Assuming for the studied complexes an averange of external factors influencing the
vector lengths M-L, then the non-rigidity of the central atoms in the crystal structures
increases in an order which is contrary to their trend to attain a maximum equidistance
(6), i.e. (9):

$$Co(III)_{LS} < Ni(II)_{HS} < Co(II)_{HS} \approx Fe(III)_{HS} < Zn(II) < Ni(II)_{LS} < Mn(II)_{HS} < Fe(II)_{HS}$$
$$< Cu(II) . \qquad\qquad (9)$$

According to this order the central atom Cu(II) is, as regards crystal structure, more
nonrigid than $Ni(II)_{LS}$ and $Ni(II)_{HS}$. This is in agreement with the higher rigidity of
octahedral Ni(II) complexes with regard to the tetragonal distortions of the coordination
polyhedron in comparison with Cu(II) complexes which is reflected by the stereochemical change of an octahedral to a square-planar coordination and usually connected with a
change of the spin state[3, 64]. The peculiarities of octahedral Cu(II) complexes are
described in several reviews[3, 65–68] in connection with the Jahn-Teller dynamic effect. The
chromophores of this central atom in the solid state are often of the type CuO_6 or CuN_6 in
which four Cu-O bonds and Cu-N bonds, respectively, have approximately in plane with
the mean length R_S and two are out-of-plane with the mean length R_L[69] (Fig. 49).
Figure 50 shows the point diagram R_L vs. R_S made up of crystal structural data of Cu(II)
compounds with chromophores CuO_6 and CuN_6 reported in[3]. In Fig. 50 (curve b) such a
point diagram is shown for Cu(II) compounds with two equidistant out-of-plane bonds
and four in-plane bonds within experimental errors on the significance level 0.05[3].
Spearman's correlation coefficients R^{22} for the data depicted in the point diagrams of
Fig. 50 are statistically significant ($R \geqq R_{crit}$; Table 7). Therefore, in both cases, the
interdependences of R_L vs. R_S can be presumed (denoted in Fig. 50 by dashed lines). In

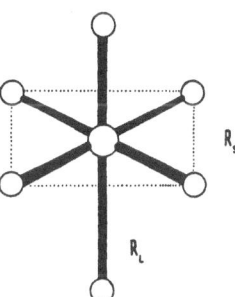

Fig. 49. Schematic representation of bonds in the plane and out of
the plane

Table 7. Statistical parameters of the correlations R_S vs. R_L for copper(II) compounds containing
CuO_6 and CuN_6 chromophores

Restriction	n	$\overline{R}_S$	$\overline{R}_L$	(R_S)	(R_L)	Reliability interval $\overline{R}_S$ [Å]	Reliability interval $\overline{R}_L$ [Å]	R^a	$R_{crit}{}^b$
		[Å]	[Å]	[Å]	[Å]				
no	111	1.99	2.56	0.06	0.33	1.97–2.01	2.46–2.66	– 0.72	0.19
2 + 4, 6, 4 + 2	27	1.99	2.52	0.07	0.34	1.94–2.04	2.32–2.72	– 0.80	0.38

[a] R, Spearsman's correlation coefficient[22]
[b] R_{crit}, critical value of R^{22}

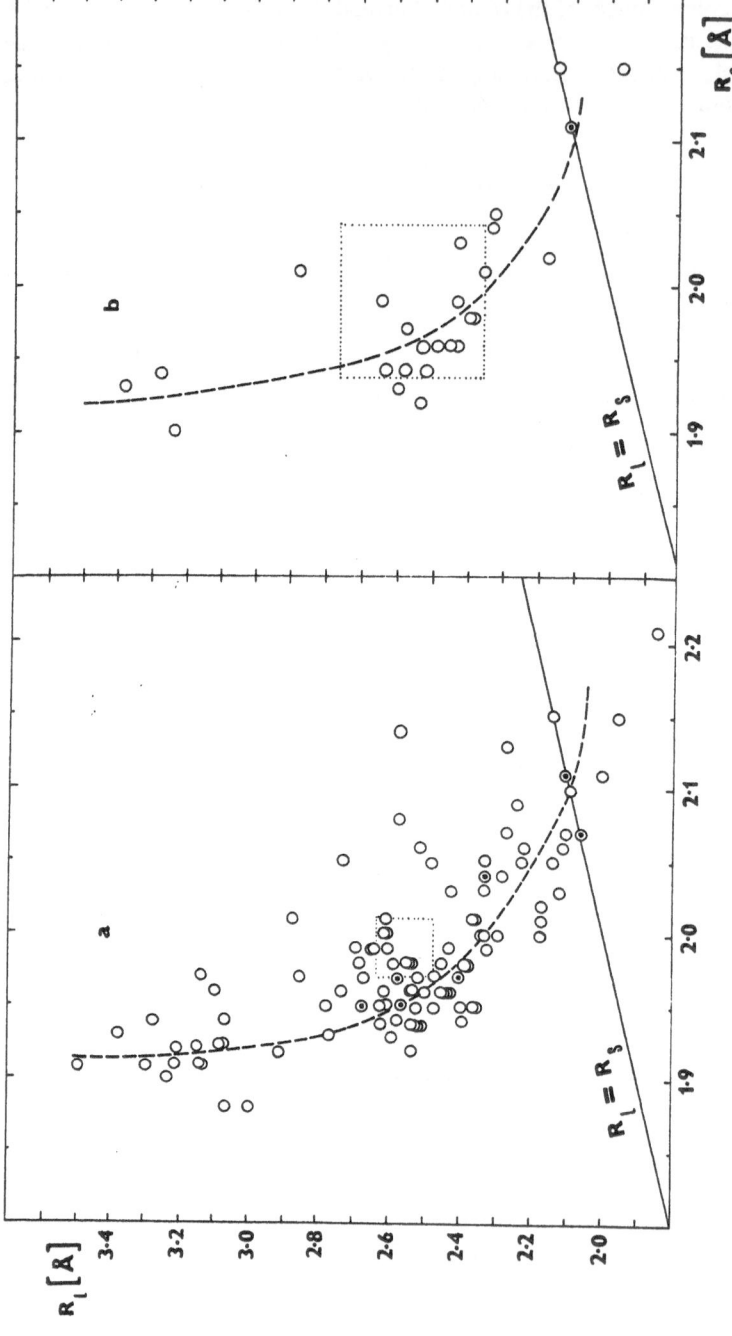

Fig. 50 a, b. Point diagrams of R_L vs. R_S of Cu(II) compounds containing CuO_6 and CuN_6 chromophores; a without conditions; b with coordination numbers 2 + 4, 6 and 4 + 2 (Ref. 3)

both diagrams, however, the points along these curves are not located at regularly distances. Assuming a normal distribution of both variables R_S and R_L then, with the probability $P = 1 - \alpha$, they are in the intervals

$$P[\overline{R}_S - z_\alpha \sigma(R_S)/\sqrt{n} \leq R_S \leq \overline{R}_S + z_\alpha \sigma(R_S)/\sqrt{n} \; ;$$

$$\overline{R}_L - z_\alpha \sigma(R_L)/\sqrt{n} \leq R_L \leq \overline{R}_L + z_\alpha \sigma(R_L)/\sqrt{n}] = 1 - \alpha$$

where $z_\alpha = 3.06$ for $\alpha = 0.005$[22)]; $\overline{R}_S$, $\overline{R}_L$, $\sigma(R_S)$ and $\sigma(R_L)$ are mean values and standard deviations calculated according to Eqs. (5) and (2); n is the number of pairs R_{S_i}, R_{L_i}. In Fig. 50 these regions with a probability of 99.5% are marked by dotted lines. From all the chromophores of CuO_6 and CuN_6 with no requirements of the bond lengths, 10% have R_S and R_L values included in this domain (Fig. 50, curve a). From the chromophores with equidistant bonds in the equatorial plane and axial direction, 52% fall into the region of probability of $P = 99.5\%$ (in Fig. 50 denoted by pointed lines), within experimental errors, i.e. with coordination numbers 2 + 4, 6 and 4 + 2. For chromophores of the CuO_6- and CuN_6-type, the elongated tetragonal-bipyramidal geometry (coordination number 4 + 2) with R_S and R_L values in the range 1.94–2.04 and 2.32–2.72 Å (Table 7) is thus the most probable. According to quantum chemical calculations[3, 70)] based on the Jahn-Teller theorem[71, 72)], such a geometry of the coordination polyhedron is energetically more advantageous for tetragonally deformed octahedral complexes Cu(II) than the compressed one.

It should be noted that in the examined complexes the most stable geometry of the coordination polyhedron is only exhibited by chromophores of the central atoms Fe(II)$_{HS}$, Fe(III)$_{HS}$ and Ni(II)$_{LS}$ with a homogeneous coordination sphere. It is interesting that for the central atoms Co(II)$_{HS}$, Co(III)$_{LS}$, Ni(II)$_{HS}$ and Cu(II) these are chromophores having both a homo- and heterogeneous coordination sphere. Both complexes with the central atoms Mn(II)$_{HS}$ and Zn(II) contain chromophores with a homogeneous sphere.

5 Abbreviations

$C_2H_4N_2(CH_2COO)_4$	ethylenediaminetetraacetate	MeTACN	2-methyl-1,4,7-triazacyclononane
$C_3H_8N_4$	malondiamidine	$Me_2Ga(N_2C_3H_3)_2$	dimethyl-bis(pyrazol-1-yl)gallate
$C_{10}H_8N_2$	2,2'-bipyridyl		
$C_{10}H_9N_2$	2,2'-bipyridilium	MO	molecular orbital
$C_{18}H_{14}N_4$	dibenzotetraaza[14]-annulene	NCD	deuteriumcyanide
		OAOH	oxamide oximato
$C_{12}H_{13}N_2O$	Schiff base	penten	N,N,N',N'-tetrakis-(2'-
$C_3N_3O_2H_2$	6-azauracilate		aminoethyl)-1,2-
chxn	1,2-diaminocyclohexane		diaminoethane
dien	diethylenetriamine	pyNO	pyridine N-oxide
DMF	dimethylformamide	quin	quinoline
3,3'-dmbpy	3,3'-dimethyl-2,2'-bipyridyl	tar	tartrate
		TPP	tetraphenylporphyrinate
EMG	ethylmethylglyoxime	TTA	thenoyltrifluoroacetone
HS	high-spin	VEP	vector equilibrium principle
en	ethylenediamine		
LS	low-spin		

6 References

1. Britton, D., Dunitz, J. D.: Acta Crystallogr. *A 29*, 362 (1973)
2. MTP Internat. Review of Science, Vol. 9, p. 37, Tobe, M. L., London, Butterworths, 1972; Muetterlies, E. J.: Inorg. Chem. *4*, 769 (1965)
3. Gažo, J. et al.: Coord. Chem. Rev. *19*, 253 (1976); *43*, 87 (1982)
4. Kitajgorodskij, A. J.: Kristallografija *2*, 456 (1957)
5. Fisher, W.: Z. Krist. *133*, 18 (1971)
6. Zorkij, P. M. et al.: Zh. Strukt. Chem. *19*, 633 (1969)
7. Loeb, A. L.: Acta Crystallogr. *15*, 219 (1962)
8. Lima-de-Faria, J.: Z. Krist. *122*, 359 (1965)
9. Zorkij, P. M., Belškij, V. K.: Sovremennije Problemy Fizičeskoj Chimii, Tom 4, Sahparonov, M. J., Akišina, P. A. (eds.), Izd. MGU, *1970*, 379
10. Loeb, A. L.: J. Solid-State Chem. *1*, 237 (1970)
11. Hellner, E. E.: Structure and Bonding *37*, 61 (1979)
12. Laves, F.: Crystal Structure and Atomic Size, in: Theory of Alloys, Cleveland, Ohio, Amer. Soc. Met. 1956
13. Loeb, A. L.: Acta Crystallogr. *17*, 179 (1964)
14. Lima-De-Faria, J., Figuerido, M. D.: J. Solid-State Chem. *16*, 7 (1976)
15. Orgel, L. E.: An Introduction to Transition-metal Chemistry, Ligand-field theory, London, Methuen; New York, John Wiley and Sons 1966
16. Franck, F. C., Kasper, J. S.: Acta Crystallogr. *11*, 184 (1958)
17. Franck, F. C., Kasper, J. S.: ibid. *12*, 483 (1959)
18. Boča, R., Pelikán, P.: Chem. Zvesti *36*, 35 (1982)
19. Boča, R.: To be published
20. Boča, R., Pelikán, P.: Inorg. Chem. *20*, 1618 (1981)
21. Ito, T., Sugimoto, M., Ito, H.: XXII.I.C.C.C., Budapest, Aug. 23–27, 1982, Abstracts of Papers *1*, 87 (1982); and private communication
22. Zacks, Sh.: The Theory of Statistical Inference, New York, Wiley 1971
23. Hald, A.: Statistical Theory with Engineering Applications, Wiley, New York 1955
24. Robinson, K., Gibbs, G. V., Ribbe, P. H.: Science *172*, 567 (1971)
25. Glowiak, T., Ciumik, Z.: Acta Crystallogr. *B 34*, 1980 (1978)
26. Taylor, D.: Aust. J. Chem. *31*, 713 (1978)
27. Daran, J. C., Jeannin, Y.: Acta Crystallogr. *B 31*, 1838 (1975)
28. Gagné, R. R., Koval, C. A., Smith, T. J.: J. Am. Chem. Soc. *99*, 8367 (1977)
29. Hair, J. N., Beattie, J. K.: Inorg. Chem. *16*, 245 (1977)
30. Figgis, B. N., Skeleton, B. W., White, A. H.: Aust. J. Chem. *31*, 57 (1978)
31. Bergendahl, T. J., Wood, J. S.: Inorg. Chem. *14*, 338 (1975)
32. Pretorius, J. A., Boeyens, J. C. A.: J. Inorg. Nucl. Chem. *40*, 1519 (1978)
33. Boeyens, J. C. A., Hancock, R. D., McDougall, G. J.: J. Afr. J. Chem. *32*, 23 (1979)
34. Kruger, G. J., Reynhardt, E. C.: Acta Crystallogr. *B 34*, 915 (1978)
35. Muto, A., Marumo, F., Saito, Y.: ibid. *B 26*, 226 (1970)
36. Solans, X. et al.: ibid. *B 35*, 2181 (1979)
37. Burštejn, I. F. et al.: Koord. Chim. *4*, 282 (1978)
38. Sato, S., Saito, Y.: Acta Crystallogr. *B 33*, 860 (1977)
39. Ohba, S., Sato, S., Saito, Y.: ibid. *B 35*, 957 (1979)
40. Harnung, S. E. et al.: Inorg. Chem. *15*, 2123 (1976)
41. Marumo, F., Utsumi, Y., Saito, Y.: Acta Crystallogr. *B 26*, 1492 (1970)
42. Kobayashi, A., Marumo, F., Saito, Y.: ibid. *B 28*, 2709 (1972)
43. Internat. Union of Pure and Applied Chem., Information Bull., No. 33, 1968, Inorg. Chem. *9*, 1 (1970)
44. Sato, S., Saito, Y.: Acta Crystallogr. *B 34*, 3352 (1978)
45. Templeton, D. H.: ibid. *B 35*, 1608 (1979)
46. Mikami, M.: ibid. *B 33*, 1485 (1977)
47. Endres, H.: ibid. *B 35*, 625 (1979)

48. Endres, H.: ibid. *B 34*, 2306 (1978)
49. Weiss, M. C., Gordon, G., Goedken, V. L.: Inorg. Chem. *16*, 305 (1977)
50. Rendle, D. F., Storr, A., Trotter, J.: J. Chem. Soc., Dalton Trans. *1975*, 176
51. Bowers, R. H., Banks, C. V., Jacobson, R. A.: Acta Crystallogr. *B 28*, 2318 (1972)
52. Schwarzenbach, D., Schmelczer, R.: ibid. *B 34*, 1827 (1978)
53. Takagi, S., Joesten, M. D., Lenhert, P. G.: ibid. *B 31*, 1968 (1975)
54. Morosin, B., Haseda, T.: ibid. *B 35*, 2856 (1979)
55. Van Ingen Schenau, A. D., Verschoor, G. C., Romers, C.: ibid. *B 30*, 1686 (1974)
56. Nestorova, J. M., Poraj-Košic, M. A., Logvinenko, V. A.: Zh. Struct. Chim. *21*, 171 (1980)
57. Stewart, J. M., Kundell, F. A., Baldwin, J. C.: The X-ray 70 System. Computer Science Center, Univ. Maryland, College Park, Maryland 1976
58. Hursthouse, M. B., New, D. B.: J. Chem. Soc., Dalton Trans. *1977*, 1082
59. Mosset, A., Bonnet, J. J., Galy, J.: Acta Crystallogr. *B 33*, 2639 (1977)
60. Gagné, R. R., Henling, M. L., Kistenmacher, T. J.: Inorg. Chem. *19*, 1226 (1980)
61. Ahmed, F. R. et al.: NRC Crystallographic Programs for the IBM/360 System. National Research Council, Ottawa, Canada 1966
62. Spaulding, D. et al.: J. Am. Chem. Soc. *96*, 982 (1974)
63. Boča, R.: Chem. Zvesti *35*, 769 (1981)
64. Jóna, E. et al.: Koord. Chim., in press
65. Robinson, K., Gibbs, G. V., Ribbe, P. H.: Science *172*, 567 (1971)
66. Hathaway, B. J. et al.: Structure and Bonding *14*, 49 (1973)
67. Burdett, J. K.: Inorg. Chem. *14*, 931 (1975)
68. Bersuker, J. B.: Coord. Chem. Rev. *14*, 357 (1975)
69. Hathaway, B. J., Hodgson, P. G.: J. Inorg. Nucl. Chem. *35*, 4071 (1973)
70. Yamatera, H.: Acta Chem. Scand. *A 33*, 107 (1979)
71. Jahn, H. A., Teller, E.: Proc. R. Soc. *A 161*, 220 (1937)
72. Jahn, H. A.: Proc. R. Soc. *A 164*, 117 (1938)

Complexing Modes of the Phosphole Moiety

François Mathey[1], Jean Fischer[2] and John H. Nelson[3]

[1] Laboratoire CNRS-SNPE, 2-8 rue H. Dunant B.P. n°28, 94320 Thiais, France
[2] Laboratoire de Cristallochimie, ERA 08, Institut Le Bel, Université Louis Pasteur,
  67070 Strasbourg, Cedex, France
[3] Department of Chemistry, University of Nevada, Reno Nevada 89557 U.S.A.

This review shows that the growing interest aroused by phospholes in coordination chemistry can be attributed to three main reasons: 1. The influence of the partial delocalization of the phosphorus lone pair upon the complexing ability of phospholes. 2. The discovery of phosphametallocenes. 3. The tremendous variety of complexing modes found for the phosphole nucleus. For a given phosphole, no less than ten different types of complexes can be obtained.

Structure and Bonding 55
© Springer-Verlag Berlin Heidelberg 1983

I. Introduction

Among the numerous reviews dealing with phospholes only two incorporated chapters comprehensively describing their complexes[1, 2]. But even the most recent[2] is now completely outdated: two thirds of the publications on these compounds have appeared since it was written. The growing interest aroused by phospholes in coordination chemistry can be attributed to three main reasons. The first reason is a question: what is the influence of the partial delocalization of the phosphorus lone pair upon the complexing ability of phospholes? This is related to the widely discussed question of phosphole aromaticity. The second reason is the discovery of phosphametallocenes. Apart from their interest as phosphorus analogs of η^5-cyclopentadienyl complexes, phosphametallocenes are the only λ^3-phosphorus-carbon heterocycles for which an extensive chemistry of electrophilic substitutions at carbon has thus far been demonstrated[1]. The third reason lies in the tremendous variety of complexing modes found for the phosphole nucleus. For a given phosphole, no less than ten different types of complexes can be obtained even if we consider only the tervalent oxidation state of phosphorus. Another less obvious, but compelling reason is that phospholes are now readily available on a large scale. These reasons suggested to us that a review on this new area would be timely.

We have excluded the widely studied dibenzophospholes from this review because they have no "free" dienic system and, thus, act only as classical 2-electron ligands. But, we have included a section on arsole and stibole complexes, and a brief comparison between azametallocenes and their phosphorus analogs in order to illustrate the influence of the heteroatom upon the properties of these systems. This review is intended to be a complete coverage of the literature up to October 1981.

II. Phosphole Complexes

II.1. Structural and Electronic Features of the Phosphole Ring

Numerous recent reviews describe the properties of the phosphole ring[2-4]. Our aim here is just to summarize the salient points which are useful for understanding the behaviour of phospholes towards transition metal derivatives.

It has already been stated that phospholes are now readily available. The various syntheses of phospholes are depicted in Scheme 1.

The first synthesis is the simplest and most general route to these systems and, since its recent optimization[7], it allows the preparation of phospholes on a large scale.

The most significant X-ray crystal structure analysis has been performed on 1-benzyl-phosphole[13]. The principal data are given in Fig. 1.

It can be seen that the three C-C bonds within the ring are sharply unequal indicating a poor electronic delocalization. Nevertheless, some delocalization is suggested by the

---

1 In more classical "aromatic" λ^3-phosphorus-carbon heterocycles there is always a competition between the electrophilic attack at carbon and at the phosphorus lone pair

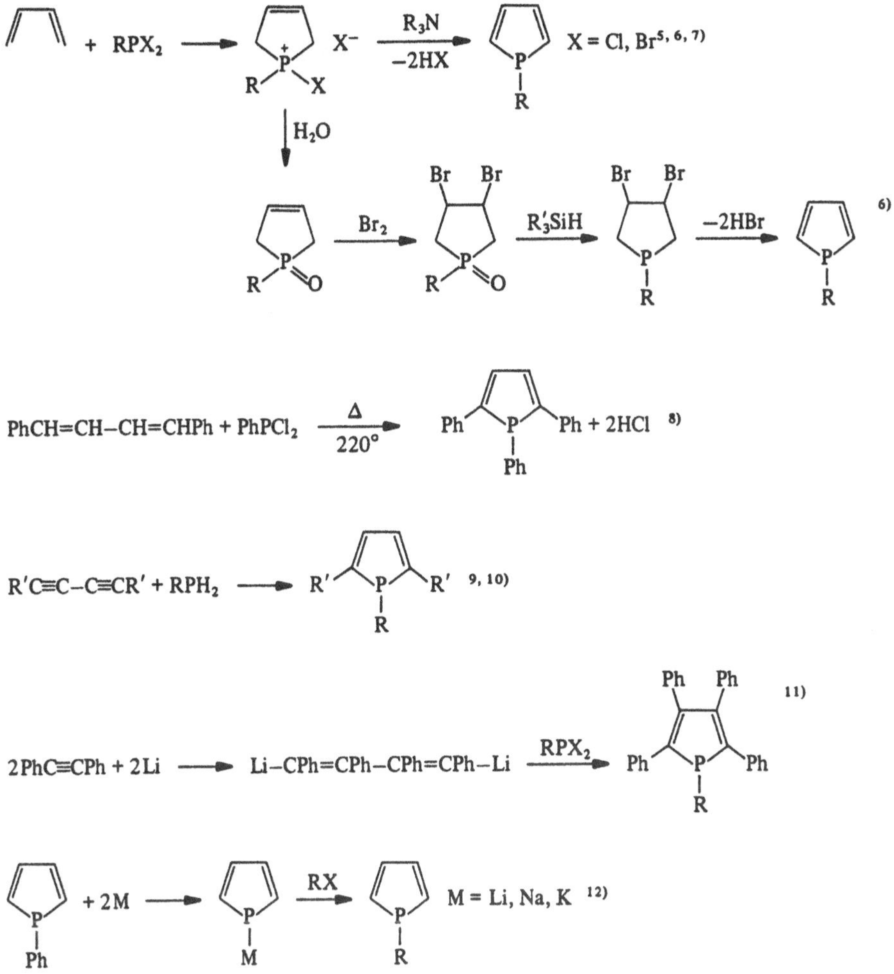

PhCH=CH–CH=CHPh + PhPCl$_2$ $\xrightarrow[220°]{\Delta}$ Ph—P(Ph)—Ph + 2HCl   [8]
(Ph on P)

R'C≡C–C≡CR' + RPH$_2$ ⟶ R'—(P,R)—R'   [9, 10]

2 PhC≡CPh + 2 Li ⟶ Li–CPh=CPh–CPh=CPh–Li $\xrightarrow{RPX_2}$ Ph—(P,R)—Ph (Ph, Ph)   [11]

(phosphole)$_{Ph}$ + 2 M ⟶ (phosphole)$_M$ $\xrightarrow{RX}$ (phosphole)$_R$ M = Li, Na, K   [12]

Scheme 1. Syntheses of phospholes

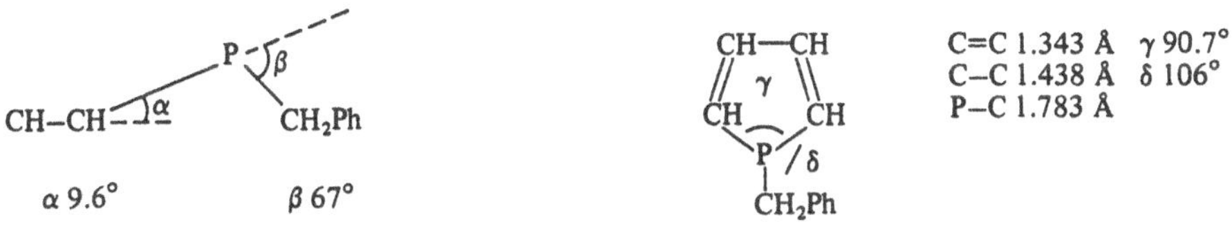

α 9.6° β 67°

CH—CH   C=C 1.343 Å γ 90.7°
CH   CH   C–C 1.438 Å δ 106°
P   CH$_2$Ph   P–C 1.783 Å

Fig. 1. Structure of 1-benzylphosphole[13]

shortening of the intracyclic P–C bond (1.78 Å vs 1.84 Å for the sum of the covalent radii). Another notable feature is the value of the intracyclic CPC angle (90.7°) which indicates that the system is very strained (normal CPC angles in phosphines ≃ 100°). As a consequence, phospholes are sterically undemanding ligands. Finally, the low phosphorus pyramidal inversion barriers of phospholes (≈ 16 kcal/mol[10]) show that their planar state is highly stabilized by electronic delocalization.

It is instructive to compare phospholes to other structurally similar phosphorus donors. The most appropriate ligands with which comparisons should be made are phenyldivinylphosphine (I) and dimethylphenylphosphine (II).

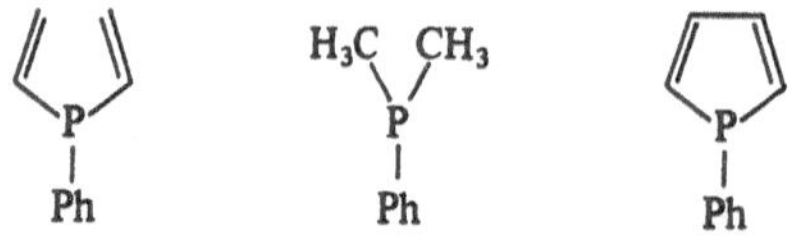

| I | II | III |

All these ligands (I–III) are pyramidal at phosphorus, but with varying degrees of "sp" hybridization. Due to their reduced phosphorus inversion barrier[10] phospholes are more planar than typical phosphines and their phosphorus lone pair possesses less s-character than does the phosphorus in either I or II. Thus from a frontier orbital point of view[14, 15], since the lone pair is likely the HOMO for all these ligands, we would anticipate that I and II might be poorer donors than III. Likewise, III is considerably less bulky than either I or II. In sum then, if cyclic conjugation is not large in III, its donor ability should approximate those of I and II.

Experimentally, the following order of donor ability has been observed:

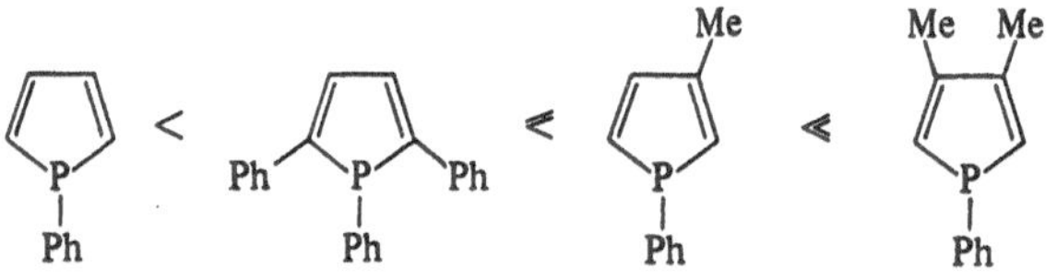

Indeed, the oxidation potential (assessed qualitatively by the ease of formation of the oxides and sulfides) of these ligands follows this same order. Exemplary data for the reaction

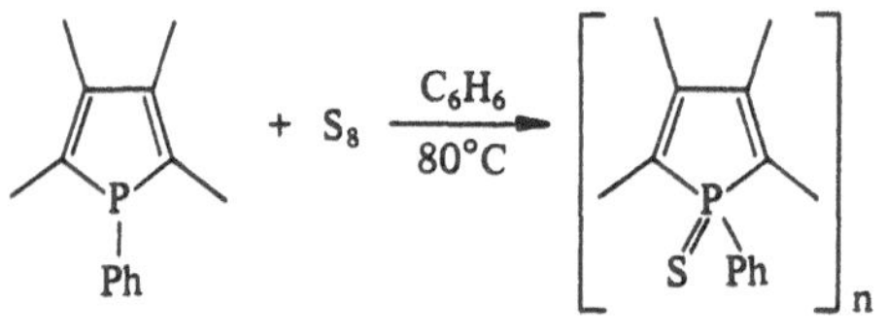

are as follows:

Phosphole[a]	Time	Yield	n	$\delta^{31}P$ (in $CDCl_3$)
PP	48 h	50%	2 endo	110.5, 62.0³ $J_{PP} = 42$
TPP	6 h	35%	1	48.6
MPP	7 h	84%	2 endo	109.5, 62.7³ $J_{PP} = 43.9$
DMPP	3 h	90%	1	46.1

[a] Data collected in our laboratory, $\delta^{31}P$ relative to 85% H_3PO_4, downfield shifts positive

This ordering is quite normal except for TPP. The electron withdrawing ability of phenyl groups on the 2- and 5-positions in concert with their increased steric hindrance would portend a low donor ability for TPP. In fact the structural data for 1-benzylphosphole[13] and 1,2,5-triphenylphosphole[16] suggest that there is less cyclic conjugation in the latter.

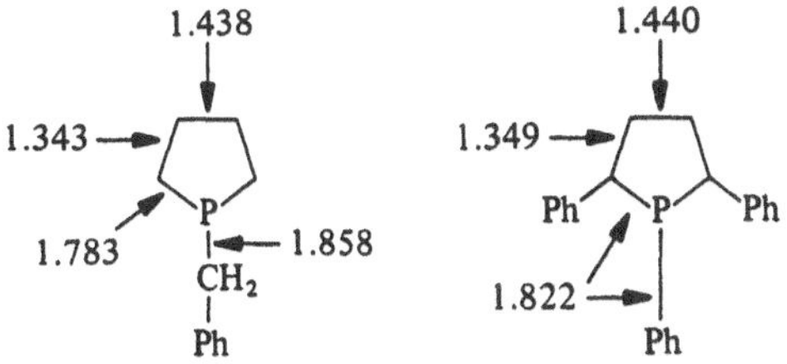

Thus, the conjugation in TPP is better represented as in IV than V, and as a result, the phenyl groups are not very effective in removing electron density from phosphorus in TPP.

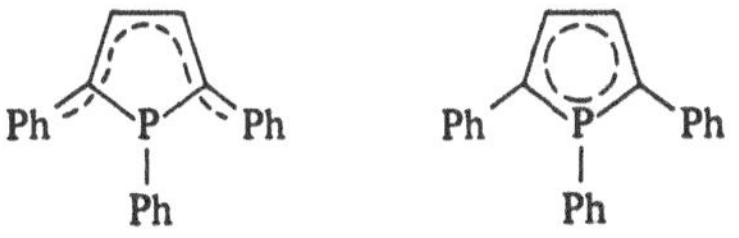

Contrary to previous arguments[17], the significantly enhanced donor ability of DMPP is likely due to the repulsive interaction between the methyl groups which places more s-character in the intracyclic P-C bonds leaving the lone pair orbital with an increased amount of p-character. Data[18] supporting these conclusions are found in Table 1.

It has been suggested[19] that increased shielding of the P-phenyl carbon and an algebraic increase in $^1J(P, C-2)$ both signify an increase in phosphole cyclic delocalization.

Table 1. ^{13}C NMR data for selected phospholes and phosphines

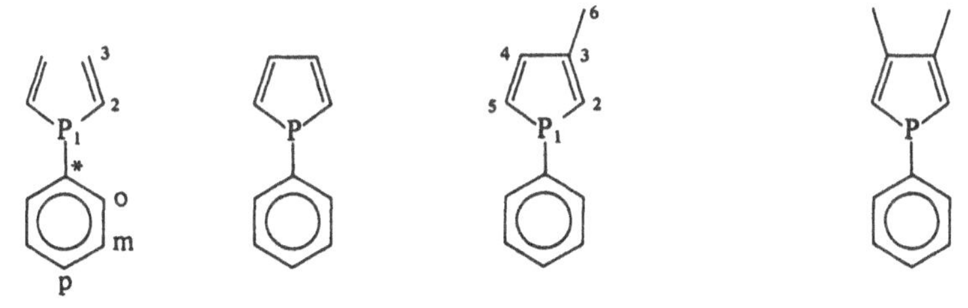

Carbon	^{13}C chemical shifts (ppm)			
2	136.95	135.11	128.56 (136.00)5	128.52
3	128.10	136.71	147.38 (140.07)4	148.75
C*	137.31	129.59	130.62	132.42
	^{13}C-^{31}P nuclear spin coupling constants (Hz)			
2	− 13.91	− 5.18	1.2(5.3)5	7.3
3	+ 23.16	+ 8.17	8.8(7.6)4	8.5
C*	9.17	− 8.63	10.5	12.2

Assuming that ^1J(P, C-2) is negative for all the phospholes, the ^{13}C NMR data suggest that the cyclic conjugation follows the order PP > MPP > DMPP. Also making the normal assumption that the absolute magnitude of $^1J_{PC}$ is proportional to the amount of s-character in the P-C bond, it is clear that both the P-phenyl and intracyclic P-C bonds increase in s-character in the order PP < MPP < DMPP.

In the ensuing chapters we will consider phospholes as two electron donors when bonding occurs to a metal by donation of the phosphorus lone pair only; one electron donors when the $> \dot{P}$: moiety bonds to a single metal center; three electron donors, when this same moiety bridges two metal centers; four electron donors, when only the diene system is coordinated; five electron donors, for the phospholyl group in phosphametallocenes; four plus two electron donors when the diene system is bound to one metal and the phosphorus behaves as a two electron donor to a second metal; four plus three electron donors when the diene system is coordinated to one metal and the $> \dot{P}$: moiety bridges two others; and five plus two electron donors when phosphametallocenes behave as two electron donors.

II.2. λ^3-Phospholes as 2-Electron Ligands

The types of complexes containing phospholes as two electron donors which have been prepared to date are summarized in Table 2. From the data in this table, it is clear that phospholes have been found to coordinate to soft metal centers, and we should classify these ligands as soft donors with some propensity towards π-acceptance. The complexes will be discussed according to the position of the metal in the periodic table.

Cr, Mo, W

Phospholes react[20, 21)] with $M(CO)_6$, THF $M(CO)_5$ or (piperidine)$_2$ $M(CO)_4$, M = Cr, Mo, W to produce the expected $LM(CO)_5$ and Cis-$L_2M(CO)_4$. No complexes of these elements in higher oxidation states have been prepared. The complexes and their infrared carbonyl stretching frequencies are listed in Table 3.

Making the normal assumption that the energy of the CO stretch is influenced by the trans ligand such that the better the sigma donor the trans ligand, the stronger the M-C bond, the weaker the CO bond and the lower the energy, v(CO) allows us to rank the donor ability of the phospholes. Using this criterion, within experimental error of at least ± 5 cm^{-1}, v(CO) for analogous complexes of Me$_2$PPh and DMPP occur at essentially the same energy implying similar metal ligand bond strengths. Likewise TPP appears to be a weaker donor towards Cr and Mo but as good a donor towards the softer W as DMPP.

The ^{31}P NMR data for the $LM(CO)_5$ complexes given in Table 3a clearly illustrate that the coordination chemical shift of the phosphole complexes is much smaller than those of similar phosphine complexes. This supports the notion of an increased potential for M-P π back-donation with phospholes and is also consistent with little phosphorus rehybridization upon coordination.

Mn, Tc, Re

Heating $Mn_2(CO)_{10}$ with 3,4-dimethyl phospholes (L) in xylene affords[23)], among other products to be discussed later, yellow $LMn_2(CO)_9$ where L occupies the axial coordination site. The infrared data for DMPP $Mn_2(CO)_9$ [v(CO), 2084, 2010, 1993, 1972, 1939 cm^{-1}] are very similar to that[24)] for $PPh_3Mn_2(CO)_9$ [v(CO), 2093, 2011, 1994, 1973, 1939 cm^{-1}] suggesting similar Mn-P bond strengths for PPh$_3$ and DMPP.

Table 2. Transition metal complexes of phospholes as two electron donors

Cr	Mo	W
LCr(CO)$_5$ III, IV	LMo(CO)$_5$ I, III, IV	LW(CO)$_5$ III, IV
Cis-L$_2$Cr(CO)$_4$ III	Cis-L$_2$Mo(CO)$_4$ III	
Mn	**Tc**	**Re**
LMn$_2$(CO)$_9$ III	NONE	L$_3$Re$_3$Cl$_9$ II, III
L$_2$Mn$_2$(CO)$_8$ III'		L$_2$Re$_3$Cl$_9$ · CH$_2$Cl$_2$ IV
Fe	**Ru**	**Os**
L$_2$Fe(CO)$_2$(O$_2$CCF$_3$)$_2$ III	L$_3$RuCl$_2$ I, IV	NONE
LL'Fe(CO)$_2$(O$_2$CCF$_3$)$_2$ III	L$_4$RuCl$_2$ III	
L$_2$Fe(CO)$_2$CS$_2$ III	L$_2$Ru(CO)$_2$Cl$_2$ II, III, IV	
LFe$_2$(CO)$_3$Cp$_2$ III	L$_3$RuCOCl$_2$ II, III	
LFe(CO)$_4$ III, IV	L$_2$L'RuCl$_2$ III	
and others	L$_4$Ru$_2$Cl$_5$ II	
	LRuCl$_3$ II	
	L$_2$RuCl$_3$ IV	
Co	**Rh**	**Ir**
L$_2$CoCl$_2$ II, III	L$_3$RhCl$_3$ I, II, III, L$_2$RhHCl$_2$ IV	L$_3$IrCl I
	L$_3$Rh(PhCO)Cl$_2$ I (L$_2$RhCl)$_2$ IV	L$_3$IrCl$_3$ II, III
	L$_5$Rh$_2$COCl$_4$ II, III	
	L$_5$Rh$_2$Cl$_4$ I, II, III	
	L$_2$RhCOCl I, II, III, IV	
	L$_3$RhCl II, III	
	LRhCl$_3$ I, IV	
Ni	**Pd**	**Pt**
[L$_3$NiCl]$^+$Cl$^-$ I	[L$_3$PdCl]$^+$Cl$^-$ I	L$_2$PtX$_2$ II, III, IV
L$_2$NiCl$_2$ II, III	L$_2$PdX$_2$ II, III, IV	L$_3$PtX$_2$ IV
LNi(CO)$_3$ IV	X = Cl, Br, I, SCN, N$_3$	X = Cl, Br, I, SnCl$_3$
		[L$_3$PtCl]$^+$Cl$^-$ III
Cu	**Ag**	**Au**
LCuCl IV	NONE	NONE
Zn	**Cd**	**Hg**
NONE	NONE	LHgX$_2$ IV
		X = Cl, Br, I

L' = PPh$_3$,

I–PP II–MPP III–DMPP IV–TPP

Table 3. Infrared data (cm^{-1}) for Cr, Mo, and W carbonyl phosphole complexes

L	Cr		Mo		W	
	LCr(CO)$_5$	CisL$_2$Cr(CO)$_4$	LMo(CO)$_5$	CisL$_2$Mo(CO)$_4$	LW(CO)$_5$	CisL$_2$W(CO)$_4$
Me$_2$PPh[22]		2003, 1880	2068[a]	2011, 1895		2008, 1883
PP			2072[a]			
DMPP	2060[a]	2005, 1920	2060[a]	2013, 1920	2067[a]	
TPP	2077[a]		2070[a]	2035, 1936	2063[a]	2029, 1920

[a] Only the highest energy ν(CO) is given

Table 3a. Coordination chemical shifts of LM(CO)$_5$[a]

M \ L	(CH$_3$)$_3$P*	(C$_6$H$_5$)$_3$P*	DMPP[20]
Cr	70.7	61.5	51.3
Mo	46.9	43.2	29.7
W	24.3	26.6	10.0

[a] Defined as δ^{31}P complex – δ^{31}P ligand, * data from: Guns, M. F., Claeys, E. G., and Van Der Kelen, G. P., J. Mol. Struct., *54*, 101 (1979)

Under similar conditions 1-n-butyl-3,4-dimethylphosphole affords L$_2$Mn$_2$(CO)$_8$ for which the infrared data suggest the unusual diequatorial coordination[23]. This clearly demonstrates the small steric bulk of the phospholes.

Both MPP and DMPP react directly with Re$_3$Cl$_9$ in ethanol at room temperature to produce purple or brown L$_3$Re$_3$Cl$_9$, whereas TPP forms the less stable L$_2$Re$_3$Cl$_9$ · CH$_2$Cl$_2$ under reflux and after recrystallization from CH$_2$Cl$_2$/ether[25]. This implies that both MPP and DMPP are better donors than TPP towards the borderline soft metal Re(III). The spectroscopic properties of these complexes are entirely conventional and very similar to those of the analogous PPh$_3$ complexes[26, 27].

Fe, Ru, Os

DMPP reacts with [(C$_5$H$_5$)Fe(CO)$_2$]$_2$ in boiling benzene to produce mainly (C$_5$H$_5$)Fe$_2$(CO)$_3$L in which the phosphole replaces one of the terminal carbonyls[28]. Complexes of type VI have been prepared[29] and found to react with trifluoroacetic acid to conveniently produce VII:

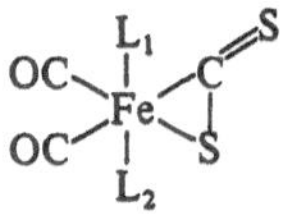

$$ \begin{array}{c} OC \\ \diagdown \\ OC \diagup \end{array} \overset{L_1}{\underset{L_2}{\overset{|}{Fe}}} \overset{S}{\underset{S}{\diagup \diagdown}} C \qquad + 2\ CF_3CO_2H \longrightarrow $$

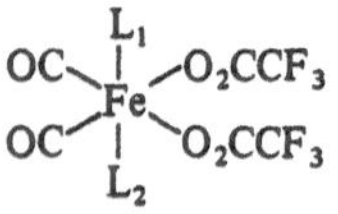

VI

$$ \begin{array}{c} OC \\ \diagdown \\ OC \diagup \end{array} \overset{L_1}{\underset{L_2}{\overset{|}{Fe}}} \overset{O_2CCF_3}{\underset{O_2CCF_3}{\diagup}} \qquad \begin{array}{l} L_1 = L_2 = DMPP \\ L_1 = DMPP, L_2 = PPh_3 \end{array} $$

VII

Table 4. $\nu(CO)$ in cm^{-1} for LFe(CO)$_4$

L	$\nu(CO)$	Ref.
TPP	2038, 1963, 1940, 1932	21
DMPP	2051, 1979, 1948, 1939	30
PPP	2066, 1992, 1961, 1953	11

Infrared data in the CO region of VII and the analogous Me$_2$PPh complex are essentially identical [L = Me$_2$PPh, $\nu(CO)$ 2050, 1997; L = DMPP, $\nu(CO)$ = 2050, 2000 cm^{-1}] once again implying similar metal ligand bond strengths for these two ligands.

Both TPP[21] and DMPP[30] react with Fe$_2$(CO)$_9$ to furnish the classical LFe(CO)$_4$ under relatively mild conditions. Comparing the CO stretching frequencies (Table 4) for the LFe(CO)$_4$ complexes, the following order of phosphole donor ability is observed: pentaphenylphosphole (PPP) < DMPP < TPP. This ordering suggests that when steric effects are at a minimum, TPP becomes a good σ-donor.

Phenyl groups in 3 and 4 position inductively reduce donor ability; in 2 and 5 position this effect is not observed as these phenyls are most likely not conjugated with the phosphorus.

An extensive series of ruthenium complexes have been prepared[25, 31]. Treatment of Ru(CO)$_2$Cl$_2$ in ethanol at room temperature with DMPP and analogs yields yellow trans L$_2$Ru(CO)$_2$Cl$_2$(VIII) as well as some L$_3$RuCOCl$_2$, demonstrating that the 3,4-dimethyl-

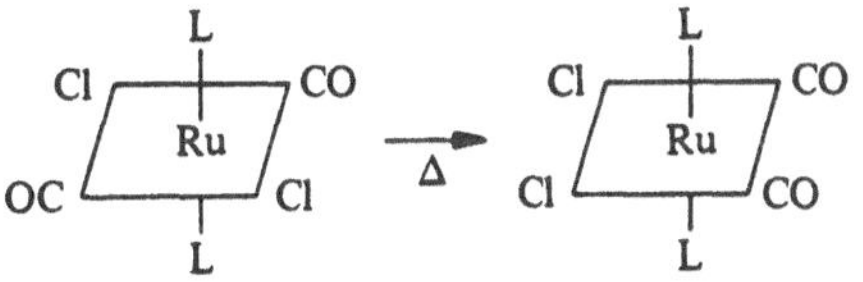

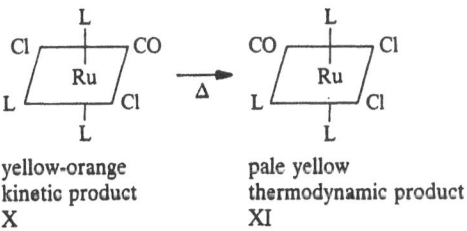

yellow
kinetic product
VIII

colorless
thermodynamic product
IX

phospholes like Me$_2$PPh[32] are sterically small and readily produce tris ligand monocarbonyl complexes. The kinetic trans isomer(VIII) is thermally isomerized to the more thermodynamically stable cis isomer(IX) in all cases[31] when L is 1-R-DMP, Me$_2$PPh, or MePPh$_2$. The thermodynamically stable isomer IX, can be prepared directly by reacting any of these ligands with Ru(CO)$_2$Cl$_2$ in refluxing 2-methoxyethanol. The tris ligand complexes X and XI are similarly prepared[25, 31, 32, 33] and a similar isomerization occurs for them.

yellow-orange
kinetic product
X

pale yellow
thermodynamic product
XI

Table 5. $\nu(CO)$ and $\nu(RuCl)$ in cm^{-1} for $L_nRu(CO)_mCl_2$

L	n	m	Structure	$\nu(CO)$	$\nu(RuCl)$
Me$_2$PPh	2	2	VIII	1996	320
	2	2	IX	2037, 1972	304
	3	1	X	1965	315, 300
	3	1	XI	1923	288, 229
DMPP	2	2	IX	2055, 1990	312, 285
	3	1	X	2000	375, 325
MPP	3	1	XI	1960	320

Infrared data (Table 5) for these series of compounds show again that Me$_2$PPh and DMPP have similar ligating ability.

The crystal structure of L$_3$RuCOCl$_2$ (L = DMPP) demonstrates the softness of this ligand in that the Ru-P bond lengths are very responsive to the nature of the trans ligand [Ru-P 2.365(1) Å and 2.356(1) Å (trans to P) and 2.411(1) Å (trans to CO)]. These distances should be compared with those found for[31, 34] (Bzl$_3$P)$_2$Ru(CO)$_2$Cl$_2$ (structure X) Ru-P = 2.398 Å, and (BzlPPh$_2$)$_2$Ru(CO)$_2$Cl$_2$ (structure IX) Ru-P = 2.40 Å. By comparison these data demonstrate that the Ru(II) phosphole bond is shorter and stronger than analogous Ru(II) phosphine bonds. Similar Ru-P bond lengths have been reported for (Ph$_3$P)$_3$RuCl$_2$[35] with Ru-P = 2.38 Å, and (Me$_2$PPh)$_2$RuCl$_2$(CO)(C$_2$H$_4$)[36] with Ru-P = 2.414 Å. The Ru-Cl bond lengths of (DMPP)$_3$RuCOCl$_2$ [2.407(1) Å and 2.423(1) Å] are similar to those found in other Ru(II) phosphine complexes: 2.387 Å in (Ph$_3$P)$_3$RuCl$_2$, 2.55 Å in (Et$_2$NCS$_2$)$_4$Ru$_3$(CO)$_3$Cl$_2$[37], 2.429(2) Å in (BzlPPh$_2$)$_2$Ru(CO)$_2$Cl$_2$, and 2.4185(1) Å in (Bzl$_3$P)$_2$Ru(CO)$_2$Cl$_2$.

The yields of the (phosphole)$_n$Ru(CO)$_m$Cl$_2$ complexes are rather low[31] because several products are formed in these reactions as the phospholes more readily replace CO than other phosphines do and the phosphole complexes are very soluble.

Reaction of (Ph$_3$P)$_4$RuCl$_2$ with 1-R-3,4-dimethylphospholes in refluxing hexane[31] produces trans L$_4$RuCl$_2$ which thermally and photochemically isomerize to the thermodynamically more stable cis-L$_4$RuCl$_2$ in high yield. The steric effect of the exocyclic substituent is demonstrated in these reactions in that 1-t-butyl-3,4-dimethylphosphole produces the purple stable pentacoordinate complex (PPh$_3$)(Bu^t-DMP)$_2$RuCl$_2$ under similar conditions. This latter complex is a configurationally and dissociatively stable pentacoordinate complex whose ^{31}P NMR spectrum displays PP coupling at room temperature in CH$_2$Cl$_2$. By way of contrast (Ph$_3$P)$_3$RuCl$_2$ is not stereochemically rigid above -90 °C in CH$_2$Cl$_2$. Dissociation of PPh$_3$ from (PPh$_3$)$_4$RuCl$_2$ is rapid and complete at room temperature but dissociation of phosphole from (phosphole)$_4$RuCl$_2$ is slow occurring over several hours at room temperature. These data offer the first definitive evidence that the 3,4-dimethylphospholes are better donors at least towards Ru(II) than PPh$_3$.

Gently heating solutions of RuCl$_3 \cdot$ 3H$_2$O in ethanol with DMPP, MPP, or PP produces[25] (DMPP)$_3$RuCl$_3$, (MPP)RuCl$_3$ and (PP)RuCl$_3$. If these reactions are performed under prolonged reflux with an excess of ligand, reduction to Ru(II) occurs affording L$_3$RuCl$_2$ with L = PP or DMPP whereas MPP only brings about partial reduction and (MPP)$_4$Ru$_2$Cl$_5$ (a mixed valence Ru(II) and Ru(III) complex) is formed. This is especially surprising since the oxidation potentials of the ligands should vary in the order

PP > MPP > DMPP, i.e., the DMPP should be the best reducing agent and PP the worst in this series. When $RuCl_3 \cdot 3 H_2O$ is refluxed with MPP in n-butanol, $(MPP)_3Ru(CO)Cl_2$ is obtained.

The pentacoordinate L_3RuCl_2 complexes, in contrast to the hexacoordinate L_4RuCl_2, seem to dissociate in solution[25] as evidenced by differing electronic spectra in the solid state and in solution. Evidently the equilibria $L_4RuCl_2 \rightleftharpoons L_3RuCl_2 \rightleftharpoons L_2RuCl_2$ favor disproportionation such that both L_4RuCl_2 and L_2RuCl_2 are generally more stable than L_3RuCl_2 when L is a phosphole. This equilibrium should be further investigated by ^{31}P NMR as these species may be very good hydrogenation catalysts.

Co, Rh, Ir

PP, MPP, and DMPP all react with $CoCl_2 \cdot 6 H_2O$ in ethanol to form green air stable crystalline solids of formula $L_2CoCl_2 \cdot 0.5 CH_2Cl_2$ after recrystallization from $CH_2Cl_2/$ ether[25]. The solvent of crystallization could not be removed from these complexes even under high vacuum. These paramagnetic complexes (μ_{eff} = ca. 4 BM) have electronic spectra which are very similar to that of $(PPh_3)_2CoCl_2$ leaving little doubt that they are normal tetrahedral Co(II) complexes[38]. These data also suggest that towards Co(II) these phospholes and PPh_3 have similar ligating abilities.

The complexes $HCo(CO)_3L$, where L is a variety of phospholenes, phospholes, and a phospholane, were compared[39] under industrial conditions for the hydroformylation of 1-octene. It was found that 3,4-dimethyl-phospholenes and phospholes give comparably high yields of linear 1-nonanol and that both classes of ligands gave more efficient and more selective catalysts than the best[40] phosphine, 9-phenyl-9-phosphabicyclononane. It was argued that the phospholes are strongly bound to cobalt, greatly increase the hydridic character of the Co-H bond, and favor the formation of $HCo(CO)_2L$ (olefin) in preference to $HCo(CO)_3$ (olefin) under the reaction conditions.

Reaction of $RhCl_3 \cdot 3 H_2O$ with PP, MPP, or DMPP in ethanol produces[25, 41] L_3RhCl_3 containing solvent of crystallization. No reduction to Rh(I) occured and no hydride species are formed. The general lack of reduction of both Ru(III) and Rh(III) by these ligands is probably the result of their strong donor ability and not a lack of reducing ability. In contrast to these reactions TPP forms $(TPP)RhCl_3$ instead of L_3RhCl_3-type complexes, probably for steric reasons. Infrared data suggest that the L_3RhCl_3 complexes possess facial rather than meridional geometry. ^1H NMR supports this assignment as only one methyl resonance is observed for $(DMPP)_3RhCl_3$ at $\delta 2.13$. With prolonged heating under reflux $(TPP)_2RhHCl_2$ is formed. This complex is an extremely active[45] homogeneous hydrogenation catalyst in benzene containing Et_3N and it hydrogenates 1-hexene at 20 °C and less than 1 atm H_2. The authors conclude that PP, MPP, and DMPP have similar donor properties toward Rh but are in some way different from TPP.

When carbon monoxide is passed through a hot solution of $RhCl_3 \cdot 3 H_2O$ in ethanol containing MPP or DMPP, $L_5Rh_2(CO)Cl_4$ are obtained[25]. Under these same conditions PP produces the benzoyl complex $(PP)_3Rh(PhCO)Cl_2 \cdot CH_2Cl_2$. The infrared spectra of the two $L_5Rh_2(CO)Cl_4$ complexes show νCO (~ 1965 cm^{-1}) and $\nu RhCl$ (280, 300) in the normal ranges and the benzoyl complex shows the ketonic νCO at 1680 cm^{-1}. Increasing the reaction times causes complete decarbonylation yielding $L_5Rh_2Cl_4$ in each case, demonstrating that all these ligands will displace carbon monoxide from rhodium. The benzoyl complex is interesting as it represents an instance of thermal metal promoted phosphorus carbon bond cleavage, a subject which will be dealt with in greater detail subsequently. All these complexes are diamagnetic and do not ionize in solution. Molecular weight data suggest that the $L_5Rh_2(CO)Cl_4$ and $L_5Rh_2Cl_4$ species are chloride bridged dimers and their diamagnetism is consistent with mixed valent Rh(I), Rh(III) complexes possessing either structure XII or XIII. These complexes

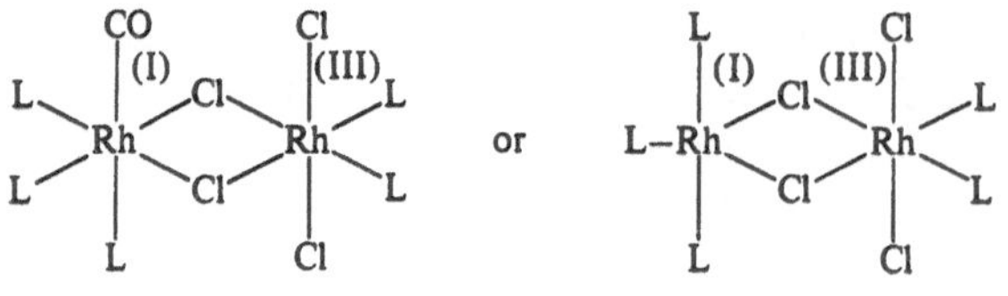

partially dissociate to monomers in solution. TPP produces the more conventional $L_2RhCOCl$ complex by reaction under carbon monoxide. Reaction of PP, MPP, and DMPP with preformed $Rh_2(CO)_4Cl_2$ also produces the rhodium (I) complexes $L_2RhCOCl$. These complexes are isomeric mixtures of the cis and trans isomer rather than the more common trans isomers and they undergo isomerization in solution as suggested by infrared data. This isomerization should be further studied by ^{31}P NMR as it suggests that these complexes are substitutionally labile and they might be good hydrogenation catalysts.

It should be noted[42] that PPh_3 reacts with $RhCl_3 \cdot 3H_2O$ to produce $(PPh_3)_3RhCl$; alkyl phosphines such as Me_3P, Et_3P, Pr_3P, and Cy_3P produce L_3RhCl_3 and mixed alkyl aryl phosphines produce L_3RhCl_2. In this respect the phospholes resemble the more basic trialkyl phosphines in their behavior toward $RhCl_3$.

Hence, PP, MPP, and DMPP have donor strengths similar to trialkylphosphines, whereas TPP resembles mixed alkyl aryl phosphines, i.e., is less basic. Comparison of $\nu(CO)$ in the $L_2RhCOCl$ systems (Table 6) with the values for similar complexes strongly indicates that these phospholes are of comparable donor ability with the tertiary alkyl and aryl phosphines. Within these phospholes, the donor ability seems to follow the order DMPP > MPP $\approx$ PP > TPP since νCO increases along this series. All have similar RhCl bond strengths as evinced by $\nu RhCl$.

Both MPP and DMPP react[25] with $IrCl_3 \cdot 3H_2O$ in refluxing ethanol to produce L_3IrCl_3, while under similar conditions PP reduces Ir(III) to Ir(I) to yield $(PP)_3IrCl$. At room temperature in ethanol all three ligands yield the green $LIrCl_3$ which could not be satisfactorily purified.

Table 6. Infrared data for trans $L_2RhCOCl$

L	CO	RhCl	Ref.
Me_3P	1954	302	43, 44
Me_2PPh	1970	305	43, 44
PPh_3	1978	not given	43, 44
DMPP	not given	310	25
MPP[a]	2070, 1970	320, 290	25
PP[a]	2060, 1970	315, 290	25
TPP	1985	320	41

[a] Isomeric mixtures of cis and trans isomers

Ni, Pd, Pt

Though[45] TPP, 1-methylphosphole, or 1-benzylphosphole do not coordinate to Ni(II), 1-benzyl-3,4-dimethylphosphole forms[46] purple L_2NiCl_2 which is sufficiently stable to be recrystallized. In solution, however, its ^1H NMR spectrum lacks any evidence of spin-spin coupling. MPP and DMPP react[47] with $NiCl_2 \cdot 6H_2O$ in refluxing ethanol to form brown L_2NiCl_2 while under the same conditions PP forms $(PP)_3NiCl_2$. The L_2NiCl_2 complexes have intermediate magnetic moments ($\mu_{eff} = 1.9$–2.12 BM) and give broad ^1H NMR spectra. In this respect they behave very much like the Me_2PPh and $MePPh_2$ complexes which LaMar[48] has shown to undergo tetrahedral-square planar interconversion in solution. The putatively pentacoordinate $(PP)_3NiCl_2$ complexes were shown by conductance measurements to be four coordinate and ionic, i.e., $[(PP)_3NiCl]^+Cl^-$. Thus, these complexes are very much like the Me_2PPh and $MePPh_2$ complexes which dissociate both in solution and slowly in the solid state and which catalyse the autoxidation of the phosphorus donor.

Refluxing $PdCl_2$ with TPP[49], MPP[47], or DMPP[47] in ethanol produces air stable orange-yellow L_2PdCl_2 complexes but under similar conditions PP produces ionic $[(PP)_3PdCl]^+Cl^-$. The complexes of the 1-substituted 3,4-dimethylphospholes are better prepared[50] by reacting the ligands with $(PhCN)_2PdCl_2$ in CH_2Cl_2 or benzene in which case, if no excess ligand is present, stable crystalline L_2PdCl_2 complexes are easily isolated. These complexes can be readily converted[50, 51] by metathesis with the appropriate NaX in $CH_3OH/CHCl_3$ solutions into L_2PdX_2 complexes (X = Br, I, N_3, SCN). Isomerization energetics obtained[52] for the isomerization cis-L_2PdX_2 = trans-L_2PdX_2 show that these phospholes stabilize the cis isomer more than either[53] Me_2PPh or $MePPh_2$ because both the Pd-P and Pd-X bonds are stronger in the cis-phosphole complexes than they are in the cis phosphine complexes (see Table 7).

Solvent effects are less dominant in the isomerization of the phosphole complexes than in the isomerizations of the phosphine complexes. Likewise changes in internal bond strengths are more important for the phosphole complexes than for the phosphine complexes. This is because both the Pd-P and Pd-X bonds are stronger and shorter in the phosphole than in the phosphine complexes suggesting that DMPP is both a better σ-donor and π-acceptor toward Pd(II) than is Me_2PPh.

Table 7. Isomerization energetics for the isomerization cisL_2PdCl_2 = trans L_2PdCl_2 in $CDCl_3$ and C_6H_6

L	ΔH (kcal/mol)	ΔS (eu)	ΔG_{303} (kcal/mol)
	$CDCl_3$		
DMPP	all cis	–	–
Me_2PPh	3.1	9.0	0.36
$MePPh_2$	4.6	20.2	-1.56
	C_6H_6		
DMPP	2.76	10.50	-0.422
Me_2PPh	all trans	–	–
$MePPh_2$	all trans	–	–

P—2.260   Cl
    Pd< 2.362
P—   Cl

P—2.241   Cl
    Pd< 2.351
P—   Cl

(Me$_2$PPh)$_2$PdCl$_2$

$\left(\text{⬠} \right)_2$PdCl$_2$ (with Ph)

In the phosphole complex the phosphole ring is planar with the two methyl groups located in the plane of the ring, despite the fact that they are well within Van der Waals contacts. There is weak residual aromaticity within the ring as is also found in the ruthenium complex (DMPP)$_3$RuCOCl$_2$ (vide supra). The intracyclic bond angle CPC of 92.7° is small and the ligand possesses essentially the same configuration as free 1-benzyl-phosphole suggesting little structural reorganization upon coordination. The same conclusion is true for (DMPP)$_3$RuCOCl$_2$. Studies of the L$_2$Pd(CNS)$_2$ complexes demonstrate that the stereoelectronic properties of the ligands (L = 1-substituted-3,4-dimethylphosphole) influence the thiocyanate bonding mode in much the same way as has been found[54] for Me$_n$PPh$_{3-n}$ and Bzl$_n$PPh$_{3-n}$ complexes. The crystal structures of two of the phosphole complexes (1-phenyl-3,4-dimethylphosphole)$_2$Pd(SCN)(NCS)[55] and (1,3,4-trimethylphosphole)$_2$Pd(SCN)$_2$[51] are illustrated schematically in Fig. 2.

As can be seen by comparing the Pd-P bond lengths in these compounds, the phosphole ligand is quite polarizable and the Pd-P bond lengths are very sensitive to the nature of the trans ligand. The residual aromaticity of the inequivalent phospholes in (DMPP)$_2$Pd(SCN)(NCS) is also quite different, further illustrating the phosphole polarizability.

Another measure of the Pd-P bond strengths is the ^{31}P NMR coordination chemical shift, $\Delta\delta^{31}$P which is given by the equation $\Delta\delta^{31}$P = A δ^{31}P(ligand) + B where $\Delta\delta^{31}$P = δ^{31}P(complex)-δ^{31}P(ligand). The L$_2$PdX$_2$ complexes[50] of both phosphines and phospholes obey this linear relationship equally well. It is clear from comparisons of the A values (which reflect phosphorus substituent effects and polarizability) and the B values (which reflect average coordination chemical shifts) that the phosphole complexes have smaller average coordination chemical shifts and larger polarizabilities than phosphines (see Table 8).

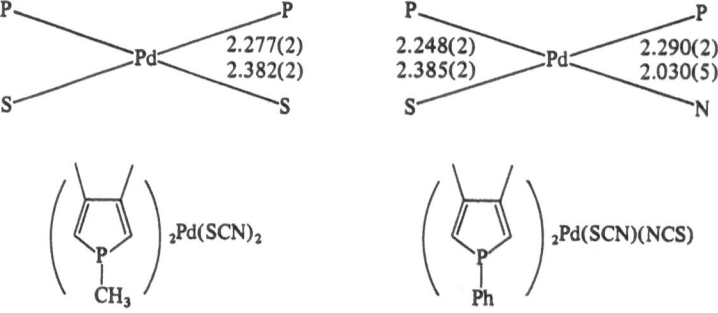

Fig. 2. Schematic structures of two phosphole palladium thiocyanate complexes

Table 8. Coordination chemical shift equations for palladium complexes

Compd	A	B	$r^{2 a}$
L = all phospholes			
Cis-L_2PdCl_2	-0.382	32.11	0.85
Cis-L_2PdBr_2	-0.459	28.19	0.65
trans-L_2PdBr_2	-0.342	21.92	0.92
Cis-$L_2Pd(N_3)_2$	-0.328	30.58	0.90
L = phosphines			
Cis-L_2PdCl_2	-0.212	38.63	0.61–0.91
trans-L_2PdCl_2	-0.304	26.79	0.69–0.96
Cis-$L_2Pd(N_3)_2$	-0.440	28.25	0.56–0.98
trans-$L_2Pd(N_3)_2$	-0.353	23.47	0.83–0.98

[a] correlation coefficients for the equation $\Delta\delta^{31}P = A\,\delta^{31}P(\text{ligand}) + B$

The smaller coordination chemical shifts for the phosphole complexes do not result from weaker Pd-P bonds, as is clear from the X-ray structural results, but rather from the existence of some π-backbonding between palladium and phosphole as already seen earlier for Cr, Mo, and W(CO)$_5$ complexes. The π-backbonding is synergistic and reinforces the Pd-P σ bond while at the same time restoring some of the phosphorus electron density removed by this σ-donation. The net result is that the drift of electron density away from the phosphole phosphorus upon coordination to palladium is less than that which occurs for phosphines. Thus the coordination chemical shifts also illustrate that the phospholes are softer and more polarizable than similar phosphines such as Me$_2$PPh.

Heating solutions of PtCl$_2$ in ethanol with PP and MPP produces[47] colorless L$_2$PtCl$_2$ while under similar conditions DMPP gave the yellow [(DMPP)$_3$PtCl]$^+$Cl$^-$ $\cdot$ 0.5 CH$_2$Cl$_2$ after recrystallization from CH$_2$Cl$_2$/ether. The latter complex was shown to be ionic by conductance measurements. Reaction[56] of either K$_2$PtCl$_4$ in ethanol/water or (PhCN)$_2$PtCl$_2$ in CH$_2$Cl$_2$ with 1-R-3,4-dimethylphospholes produces cis-L$_2$PtCl$_2$. Metathesis of these complexes with the appropriate NaX in ethanol/water produces cis-L$_2$PtBrCl, cis-L$_2$PtBr$_2$, and cis- and trans-L$_2$PtI$_2$. All these complexes are pale yellow to orange in color. These complexes, like the analogous palladium complexes, have been completely characterized by infrared, conductance, and NMR spectroscopy including ^{195}Pt NMR.

Addition of excess phosphole to CH$_2$Cl$_2$, CDCl$_3$, or CH$_3$OH solutions of either L$_2$PdX$_2$ or L$_2$PtX$_2$ causes a color change from yellow to red and the resulting solutions show non-electrolyte behavior (no measurable conductance). At 25 °C the ^{31}P NMR of these solutions indicate rapid inter- and intra-molecular equilibria as only broad resonances are observed. Variable temperature ^{31}P spectra for the less labile platinum complexes indicate the existence of two dynamic processes. The first is described by the equilibrium: L$_2$PtX$_2$ + L = L$_3$PtX$_2$, and the second is the permutational isomerization of the pentacoordinate L$_3$PtX$_2$ species. The thermodynamic parameters given in Table 9 have been obtained.

These data indicate that the formation of L$_3$PtX$_2$ is enthalpy favored and entropy disfavored. The relative thermodynamic stability of the L$_3$PtX$_2$ complexes is a function of ligand steric bulk; the smaller ligand gives the greater stability. The stereochemical

Table 9. Equilibrium thermodynamics for the reaction:

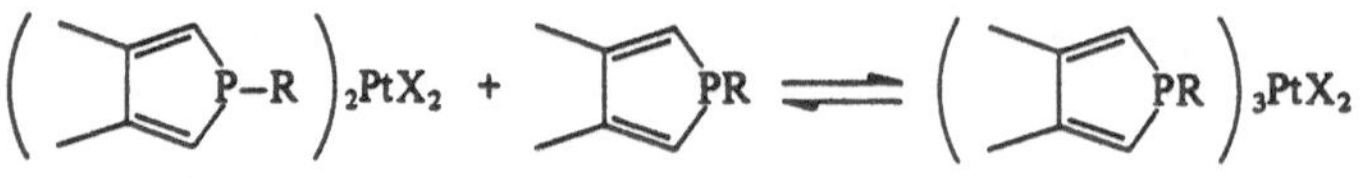

$$\left(\begin{array}{c} \\ \end{array} P\!-\!R \right)_2 PtX_2 \; + \; \begin{array}{c} \\ \end{array} PR \; \rightleftharpoons \; \left(\begin{array}{c} \\ \end{array} PR \right)_3 PtX_2$$

R	X	ΔG_{300}	ΔH (kcal/mol)	ΔS (eu)	T intra (K) [a]	T inter (K)
t-C$_4$H$_9$	Cl	1.28	− 3.0	− 14.2	~ 293	~ 318
t-C$_4$H$_9$	Br	1.65	− 4.1	− 19.0	~ 303	~ 333
C$_6$H$_5$	Cl	0.69	− 2.4	− 10.2	~ 263	~ 303
C$_6$H$_5$	Br	1.53	− 2.6	− 13.2	~ 270	~ 303

[a] Approximate coalescence temperatures for intramolecular exchange within L$_3$PtX$_2$ (Tintra) and inter molecular exchange between L$_3$PtX$_2$, L$_2$PtX$_2$, and L (Tinter)

Table 10. Coordination chemical shift equations for L$_2$PtX$_2$ and L$_3$PtX$_2$ L = 1-R-3,4-dimethylphospholes

Compound	A	B	R[a]
cis-L$_2$PtCl$_2$	− 0.110	14.06	0.66
cis-L$_2$PtBr$_2$	− 0.411	11.37	0.97
cis-L$_2$PtI$_2$	− 0.283	9.35	0.95
L$_3$PtCl$_2$			
triplet	− 0.345	12.38	0.91
doublet	− 0.779	39.61	0.83
L$_3$PtBr$_2$			
triplet	− 0.380	13.30	0.92
doublet	− 0.733	35.79	0.82
L$_3$PtI$_2$			
triplet	− 0.346	11.41	0.99
doublet	− 0.449	35.41	0.32

[a] Correlation coefficient in the equation $\Delta\delta^{31}P = A\,\delta^{31}P(\text{ligand}) + B$

rigidity of the L$_3$PtX$_2$ complexes is proportional to ligand steric bulk; the larger the ligand the more rigid the L$_3$PtX$_2$ complex. The only other ligands[57] for which L$_3$PtX$_2$ complexes have either been detected in solution or isolated are Me$_2$PPh, MePPh$_2$, Me$_3$P, Me$_2$PBzl, and dibenzophospholes.

The ^{31}P coordination chemical shifts for both the L$_2$PtX$_2$ and L$_3$PtX$_2$ complexes (Table 10) are related to the free ligand chemical shifts in a fashion analogous to those found for the palladium complexes. Likewise, for platinum phosphole complexes the coordination chemical shifts are less than those found for similar platinum phosphine complexes also indicating some platinum phosphole back donation.

It is also found (Table 11) that $^1J_{PtP}$ is smaller for phosphole complexes than for similar phosphine complexes demonstrating that there is less s-character in the platinum phosphole bond than in the platinum phosphine bond. Since the platinum s-orbital character is probably constant, this signifies less phosphorus s-character in the phosphole

Table 11. ^{31}P and ^{195}Pt data for L_2PtX_2 complexes

Complex	$\delta^{31}P$ ligand	$\delta^{31}P$ complex	$\Delta\delta^{31}P$	$^{1}J_{PtP}$	$\delta^{195}Pt^a$
cis-(DMPP)$_2$PtCl$_2$	−2.5	8.1	10.6	3345	50
cis-(DMPP)$_2$PtBr$_2$	−2.5	7.9	10.4	3268	−272
cis-(DMPP)$_2$PtI$_2$	−2.5	5.6	8.1	3125	−343
cis-(Me$_2$PPh)$_2$PtCl$_2$	−46.9	−15.8	31.1	3550	130
cis-(Me$_2$PPh)$_2$PtBr$_2$	−46.9	−15.8	31.1	3506	−113
cis-(Me$_2$PPh)$_2$PtI$_2$	−46.9	−18.0	28.9	3374	−149

^{a} Relative to 21.4 MHz with TMS at exactly 100 MHz

than in the phosphine complexes. The greater amount of phosphole p-character is consistent with the greater electronegativity of the phosphole phosphorus.

Complete $^{13}C\{^{1}H\}$ NMR data have been obtained for both the palladium and platinum complexes of the 1-R-3,4-dimethylphospholes and support the conclusions already made regarding the nature of the phosphole metal bond.

Cu, Ag, Au

The only report to date regarding the reaction of these metals with phospholes concerns the reactions of TPP with CuCl$_2$ and CuCl in hot methanol producing[45] either yellow or colorless diamagnetic TPP CuCl, respectively. In this same article AgCl was not found to react with TPP. In this respect TPP is not strictly analogous with PPh$_3$ as the latter[42] forms a family of copper and silver complexes with variable stoichiometry.

Zn, Cd, Hg

TPP reacts[49] with HgCl$_2$ to form the colorless chloride bridged dimer [(TPP)HgCl$_2$]$_2$ in which[58] the Hg-P bond [2.438(10) Å] is slightly longer than that found for [(PPh$_3$)HgCl$_2$]$_2$ [2.406(7) Å]. The HgCl bridging bond distances also differ for the two compounds: [2.623(8) and 2.658(8) Å] in [(PPh$_3$)HgCl$_2$]$_2$ and [2.542(13) and 2.747(14) Å] in [(TPP)HgCl$_2$]$_2$. The thermodynamics[59] of metal ligand bond formation, determined by solution calorimetry, for interaction of phosphorus donors with HgX$_2$ (Table 12) show that TPP forms weaker bonds to mercury than either PPh$_3$ or MePPh$_2$.

Table 12. Heats of Interaction Between HgX$_2$ and Phosphorus Ligands

Ligand	HgCl$_2$ −ΔH (Kj/mol)	HgBr$_2$ −ΔH (Kj/mol)	HgI$_2$ −ΔH (Kj/mol)
PPh$_3$	75 ± 4	77 ± 4	65 ± 3
MePPh$_2$	92 ± 2	89 ± 2	82 ± 2
TPP	~67	~61	no reaction

II.3. λ^3-Phospholes as 4-Electron Ligands

The earliest studies of the coordination chemistry of phospholes involved reactions with metal carbonyls. Braye et al.[11] found that 1,2,3,4,5-pentaphenylphosphole (PPP) reacts with $Fe_3(CO)_{12}$ to form a mixture of complexes among which there is one with a PPP ring η^4-bonded to $Fe(CO)_3$. These types of complexes must be very unstable and very difficult to prepare due to the high reactivity of the lone pair of the phosphorus atom. In order to obtain them it is necessary to deactivate the phosphorus atom by electron withdrawing substituents and to prevent complexation through the lone pair by steric hindrance. These facts explain why only two η^4-phosphole complexes have been found, both of them with 2,5-diphenyl substituted phospholes (PPP and TPP[21]), and why no unambiguous characterization of the bonding mode has been established.

II.4. λ^3-Phospholes as 6-Electron Ligands

The first complex in which the phosphole ring acts as a tridentate ligand, via the phosphorus lone pair and the dienic system was discovered by Braye[11]. When reacting PPP with $Fe_3(CO)_{12}$, complex XIV was formed among other products.

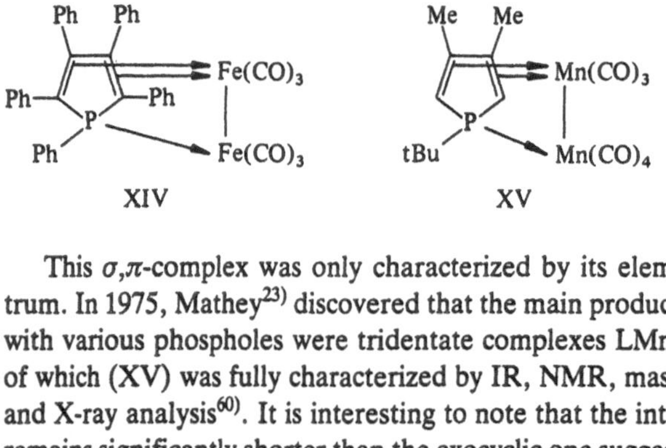

XIV   XV

This σ,π-complex was only characterized by its elemental analysis and its IR spectrum. In 1975, Mathey[23] discovered that the main products of the reaction of $Mn_2(CO)_{10}$ with various phospholes were tridentate complexes $LMn_2(CO)_7$ analogous to XIV, one of which (XV) was fully characterized by IR, NMR, mass spectrum, elemental analysis, and X-ray analysis[60]. It is interesting to note that the intracyclic P-C bond length in XV remains significantly shorter than the exocyclic one suggesting that the π-complexation of the dienic system could be responsible for this shortening instead of the classically invoked cyclic delocalization. In 1977, Mathey[30] reinvestigated the complexation of phospholes with various iron carbonyls. Using DMPP and $Fe_3(CO)_{12}$ in boiling toluene, a new complex (XVI) was obtained without an iron-iron bond (characterized by elemental analysis, mass spectrum, NMR, and IR spectroscopies).

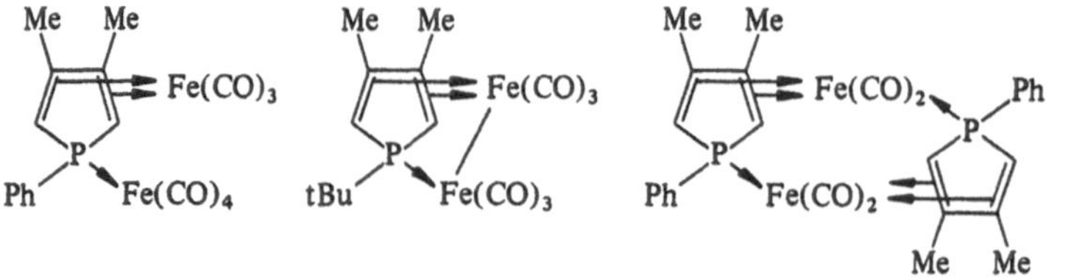

XVI   XVII   XVIII

Using the more bulky 1-t-butyl-3,4-dimethylphosphole and the same iron carbonyl, the main product was complex XVII analogous to Braye's complex XIV. Thus the more bulky the P-substituent on the phosphole ring, the more likely the formation of the metal-metal bond.

When heated at 150 °C in xylene for 24 h with an excess of phospholes, Santini[61] showed that complexes of type XVI gave sandwich complexes of type XVIII with a head-to-tail configuration which were fully characterized including an X-ray crystal structure analysis. Depending on the starting materials, a number of mixed homo- and hetero-bimetallic sandwiches could be obtained. On the basis of structural and NMR data, it appears that there is a strong through-space interaction between the non-bonded iron and phosphorus atoms in these complexes.

Finally, Mathey[62] showed that 5-phenyl-1,2,3,4,6,7,8,9-octahydrodibenzophos-phole(XIX) reacts with $Fe_3(CO)_{12}$ to form a mixture of complexes among which, one (XX) has the same structure as complex (XVI); this complex was characterized by mass spectrum and NMR.

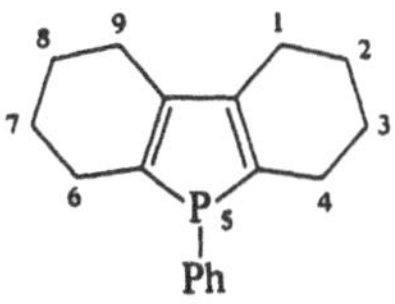

XIX 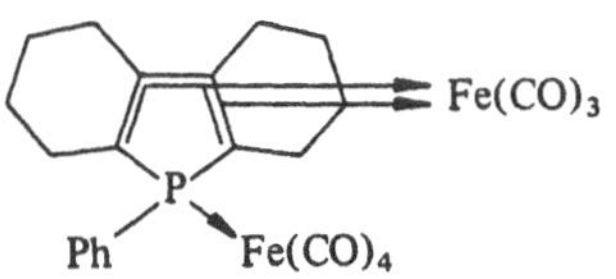   XX

Table 13 summarizes the geometrical data from the X-ray structures for the 6-electron phosphole complexes. In each structure the ring carbon atoms are co-planar within experimental error, the P atom lies out of the $C_2 \rightarrow C_5$ mean plane being 0.194 to 0.644 Å away from the metal atom, leading to a folding of the ring around the C_2-C_5 axis by 9.0 to 29.4°. No significant correlation can be made between the folding angle α and bond angles or distances within the ring. The P-C and the C-C bonds do not change and the C-P-C angle changes are not related. Perhaps there is a correlation between α and the P-C-C and C-C-C angles, but statistically these differences are not significant.

II.5. Reactions of λ^3 Coordinated Phospholes

Santini[20] has shown that, under UV-irradiation, cis-$L_2M(CO)_4$ complexes (M = Cr, Mo, W) undergo [2 + 2] dimerizations leading to complexes such as (XXI). In the case of 3,4-

Table 13. Geometrical data of coordinated λ^3/6-electron phospholes

Complex	Metal-Ω^a P-Ω	a^b	P-C	C-C	C-P-C	P-C-C	C-C-C	Ref.
XV	1.777	0.271	11.8	1.778	1.38 → 1.42	87.7	113.3	111.8	60
c	1.671	0.194	9.0	1.773	1.426	90.6	112.0	112.2	60
XVIII	1.671	0.644	29.4	1.786	1.429	85.6	111.0	110.5	61

[a] M-Ω is the distance from the metal to the $C_2C_3C_4C_5$ mean plane; [b] see Fig. 1

[c] Complex 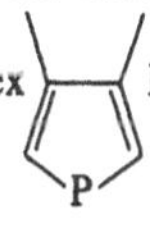 $Mn_3(CO)_{11}$ described in Sect. III.6.

dimethyl-substituted phospholes, these complexes spontaneously rearrange to give the [4 + 2] exo-dimers XXII.

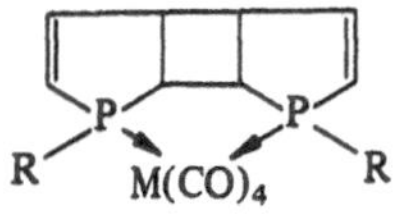

XXI 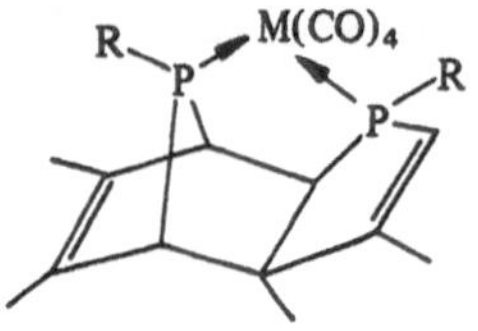   XXII

The X-ray crystal structure of complexes such as XXII shows that the bridge phosphorus is much more strained than that of classical, analogous endo-phosphole dimers. When complex XXII (M = Mo) is treated with sulfur in benzene at 50 °C, the exo-dimeric phosphole sulfide (XXIII) is obtained.

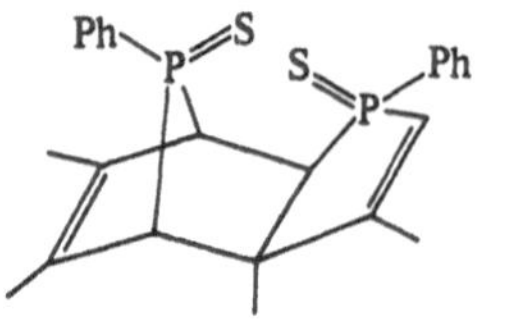

XXIII 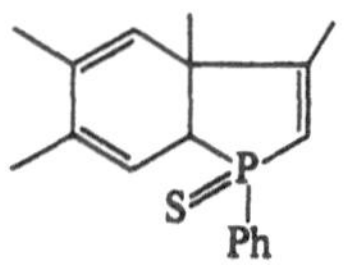   XXIV

At higher temperatures in boiling toluene, XXIII loses "RP = S" and the dihydrophosphindole sulfide (XXIV) is obtained. It has also been possible to trap the phosphinidene sulfide by reaction with 2,3-dimethylbutadiene.

It is quite certain that numerous other reactions of P-coordinated phospholes will be studied in the future. Indeed the P-complexation induces sharp changes in the reactivity of the phosphole dienic system which may lead to the discovery of new types of structures.

II.6. λ^5-Phospholes as Ligands

II.6. a. Phosphole Oxide, Sulfide, and Selenide σ-Complexes

The donor character of the oxide (TPPO), sulfide (TPPS), and selenide (TPPSe) toward transition metal halides has received considerable attention. TPPO does not react with the dichlorides of iron, nickel, or copper; with Co(II) chloride hexahydrate, a complex tentatively formulated as (TPPO)$_2$CoCl$_2$ is formed[45]. With various transition metals in higher oxidation states, stable complexes are formed; for example, TPPO reacts with Fe(III) chloride to give (TPPO)$_2$FeCl$_3$ · 0.5 CH$_2$Cl$_2$ and with Th(IV) and U(IV) to give (TPPO)$_2$MCl$_4$ (M = Th, U) and (TPPO)$_3$ThCl$_4$.

With Nb(V) and Ta(V), TPPO, TPPS, and TPPSe give a number of adducts with the general formula LMX$_5$ (L = ligand, M = metal, X = halogen)[45]. TPPS reacts with Rh$_2$Cl$_2$(CO)$_4$ to form (TPPS)RhCl(CO)$_2$ in which apparently the Rh(I) atoms are square-planar[41]. Under the same conditions TPPO does not give σ-bonded complexes. TPPO, TPPS, and TPPSe react with Ru(III) chlorides yielding adducts of the form LRuCl$_3$;

unlike Rh(III), no reduction of Ru(III) occurs[25] and no cleavage of the P = X (X = O, S, Se) bond occurs as has been observed for phosphole sulfides with iron carbonyls[11].

The Ru(II) adduct can be formed directly from the blue Ru(II) methanol solution[64, 65] to give $(TPPO)_3RuCl_2 \cdot EtOH$ in which two of the TPPO ligands are σ-bonded and one is π-bonded to the Ru atom.

Infrared examination of various adducts showed that the P = O, P = S, and P = Se stretching vibrations are shifted to lower energy on coordination and this confirms that the ligands are linked to the metal atoms through the O, S, or Se atoms. From the examination of all these data it appears that TPPO, TPPS, and TPPSe are poorer ligands than corresponding classical phosphine oxides, sulfides, and selenides probably for steric reasons.

Unfortunately, no X-ray structure is yet available to confirm this assessment.

II.6. b. Phosphole Oxide, Sulfide, Selenide, and Phospholium Salt π-Complexes

As with λ^3-phospholes, it was also Braye et al.[11] who described the first λ^5-η^4-bonded phosphole derivative. Indeed PPPO reacts with $Fe(CO)_5$ to give a η^4-complex (XXV).

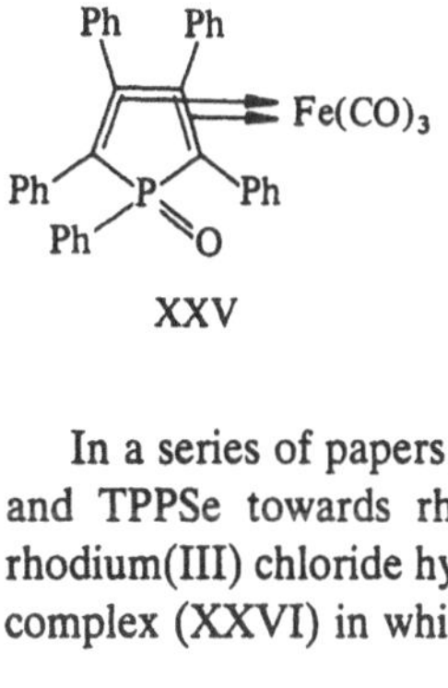

XXV

In a series of papers, Holah et al.[41, 25] studied the donor character of TPPO, TPPS, and TPPSe towards rhodium, ruthenium, and rhenium. Treatment of TPPO with rhodium(III) chloride hydrate in ethanol causes reduction to Rh(I) and gives the dimeric complex (XXVI) in which TPPO is η^4-bonded to the metal.

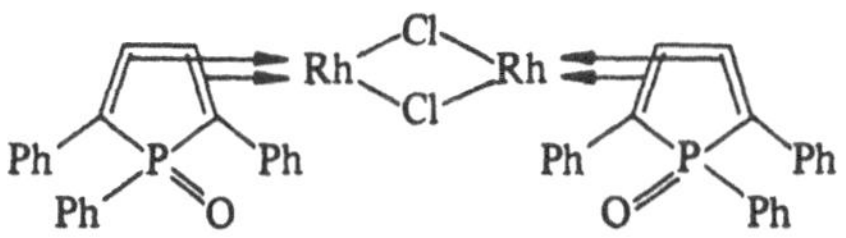

XXVI 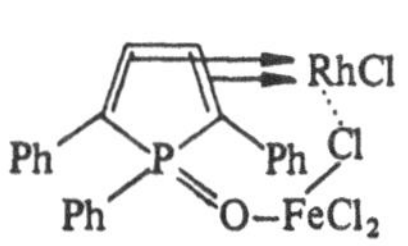   XXVII

The dimer (XXVI) yields the $FeCl_3$ complex (XXVII) containing four coordinate rhodium, the fourth chlorine atom bridging rhodium and iron. XXVI also reacts with o-phenanthroline, pyridine, and carbon monoxide to give complexes of the type $(TPPO)RhCl \cdot L$ or $(TPPO)RhCl \cdot L_2$ in which TPPO remains η^4-bonded to the metal.

The first fully characterized η^4-coordinated phosphole oxides have been described by Barrow et al.[66]; $(\eta^5$-$C_5H_5)Co(PF_3)_2$ reacts with hexafluorobut-2-yne to give $(C_5H_5)Co(C_4F_6)_2PF_3$ which undergoes hydrolysis to XXVIII and XXIX.

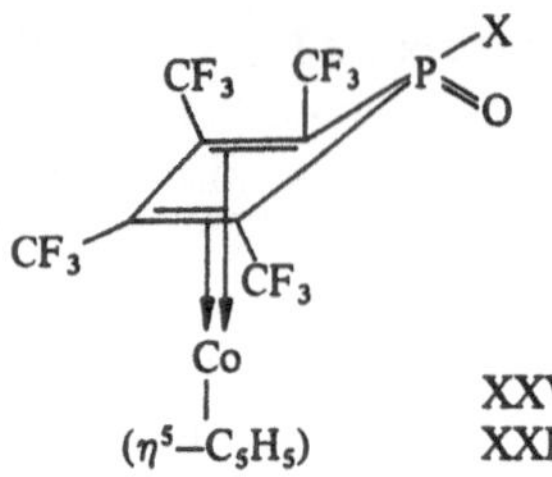

CF₃ CF₃ P—X / O ... Co (η⁵–C₅H₅)

XXVIII X = F
XXIX   X = OH

The crystal structure of **XXIX** proved that there are no interactions between the phosphorus or the oxygen atoms of the − P(O)OH group and the metal atom. The carbon atoms of the phosphole ring are co-planar within experimental error[67].

More recently, Yasufuku et al.[68] obtained η^4-coordinated phosphole oxides by reaction of a cobaltacyclopentadiene ring with phosphites:

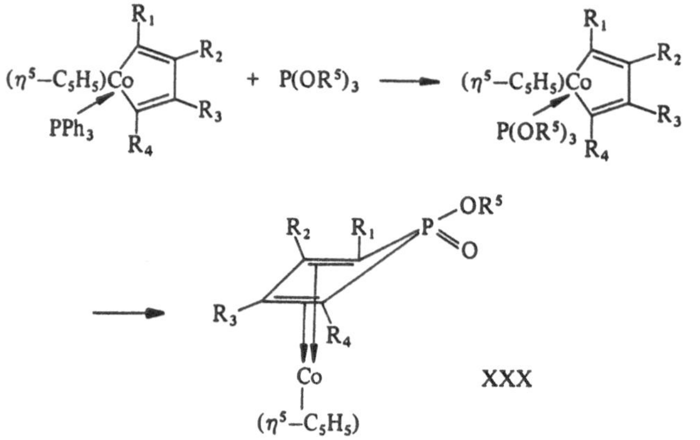

$(\eta^5–C_5H_5)Co$ ... R_1 R_2 R_3 R_4 PPh₃ + P(OR⁵)₃ ⟶ $(\eta^5–C_5H_5)Co$ ... R_1 R_2 R_3 R_4 P(OR⁵)₃

⟶ R_2 R_1 R_3 R_4 P—OR⁵ / O ... Co (η⁵–C₅H₅)   **XXX**

The molecular structures of **XXIX** and **XXX** (with $R_1 = R_2 = R_3 = R_4 = C_6H_5$ and R_5 = OCH₃) are similar.

Lindner et al.[69] described a completely different way of obtaining η^4-complexed phosphole rings. Under catalytic conditions (Raney-nickel) and with NiS elimination, the sulfur atom of the six-membered thiaphosphamanganabicyclo[2,2,1]heptadiene is eliminated.

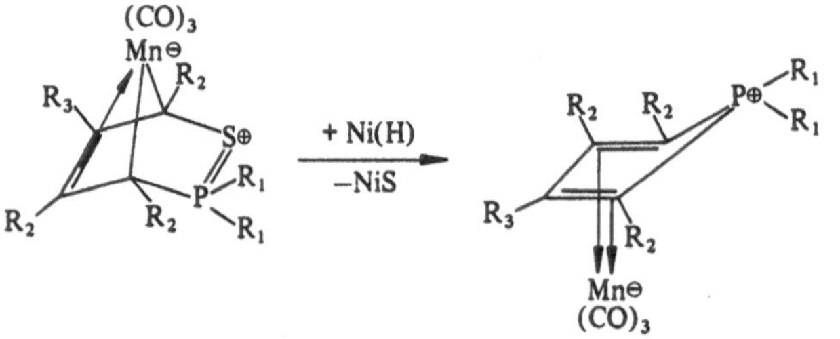

(CO)₃ Mn⊖ R_3 R_2 S⊕ R_1 P R_1 R_2 R_2 R_1 + Ni(H) / −NiS ⟶ R_2 R_2 R_3 R_2 P⊕ R_1 R_1 Mn⊖ (CO)₃

XXXI

Complex **XXXI** has been fully characterized and its crystal structure with R_1 = CH₃, $R_2 = R_3 = CO_2CH_3$ again shows an η^4-bonded phosphole ring with no Mn-P interaction. Phosphorus bears a positive charge and thus the phosphole ring is cationic.

Deschamps[70] showed that the phosphorus atoms of 3,3′,4,4′-tetramethyl-1,1′-diphosphaferrocene (DPF) are electrophilic. The reaction of DPF at − 80 °C in THF with two equivalents of t-BuLi, followed by addition of three equivalents of CH_3I, gives a stable, water-soluble, paramagnetic monocation (XXXII) which has a bis-(η^4-diene)iron structure as definitively proved by X-ray structural analysis.

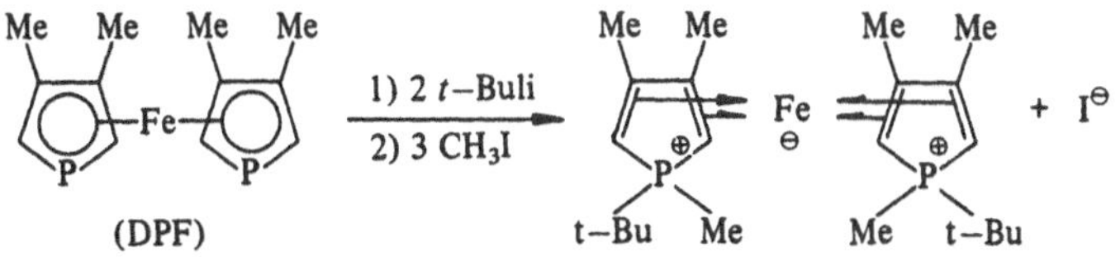

The very long Fe...P distances preclude the existence of bonds between iron and phosphorus and, thus, the phosphole rings act only as η^4-ligands by their dienic systems. The stability of cation XXXII decreases with the steric bulk of phosphorus substituents. This type of compound is immediately destroyed by aqueous acids to give the corresponding free phospholium salts.

The intermediate complex having an (η^5-phospholyl)-(η^4-phospholium)structure can be stabilized by replacing one alkyl P-substituent by an acyl as shown by Deschamps et al.[71]. The crystal structure of the stable λ^3-λ^5-zwitterion XXXIII has been determined.

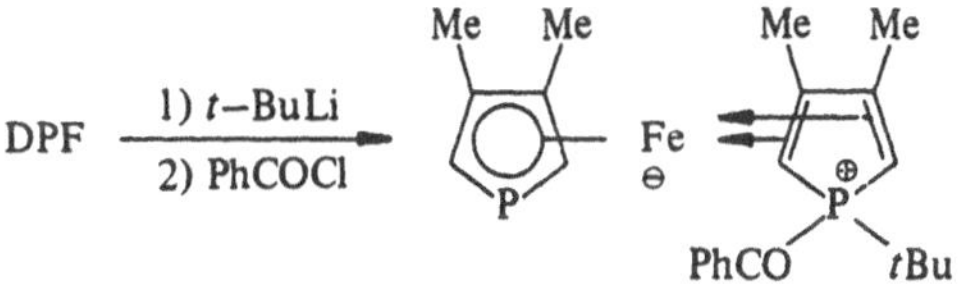

In XXXIII iron is sandwiched between the phospholyl and phospholium rings. The role of the benzoyl group is to hinder attack of the second phosphorus atom.

The thermolysis of XXXIII is very interesting. It yields DPF (as expected on the basis of mass spectral data) through a reductive elimination at phosphorus. It also yields 1-t-butyl-3,4-dimethylphosphole, probably after an initial shift of the benzoyl group from phosphorus to iron. Finally it also yields, quite unexpectedly, 1′-phenyl-3,3′,4,4′-tetramethyl-1-phosphaferrocene.

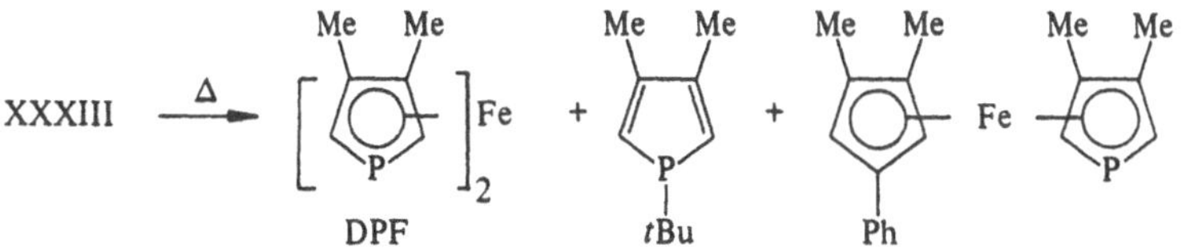

The mechanism of the phosphaferrocene formation is unknown but it almost certainly parallels the following sequence which has been established with free phospholes[63].

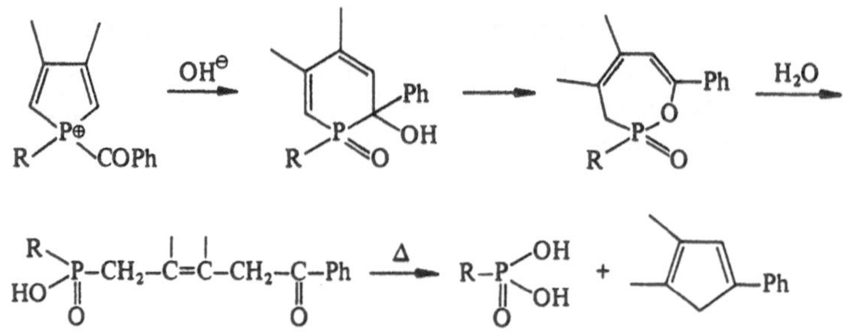

During the nucleophilic attack of DPF by phenyllithium, a spontaneous oxidation occured to give the stable diamagnetic (η^4-phospholium)(η^5-1,1-dioxophosphole) zwitterion XXXIV:

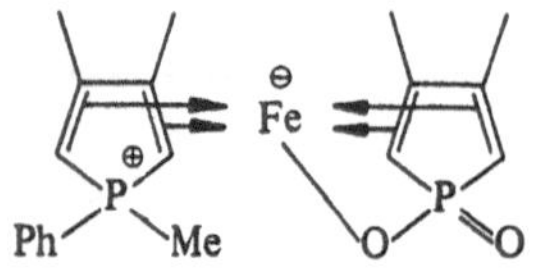

XXXIV

In contrast to the other π-complexed phosphole oxides described above, the crystal structure showed that one heterocycle is an η^4-diene and the other a 5-electron donor. One oxygen atom of the $-$ P(O)O group is bonded to iron.

Table 14 collects all structural data for η^4- and η^5-bonded λ^5-phospholes. Within experimental error, all dienic systems are planar. No correlations are apparent among geometric data such as bond distances, bond angles and the substituents of the carbon atoms C2 to C5. On the contrary, the folding angle α for η^4-phospholes varies considerably (20.8 to 34.3°) showing that this angle is very sensitive to the P-substituents and that rotation around the C2-C5 axis does not require much energy. In turn, the value of α is correlated with the P-Ω distance.

For XXXIV, the folding angle decreases quite logically to 10.4° leading to an unusually short P-Ω distance of 0.231 Å. This decrease is related to the bonding of the iron to an oxygen of the $-$ P(O)O group which, in turn, leads to an increase of the Fe-Ω distance.

III. Phospholyl Complexes

III.1. Main Features of the Phospholyl Anions

The phospholyl anions were discovered by Braye[12] and are readily prepared through alkali metal cleavage of the exocyclic phosphorus-carbon bond of phospholes:

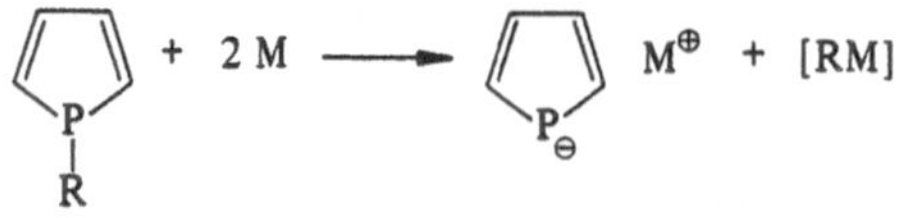

R = Ph, PhCH$_2$, Me
M = Li, Na, K

Table 14. λ^5-phospholes as 4- and 5-electron ligands

Complex	Metal $-\Omega^a$	$P-\Omega$	α^b	P-C exo	P-C intra	C2-C3	C3-C4	C4-C5	C-P-C^c	P-C-C^c	C-C-C^c	Ref.
XXIX	1.583	0.686	32.2	–	1.766	1.43	1.42	1.43	86.5	109.8	110.4	67
XXX	1.594	0.632	30.3	–	1.768	1.44	1.44	1.43	89.9^d	108.0	111.5	68
XXXI	1.710^d	0.670	32.0	1.767	1.769	1.44	1.40	1.44	88.4	108.3	111.4	69
XXXII	1.637	0.630	30.6	1.791 / 1.832	1.757	1.41	1.41	1.44	90.0	107.3	111.9	70
XXXIII	1.629	0.439	20.8	1.858 / 1.876	1.754	1.41	1.41	1.43	90.5	109.8	112.3	71
XXXIV^4	1.645	0.706	34.3	1.783 / 1.829	1.763	1.43	1.43	1.43	89.4	106.8	111.4	70
XXXIV^5	1.682	0.231	10.4	–	1.793	1.39	1.44	1.40	89.1	112.1	112.6	70

a Ω is the mean plane C2-C3-C4-C5 (the metal and phosphorus nuclei lie on opposite sides of this plane)
b α is the folding angle around the C2-C5 axis (see Fig. 1)
c Mean values
d Calculated from atomic coordinates given in original references

Independent of the nature of the R substituent, the cleavage always takes place quantitatively on the exocyclic bond in spite of the high strain of the phosphole ring. This simple fact already suggests that phospholyl anions are stabilized by electronic delocalization in the same way as the isoelectronic thiophenes. Two theoretical studies[72, 73] strongly support this hypothesis even though both of them rely upon guessed geometries: in the first case[72], the geometry is derived from that of 1-benzylphosphole by just flattening the ring and removing the benzyl substituent; in the second case [73], the geometry is that of the phospholyl ring in its manganese-tricarbonyl π-complex. The most notable difference between these two geometries concerns the respective lengths of the formal single and double carbon-carbon bonds within the ring: they are unequal in the first case (1.438 and 1.343 Å) and equal in the second case (1.420 Å). It would be necessary to perform an optimization of the geometry in order to obtain more reliable results.

Two other types of results equally suggest a strong electronic delocalization within the phospholyl anions. First, Quin[74] has found that the ^{31}P NMR signals of these anions are located at much lower field than those of more classical phosphide anions. The available data from Quin and from our own laboratory are collected in Table 15.

The recorded ^{31}P shifts fall in the upper part of the range normally associated with neutral dicoordinated phosphorus compounds; for comparison, in compound XXXV the

Table 15. ^{31}P NMR data for various phospholyl anions

Anion	M	R	R'	Solvent	δ^{31}P[a]	Ref.
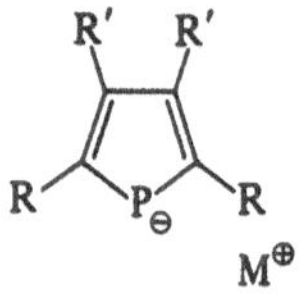	Li	H	H	THF	+ 76	our laboratory
	Li	H	Me	THF	+ 54	our laboratory
	Li	Ph	H	THF	+ 78	our laboratory
	Li	Ph	Ph	THF	+ 97	our laboratory
	K	Ph	Ph	THF	+ 96	our laboratory
XXXIX 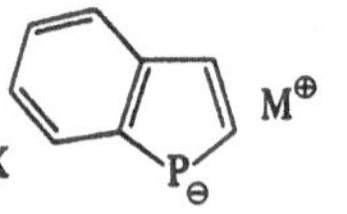	K			THF	+ 73.3	74
XXXVIII 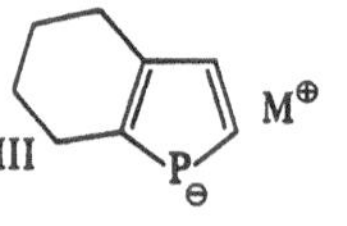	Li			THF	+ 40	our laboratory
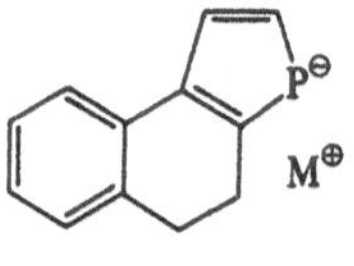	K			THF	+ 81.7	74
$R_2P^{\ominus}M^{\oplus}$	K	Ph			− 12.4	data cited in ref. 74
	K	iPr		DME/benzene	+ 23.2	data cited in ref. 74

[a] In every case except the last two, Ph M is also present in the solution. δ in ppm from 85% H_3PO_4 as external reference; δ positive for downfield shifts

^{31}P resonance appears at + 77.2 ppm[75]. This observation suggests significant contributions of canonical structures such as **XXXVI** and **XXXVII** to the representations of phospholyl anions.

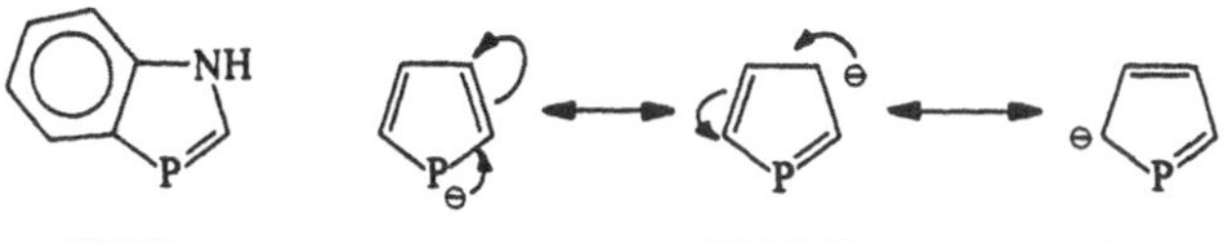

XXXV   XXXVI   XXXVII

A second striking observation was made by Quin[74]: phospholyl anions do not react with absolute ethanol in contrast to classical phosphide anions. This indicates a much reduced basicity for the phospholyl species, once again in agreement with a strong aromaticity.

Finally it is interesting to note that phosphindolyl anion **XXXIX** does react with absolute ethanol and has a ^{31}P resonance at higher field than its phospholyl counterpart **XXXVIII** (see Table 15). The adverse effect of the benzo annelation upon the aromaticity of the phospholyl ring is thus well established.

III.2. Phospholyls as 1-Electron Ligands

Phosphides are known to act in most cases as 3-electron bridging ligands. In some rare cases, however, when they are substituted by bulky and/or electron withdrawing groups such as CF_3 or C_6F_5, phosphides can also give terminal complexes in which they act as 1-electron donors. This possibility also exists with phospholyls. Thus far, only compounds **XL a–e** derived from the bulky tetraphenylphospholyl are known. They were prepared according to the two following routes:

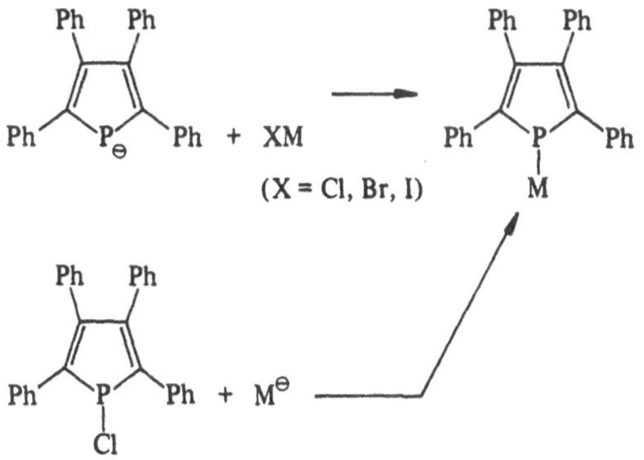

(X = Cl, Br, I)   M

a M = CpFe(CO)$_2$[76, 77]
b M = Mn(CO)$_5$[77]
c M = Re(CO)$_5$[77]
d M = CpMo(CO)$_3$[77]
e M = CpW(CO)$_3$[77]

Only the first route can be generalized since 2,5-diphenyl- and 2,3,4,5-tetraphenyl- are the only two 1-halophospholes known at the present time.

The stability of these species (**XL a** can be sublimed in vacuo at 180 °C without decomposition[76]) is certainly due, in large part, to the bulkiness of the substituted phosphole ring. This bulkiness is well evidenced by the IR spectra of **XL b** and **XL c** which show four strong and well resolved metal carbonyl stretching modes[77] indicating that the formal C_{4v} symmetry is lost together with the usual degeneracy of the E mode.

Nevertheless, an extra stabilization due to a slight delocalization of the phosphorus lone pair within the phosphole ring is not completely excluded. Additional work with less substituted phospholyls is necessary to clarify this possibility.

The chemistry performed with complexes XL a–e is also very limited if we exclude the study of their thermal decomposition. Complex XL a is readily oxidized by air to yield the corresponding P-oxide[76] indicating that the lone pair on phosphorus has kept its usual reactivity. Since the P = O IR absorption occurs at a lower frequency than normal, a P-O-Fe interaction is perhaps indicated. Finally, Abel[77] has shown that the P-M bond of XL was readily cleaved by chlorine and bromine yielding the 1-halophosphole and the organometallic halide.

III.3. Phospholyls as 3-Electron Ligands

Like ordinary phosphides, phospholyls are also able to act as 3-electron bridging ligands. Curiously, only two complexes belonging to this class appear to be described in the literature[77]. This may reflect the lack of special interest in these classical types of complexes. These species were prepared by controlled thermal decomposition of XL b and XL c:

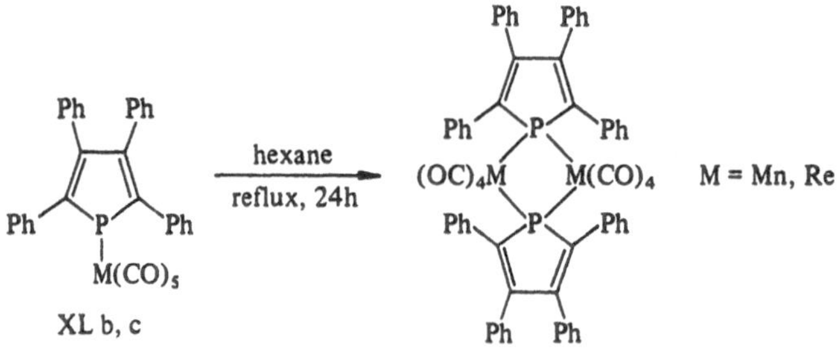

Higher temperatures cause the loss of an additional CO to produce the corresponding η^5-complexes. In our opinion, the sterically more compact and electronically more releasing 3,4-dimethylphospholyl would be more adapted to act as 3-electron donors than the tetraphenylphospholyl used here.

III.4. Phospholyls as 5-Electron Ligands – Synthesis and Properties of Phosphametallocenes

In the strict sense of the word, phosphametallocenes are the compounds resulting from the replacement by phosphorus of a CH unit in one or more cyclopentadienyl rings of any given metallocene (formula XLI or XLII). By extension and for greater convenience,

XLI   XLII   XLIII

we have given this name to any type of compound in which a phospholyl acts as an η^5 ligand (formula XLIII).

The initial impetus for the research in this field came from the observation that it was impossible to directly functionalize by electrophilic attack (or by any other means) any theoretically aromatic carbon-phosphorus heterocycle, thus precluding the development of an interesting new chemistry. For exemple, up to now, no electrophilic attack at carbon was recorded for free phospholyl anions or for phosphorins XLIV. Such is not the case with arsenic, as Ashe[78]

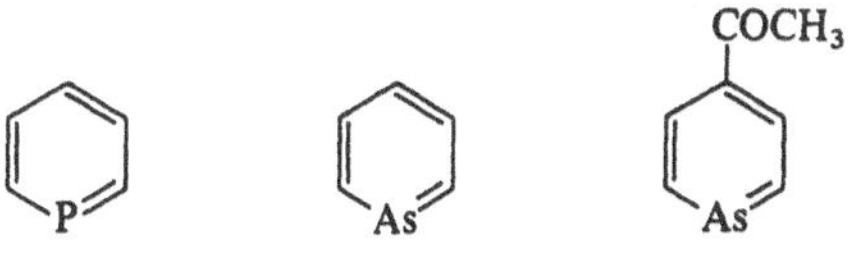

XLIV   XLV   XLVI

has demonstrated that it is possible to directly functionalize arsenin (XLV) by reaction with acetyl chloride and $AlCl_3$ to give mainly XLVI. Thus it was obvious that the failures recorded with phosphorus heterocycles were due to the very high nucleophilicity of tervalent phosphorus. In the case of free phospholyl anions, the localization of a strong negative charge on phosphorus[72] certainly has a very adverse effect. Thus it was hoped, that, in phosphametallocenes where the negative charge is "neutralized", the nucleophilicity of phosphorus would decrease to such a level that electrophilic substitutions at carbon would become feasible. Such is indeed the case, and phosphametallocenes were the first and remain, at present, the only carbon-phosphorus heterocycles for which an extensive aromatic chemistry can be developed. This is one of the reasons why they deserve special attention.

III.4. a. Synthesis

The various syntheses of phosphametallocenes can be divided into four groups according to the method chosen for building the phosphorus-phenyl bond.

The method first discovered relies upon the reaction of 1-phenylphospholes with organometallic reagents which are able to cleave the phosphorus-metal bond. Obviously, this method is not general even though the exocyclic phosphorus-carbon bond of phospholes appears to be weaker than usual. Nevertheless, when it works, this synthesis is, by far, the most convenient because of the availability of the starting phospholes. Up to now, only phosphacymantrenes[2] (η^5-phospholyl-tricarbonyl-manganese) and phosphaferrocenes have been made by this procedure. Indeed, the direct reaction of decacarbonyldimanganese with 1-phenylphospholes in boiling xylene has been shown to give phosphacymantrenes, sometimes in reasonable yields[79]:

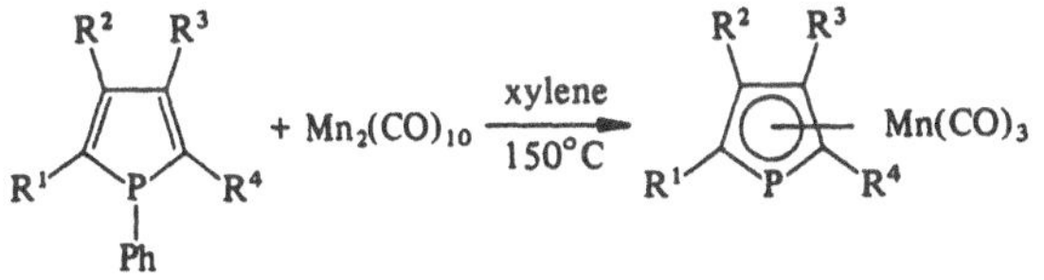

$R^1 = R^2 = R^3 = R^4 = H$ 50%[79]
$R^1 = R^3 = R^4 = H; R^2 = Me$ 40%[79]
$R^1 = R^4 = H; R^2 = R^3 = Me$ 60–80%[79]
$R^1 = R^4 = Ph; R^2 = R^3 = H$ 30%[80]
$R^1-R^2 = R^3-R^4 = (CH_2)_4$ 10%[62]

---

2 This name is built from the name cymantrene given by Cais to cyclopentadienyl-*manganese*-*tricarbonyl*

The benzo annellated compound has also been made in this way from 1-phenylphos-phindole by Nief[116].

Similarly, the reaction of dicyclopentadienyl-tetracarbonyl-diiron with 1-phenylphos-pholes yields phosphaferrocenes, although in lower yields[28, 81]:

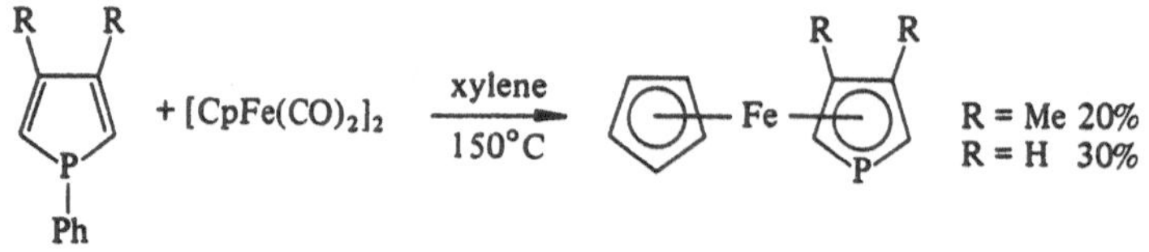

$$\text{(phosphole with R, R, P-Ph)} + [CpFe(CO)_2]_2 \xrightarrow[150°C]{\text{xylene}} \text{(phosphaferrocene)}\quad \begin{array}{l} R = Me\ 20\% \\ R = H\ 30\% \end{array}$$

Possible mechanisms for these reactions are depicted in Scheme 2. Both rely on the expulsion of a phenyl group from a transient phosphole radical such as XLVII or IL.

Phosphole radicals are known to be more readily made and more stable than analog-ous phosphine radicals[82]. This is probably due to the delocalization of the unpaired electron over the ring. On the other hand, complexes such as XLVIII are isolated from the reaction mixture and their mass spectra contain a medium-intensity peak correspond-ing to the final phosphaferrocene; however, their thermolysis in the pure state does not produce phosphaferrocenes in detectable amounts. Phenyl-substituted phosphafer-rocenes are always among the by-products of the synthesis of phosphaferrocenes. No satisfactory explanation of this fact (the formation of phenyl substituted phosphafer-rocene by radical arylation as initially proposed is not very likely since ferrocene cannot be arylated in this way[83] could be found until the very recent discovery that, at around

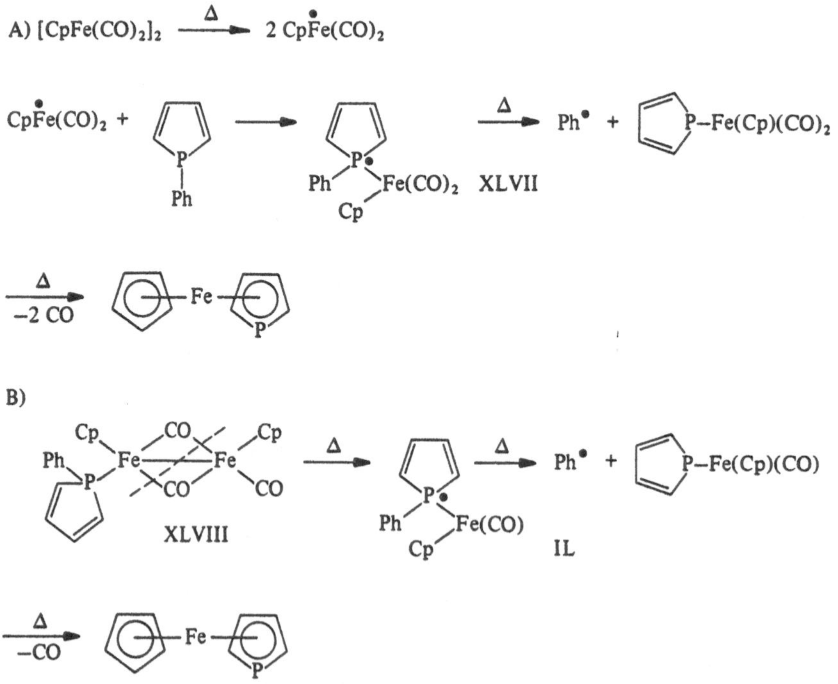

A) $[CpFe(CO)_2]_2 \xrightarrow{\Delta} 2\ Cp\dot{F}e(CO)_2$

$Cp\dot{F}e(CO)_2\ +$ (phosphole with P-Ph) $\longrightarrow$ (intermediate, Ph, Fe(CO)$_2$, Cp) **XLVII** $\xrightarrow{\Delta}$ $Ph^\bullet\ +$ (ring) P–$Fe(Cp)(CO)_2$

$\xrightarrow[-2\ CO]{\Delta}$ (phosphaferrocene, Cp–Fe–P)

B)

(structure: Cp, CO, Cp, Ph, Fe—Fe, CO, CO, P) **XLVIII** $\xrightarrow{\Delta}$ (Ph, P, Fe(CO), Cp) **IL** $\xrightarrow{\Delta}$ $Ph^\bullet\ +$ (ring) P–$Fe(Cp)(CO)$

$\xrightarrow[-CO]{\Delta}$ (phosphaferrocene, Cp–Fe–P)

Scheme 2. Possible mechanisms for the synthesis of phosphaferrocenes

150 °C, 1-phenylphospholes experience a 1,5-shift of the phenyl substituent to yield the isomeric 2H-phospholes[84], thus providing a logical answer to the problem:

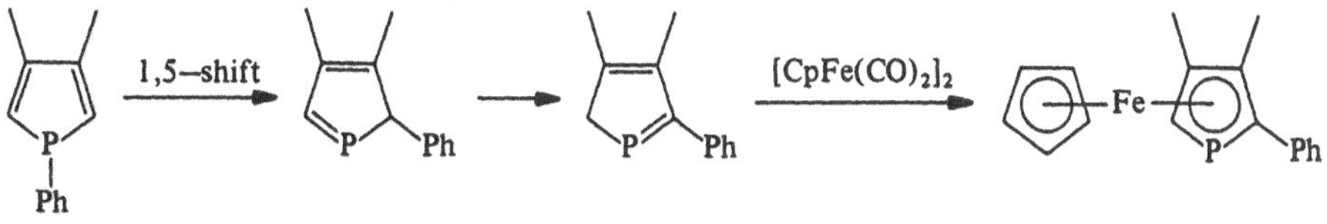

In order to avoid the presence of these phenyl-substituted phosphaferrocenes, which lowers the yield of the desired products and causes inconveniences for purification, 1-phenylphospholes have been successfully replaced by 1-tertbutylphospholes[85]:

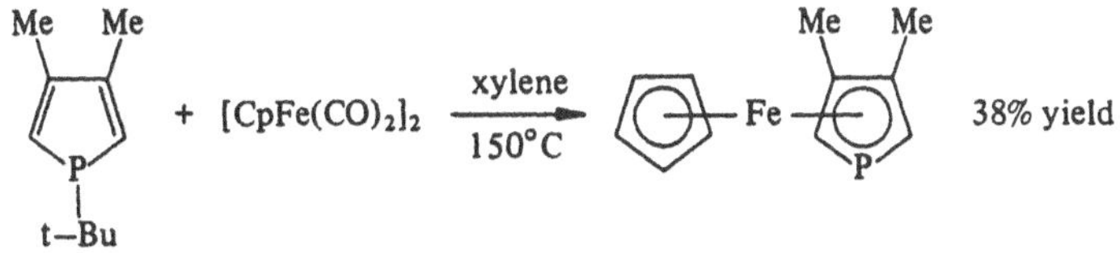

The second and most general method for the synthesis of phosphametallocenes involves the reaction of phospholyl anions with organometallic halides, yielding directly either the desired η^5-complexes or transient η^1- or μ^2-complexes which are the precursors of the desired products. Good examples of this last approach were described by Abel[86]

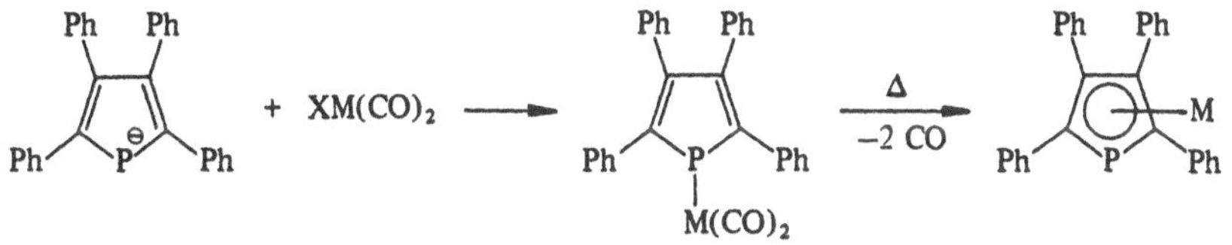

XL a–c

M = Mn(CO)$_3$, refluxing decane/20 min/49%; Re(CO)$_3$, 120 °C/1 h in vacuo/55%; CpFe, refluxing xylene/60 h/35%

Since alkali metal phospholides are reducing and nucleophilic reagents[3], very often they just reduce the metallic halide without coordinating to the metal center, or they destroy the final phosphametallocene by nucleophilic attack; this drawback can be avoided by metathetically replacing the alkali metal with another less electropositive metal. Two good examples are recorded in the literature: the synthesis of 1,1'-diphosphaferrocenes[87] and the synthesis of phosphazirconocene dichlorides[88]:

---

3 The observed side-reactions can be also due to the presence of the phenyl anion as a by-product in the synthesis of the phospholyl anions; De Lauzon has shown that a stoichiometric amount of AlCl$_3$ could selectively destroy phenyllithium without reacting with phospholyllithium (unpublished results)

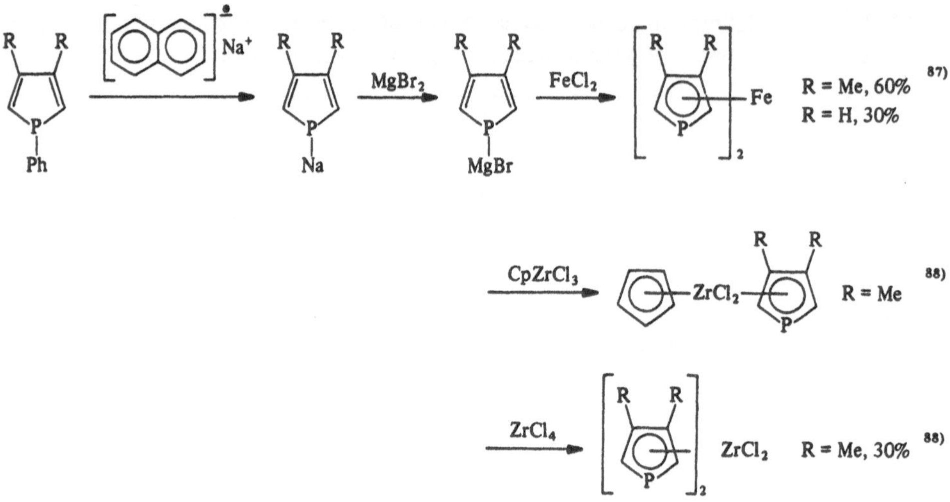

In both cases, the direct reaction of the phospholyl sodium with the metal halide gives no tractable products. Another way to avoid this drawback would be to use 1-silylphospholes as in Abels'[86] synthesis of 2,3,4,5-tetraphenylarsacymantrene. However, weakly substituted 1-silylphospholes are unknown up to now.

The third synthesis of phosphametallocenes relies on the reaction of a 1,1'-biphospholyl with a metal-metal bonded organometallic reagent. This method was first used by Abel[86] in another synthesis of 2,3,4,5-tetraphenylphosphacymantrene:

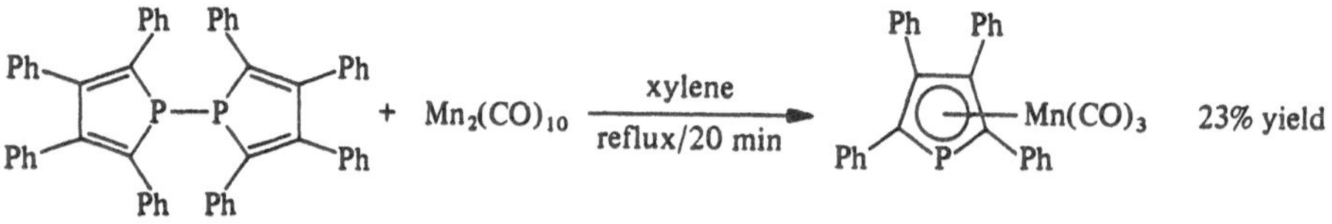

The discovery of the first known (η^5-phospholyl) dicarbonylcobalt[89] is a much more significant illustration of this approach:

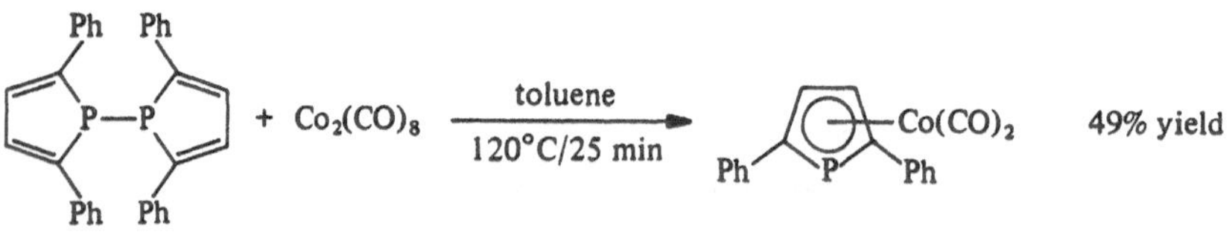

This route is milder and more general than the first one (the cleavage of the P-P bond is easier than the cleavage of the P-Ph bond) and has the additional advantage of being in a strictly neutral medium. In our opinion, this would be the best method if 1,1'-biphospholyls were readily available. Unfortunately, only heavily phenyl-substituted 1,1'-biphospholyls are presently accessible.

The principle of the fourth and last method has already been described in Sect. III.2. The phosphorus-metal bond is first created by reacting a 1-halophosphole with an organometallic anion, thus yielding an η^1-phospholyl complex which is subsequently thermolyzed (as seen before in the discussion of the second method) to give the phosphametallocene. The drawback of this method is the same as that of the preceding one: only heavily phenyl-substituted 1-halophospholes are available at the present time.

III.4. b. Physical Properties

Many types of phosphametallocenes have been characterized by X-ray structural analysis i.e. phosphacymantrenes[90], phosphaferrocenes[81], and 1,1'-diphosphaferrocenes[87]. An inspection of the data shows that the structures of metallocenes are disturbed only to a limited extent by the introduction of phosphorus into the cyclopentadienyl ring. The perturbation is restricted to the immediate vicinity of the phosphorus atom. A comparison between cymantrene and phosphacymantrene is made in Table 16; data on phosphaferrocene are given in Table 17.

Table 16. Structural comparison between cymantrene and its phospha analog[a]

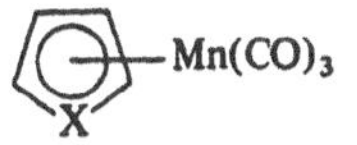

	Mn-C (ring)	Mn-C (carbonyl)	C-C (ring)	C-O	Mn-plane
X = CH	2.165	1.80	1.42	1.15	1.80
X = P[b]	2.169	1.794	1.42	1.146	1.757

[a] Data from ref. 90; distances in Å
[b] 2-benzoyl-3,4-dimethyl substituted

Table 17. Structural features of 3,4-dimethylphosphaferrocene; comparison with (1-benzylphosphole)

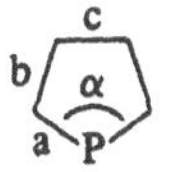

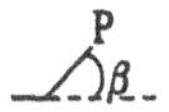

a 1.763(1.783)	Fe...C_4P	1.625
b 1.405(1.343)		
c 1.414(1.438)	Fe...C_5H_5	1.655
α 88.4°(90.7°)		

β 1.9°(9.6°); dihedral angle between the rings 3.18° eclipsed conformation

Data on phosphaferrocene from Ref. 81; data on (1-benzylphosphole) from Ref. 13; distances in Å

It can be seen that the only notable differences are the following: a) the metal is closer to the C_4P planes than to the C_5H_5 planes. This does not mean that the metal-phospholyl bond is stronger than the metal-cyclopentadienyl bond; in fact, the C_α-C_α' distances in the phospholyl rings are greater than the corresponding distances in the cyclopentadienyl rings while the metal-C (ring) distances remain the same. Thus, the metal is necessarily closer to the phospholyl planes. b) The phospholyl rings are slightly puckered contrary to the cyclopentadienyl rings. This is due to the fact that phosphorus is larger than the CH group that it replaces. Even though phosphorus is farther from the metal-ring axis than the ring carbons, planarity would require a shorter M-P distance than observed, which is practically equal to the sum of the covalent radii. This bulkiness is also responsible for the appearance of a non-zero dihedral angle between the two rings in phosphaferrocenes. More notable differences appear when the π-complexed phospholyl ligand is compared with free phospholes (Table 17). An almost complete equalization of the three ring C-C bonds occurs when passing from phosphole to phospholyl and the puckering of the ring around the C_α-C_α' axis is sharply reduced. These data reflect the increase of the electronic delocalization (aromaticity) which occurs upon complexation of the phosphole ring.

A very notable result has been obtained during an X-X study of phosphaferrocene[91]. The electronic density map shows a localized lone pair on phosphorus which is practically in the plane of the C_4P ring. This means that phosphorus behaves as a two-coordinated center even though the Fe-P distance appears to be in the normal range. A combined study of phosphacymantrene using UV photoelectron spectroscopy and E.H.T. calculations[73] has led to the same conclusion. The HOMO is not localized on phosphorus as usual in ordinary P^{III} derivatives but is mainly on the ring carbons. The n-orbital associated with the lone pair is only the fourth highest occupied orbital. The LUMO is mainly localized on phosphorus whereas it is mainly centered on manganese in cymantrene. This means that phosphorus has lost its nucleophilicity and has acquired some electrophilicity. It should, thus, behave as a normal P^{II} compound. This also indicates that electrophilic substitutions on the ring carbons appear feasible. Two interesting results have also been obtained during an IR-Raman study of phosphacymantrene[92]. The ring-metal bond is weaker in phosphacymantrene than in cymantrene: the force constants are approximately equal to 2.6 mdyne $Å^{-1}$ and 3.2 mdyne $Å^{-1}$, respectively. Moreover, the phospholyl ring appears to be a weaker donor than the cyclopentadienyl ring as monitored by the increase in frequencies of the $\nu(CO)$ stretching modes and the decrease of the frequencies of the $\nu(Mn-CO)$ modes. This again suggests that phosphorus behaves as an electrophilic center.

From the numerous NMR data on phosphametallocenes which are recorded in the literature, it is possible to extract two characteristic spectral features. First, the phosphametallocenes, contrary to most of the complexes of phosphorus ligands, have a ^{31}P resonance at high field. Table 18 gives some representative data. Thus, from this point of view, phosphametallocenes are very different from the classical P^{II} compounds with their characteristic ^{31}P resonances at very low field. There is no obvious explanation of this phenomenon. Just one remark may be made. In a normal 1-electron phosphide complex, the $\widehat{CPM}$ angle probably falls in the range of 120–130°; in a phosphametallocene, the $\widehat{CPM}$ angle is approximately 60° (e.g. 61° in a phosphacymantrene[90]). If a comparison is made with ordinary phosphines, an explanation can be suggested: A normal acyclic phosphine has a $\widehat{CPM}$ angle around 100° and a ^{31}P resonance around −10 ppm; phosphirane, with a very small intracyclic angle ($\sim 47.4°$[93]), has a ^{31}P resonance at − 341 ppm[93]. This comparison suggests that the upfield shift of the phosphametallocene ^{31}P resonances may have, at least in part, a geometrical origin. The same phenomenon is also observed when comparing the chromium carbonyl complexes of 2,4,6-triphenylphosphorin. A tremendous upfield shift of the ^{31}P resonance is observed when passing from the 2-electron $Cr(CO)_5$ complex L[94] to the 6-electron $Cr(CO)_3$ π-aromatic complex LI[95].

Table 18. ^{31}P chemical shifts of some representative phosphametallocenes[a]

Ring	M = Mn(CO)$_3$	Fe(CP)	1/2 Fe	Co(CO)$_2$
 (phosphole ring, P)	− 25.6[90]	− 67.5[81]	− 59[87]	−
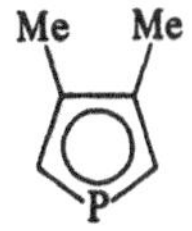 (Me Me ring, P)	− 46.6[90]	− 83.5[81]	− 72[87]	−
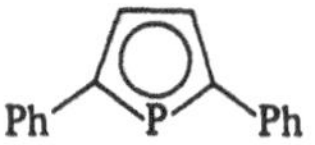 (Ph P Ph ring)	− 30.1[80]	−	− 63.6[87]	− 13.8[89]

[a] δ in ppm, positive for downfield shifts; ref. external H$_3$PO$_4$; CDCl$_3$ solvent except the cobalt compound measured in C$_6$D$_6$

Table 19. ^{13}C-^{31}P coupling constants in some potentially aromatic phosphorus heterocycles[a]

Heterocycle	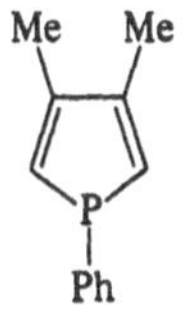 (Me Me, P, Ph)	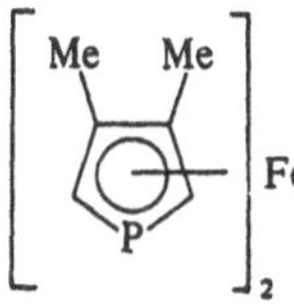 [(Me Me, P)]$_2$ Fe	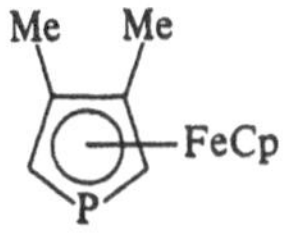 (Me Me, P)−FeCp	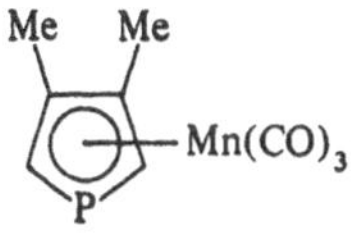 (Me Me, P)−Mn(CO)$_3$	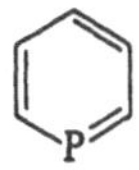 (phosphorin, P)
$^1J(C\text{-}P)$	7.3	61.6	61	62	53
$^2J(C\text{-}P)$	8.5	7.5	7.2	7.7	14
$^3J(C\text{-}P)$	3.7	0	0	0	22

[a] Coupling constants in Hz; all the data from our laboratory except for the phosphorin (Ref. 96)

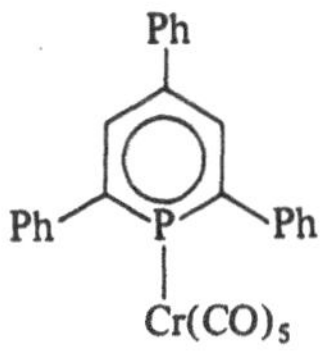

(structure L) Ph ... Ph−P−Ph, Cr(CO)$_5$ $\widehat{CPCr}$ 128° $\delta^{31}P$ 197.8 ppm

L

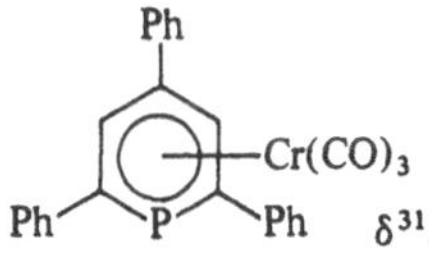

(structure LI) Ph ... Ph−P−Ph, −Cr(CO)$_3$ $\widehat{CPCr}$ 62° $\delta^{31}P$ 4.3 ppm

LI

Whatever the actual origin of this feature, it allows one to very easily recognize, with the aid of ^{31}P NMR, the presence of a phosphametallocene in a mixture of phosphole and phospholyl complexes.

Another spectral feature of phosphametallocenes which deserves some attention has been found in their ^{13}C NMR spectra. A very high $^1J(C\text{-}P)$ coupling constant is always observed for these compounds (Table 19) contrary to what is observed with phospholes

and ordinary phosphines. In fact, this high coupling constant is characteristic of methy-
lenephosphines (e.g. for mesityl-diphenylmethylene-phosphine: $^1J(C = P) = 43.5$ Hz[97]).
Thus, in this way again, phosphametallocenes resemble two-coordinated phosphorus
compounds.

To close this section, it should be noted that the ^{1}H NMR spectra of phosphacymantrene oriented in nematic phases have been investigated[98] and that an order-disorder
phase transition at 110 K has been put in evidence for the same molecule by calorimetry,
X-ray diffraction, and Raman diffusion[99].

III.4. c. Chemical Properties

The phosphorus of phosphametallocenes has almost completely lost its classical nucleophilicity. Indeed, phosphacymantrenes do not react with iodine in boiling CCl_4, with
pure methyl iodide at room temperature, or with benzyl bromide in refluxing toluene[90];
they can be dissolved without protonation in pure trifluoroacetic acid, while even cymantrene is protonated at manganese under such conditions. Phosphaferrocenes are slightly
more reactive. For example, 3,3',4,4'-tetramethyl-1,1'-diphosphaferrocene reacts with
pure benzyl bromide at 80 °C to yield the corresponding uncomplexed phospholium
salt[100]:

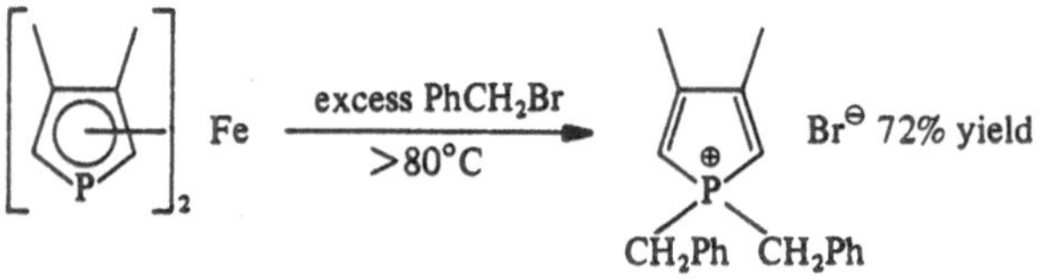

This reaction provides a mild way to decomplex the phosphole ring. The much
reduced nucleophilicity of phosphorus consequently does not interfere with the reactions
of electrophiles at carbon. It is thus possible to acylate phosphacymantrenes and phosphaferrocenes. Some examples are given below:

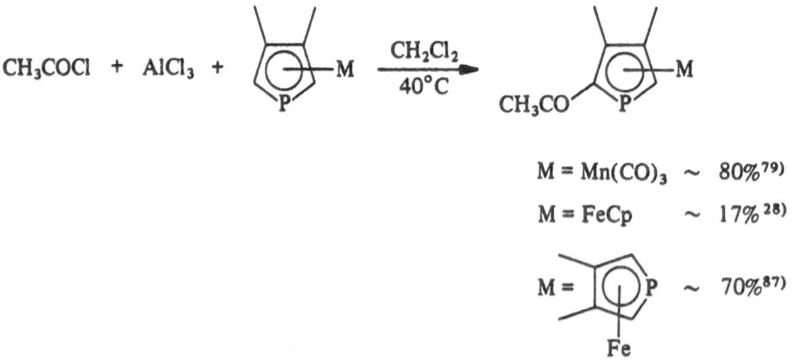

$$M = Mn(CO)_3 \sim 80\% [79]$$
$$M = FeCp \sim 17\% [28]$$

$$\sim 70\% [87]$$

The low yield obtained with monophosphaferrocenes is not due to a lack of reactivity
but to a lack of stability under the reaction conditions. However, it is clear that a given
phosphametallocene is always less reactive than the corresponding metallocene, due to
the overall electron withdrawing effect of phosphorus.

1,1'-Diphosphaferrocenes can also be diacylated if an excess of acyl halide and aluminum chloride is used:

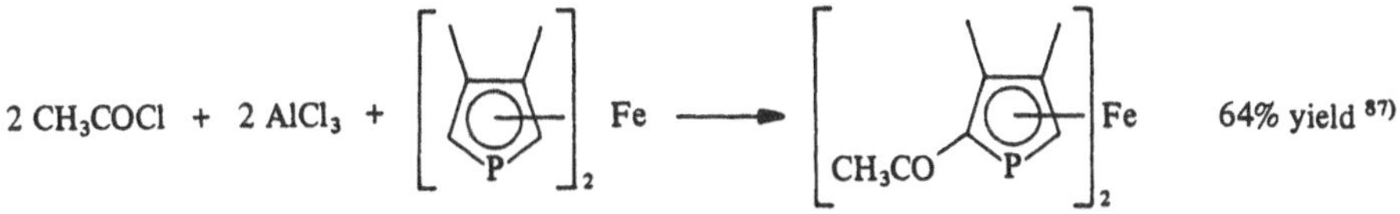

$$2 \text{ CH}_3\text{COCl} + 2 \text{ AlCl}_3 + \left[\begin{array}{c}\text{P}\end{array}\right]_2 \text{Fe} \longrightarrow \left[\begin{array}{c}\text{CH}_3\text{CO} \quad \text{P}\end{array}\right]_2 \text{Fe} \qquad 64\% \text{ yield } [87]$$

Since the two rings become chiral, the diacetylated product is a mixture of two separable diastereoisomers.

The variation of reactivity towards electrophiles follows the same pattern for metallocenes and phosphametallocenes. Thus, phosphacymantrenes are much less reactive than phosphaferrocenes. A good illustration is given by the Vilsmeier formylation. Both mono- and diphosphaferrocenes are formylated in this way whereas phosphacymantrene does not react:

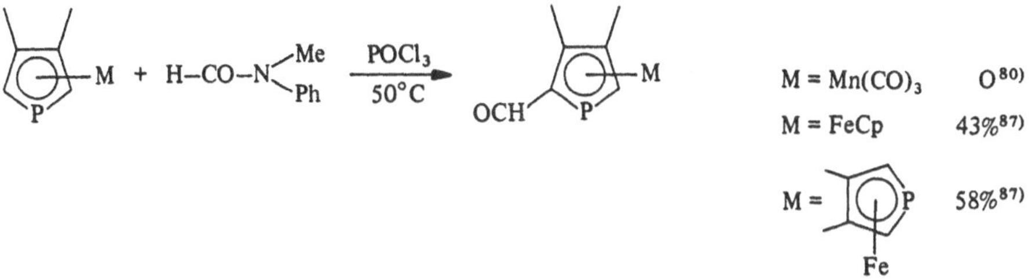

$$\text{M} + \text{H-CO-N}\overset{\text{Me}}{\underset{\text{Ph}}{}} \xrightarrow[50°C]{\text{POCl}_3} \text{M}$$

M = Mn(CO)$_3$ O [80]

M = FeCp   43% [87]

M =   58% [87]

The high reactivity of 1,1'-diphosphaferrocenes is also well exemplified by the following carboxylation reaction:

$$\left[\begin{array}{c}\text{P}\end{array}\right]_2 \text{Fe} + \text{ClCOOEt} \xrightarrow[\text{CS}_2,-10°C]{\text{AlCl}_3} \text{Fe} \qquad 59\% \text{ yield } [100]$$

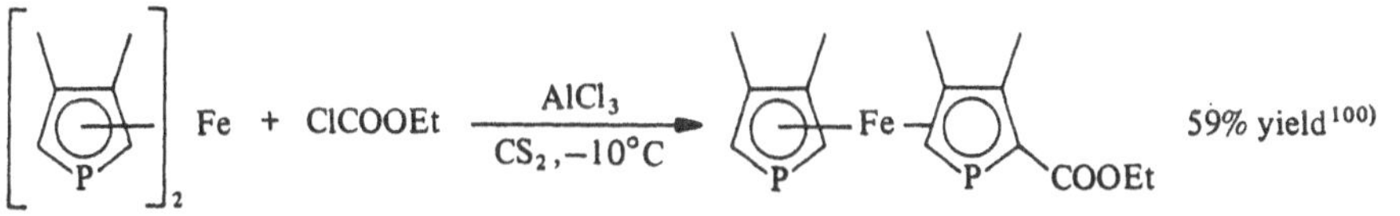

In the presence of aluminum chloride, chloroformates are known to decompose readily to give carbon dioxide and alkyl chlorides. The success of this carboxylation reaction means that the electrophilic attack on phosphaferrocene takes place faster than the decarboxylation of ethyl chloroformate.

A number of classical transformations have been performed on the various functional derivatives thus obtained; some of them are summarized below:

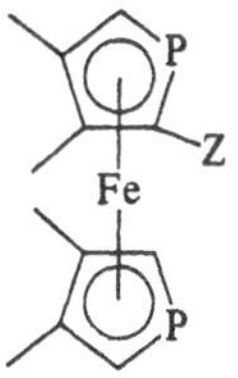

Z		Z'
-C(O)R	$\xrightarrow{\text{BH}_4^{\ominus}}$	-CH(OH)R [80, 87]
-CHO	$\xrightarrow{\text{LiAlH}_4}$	-CH$_2$OH [87]
-CHO	$\xrightarrow{\text{NH}_2\text{OH}}$	-CH=NOH $\xrightarrow{-\text{H}_2\text{O}}$ CN [100]
-C(O)R	$\xrightarrow{\text{AlH}_3}$	-CH$_2$R [80, 87]
-C(O)R	$\xrightarrow{\text{R'MgX}}$	-C(OH)RR' [87]
-C(O)OR	$\xrightarrow{\text{H}^{\oplus}}$	-C(O)OH [100]

Whereas, there is an almost complete analogy between metallocenes and phosphametallocenes in terms of electrophilic substitutions, sharp differences appear with nucleophilic reagents. It has never been possible to observe C-metalations with phosphametallocenes. In all cases, nucleophilic attack takes place at phosphorus. This is quite logical since it must be remembered that the LUMO is mainly localized on phosphorus, at least in the case of phosphacymantrenes. Phosphacymantrenes are so sensitive to nucleophilic media that they are destroyed even by sodium cyanide in ethanol[90], and reaction of butyllithium directly results in a high yield of 1-butylphospholes[90]:

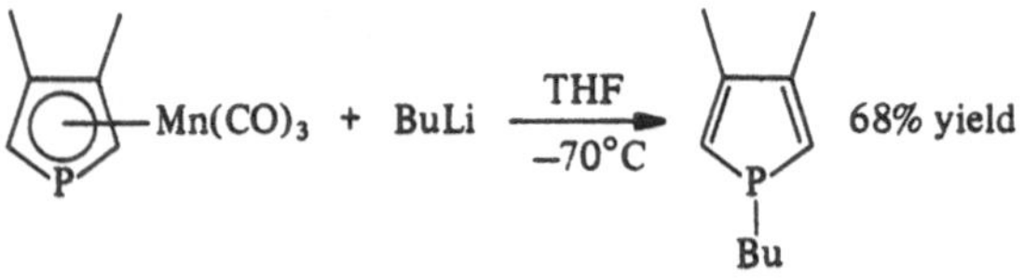

In the case of 1,1'-diphosphaferrocenes, the sensitivity to nucleophilic media decreases to such a level that it becomes possible to trap the transient anions resulting from alkyllithium attack at phosphorus. The ^{31}P NMR data suggest that the negative charge is mainly localized on iron in these anions[70].

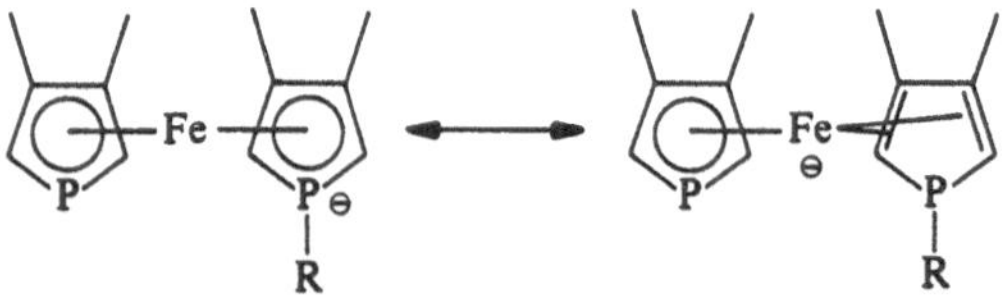

Nevertheless, all the products so far isolated following reaction of these anions with an electrophilic reagent, result from an apparent attack of the electrophile at phosphorus; however, an initial attack at iron followed by a shift from iron to phosphorus is not at all excluded. These products are described in Sect. II.6.b.

As it can be seen, a fascinating new chemistry is at hand. More conventional are some carbonyl exchanges with other ligands (PR_3, $P(OR)_3$) which have been performed on phosphacymantrenes[86, 101]:

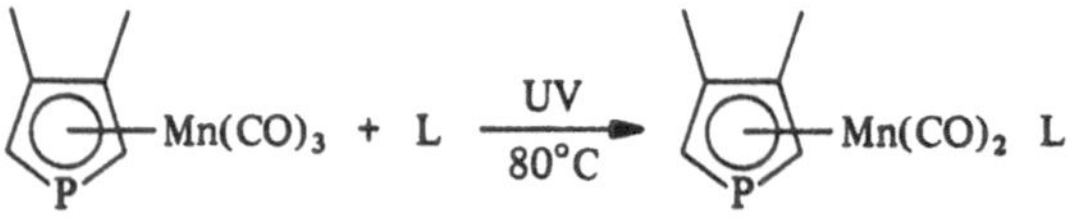

With phosphites, disubstituted products are also obtained.

III.5. Phospholyls as (5 + 2)-Electron Ligands – Complexes of Phosphametallocenes

Since phosphametallocenes still have a localized lone pair on phosphorus[91], they can behave as two electron phosphorus donors towards other metallic centers. This possibility has been demonstrated for phosphacymantrenes[80], phosphaferrocenes[85], and 1,1'-diphosphaferrocenes[100] with $Fe(CO)_4$ as the acceptor group:

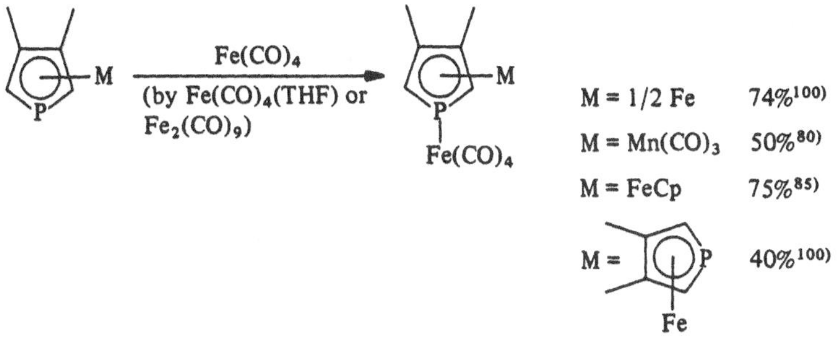

$$M = 1/2\ Fe \qquad 74\%^{100)}$$
$$M = Mn(CO)_3 \qquad 50\%^{80)}$$
$$M = FeCp \qquad 75\%^{85)}$$
$$M = \qquad 40\%^{100)}$$

The structure of the phosphaferrocene complex has been studied by X-Ray crystal structure analysis[102]. Apparently, the phospholyl nucleus remains fully aromatic; the P-Fe(CO)$_4$ bond is especially short [2.211(1) Å] indicating that phosphametallocenes are good ligands towards metals in their low oxidation states. The IR data indicate that they behave mainly as π-acceptors and are closer to phosphites than to phosphines from that standpoint. Dramatic changes are observed in the ^{31}P and ^{13}C NMR spectra upon complexation of the phosphorus atom. A very strong downfield shift (~ 100 ppm) of the ^{31}P resonance, a shielding of the α carbons by more than 10 ppm and a decrease in ^{1}J(C-P) from around 60 Hz to almost 0 Hz, all occur upon complexation. The shielding of the α carbons might be associated with the π-acceptor properties of the system.

Obviously, there is much to be done in this area, and additional work with other metals is currently in progress in our laboratory.

III.6. Phospholyls as (4 + 3)-Electron Ligands

Only one complex in which a phospholyl ring acts as a (4 + 3) electron donor through its diene and its phosphorus atom taken separately, is structurally characterized. It has been prepared from a (1-phenylphosphole)-Mn$_2$(CO)$_7$ complex, in which the diene and the phosphorus atom are already complexed, (see Sect. II.4.) by cleavage of the phosphorus-phenyl bond with manganese carbonyl under UV at low temperature[23]:

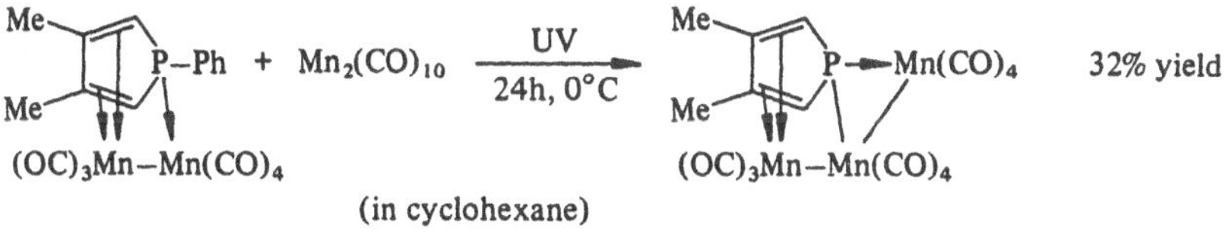

It was initially formulated without an additional Mn-Mn bond and its correct formulation was only established later by X-ray crystal structure analysis[60]. In this complex, the phospholyl ring is clearly not aromatic: the folding of the ring around the C_α-$C_{\alpha'}$ axis is as important as in free phospholes; the ^{31}P resonance is at very low field (+ 74 ppm) and the protons of the ring are more shielded than in the corresponding phosphacymantrene (3.01 vs 4.38 ppm). From another viewpoint, a notable decrease of the phosphorus − Mn(CO)$_3$ distance occurs when passing from the phosphole to the phospholyl complex [2.573(1) vs 2.468(2) Å, respectively]. This distance is only slightly longer than the longest phosphorus − Mn(CO)$_4$ bond in the phospholyl complex [2.366(2) Å]. Thus, there is clearly an interaction between the manganese atom which is π-bonded to the diene and the phosphorus atom. This explains why this complex has a great tendency to give the corresponding phosphacymantrene upon heating. Another interest of this compound lies in the occurence of a rare Mn-Mn-Mn chain arrangement.

Only one other complex in which the phospholyl ligand is supposed to act as a (4 + 3) electron donor is described in the literature. It is a by-product of the high temperature synthesis of phosphacymantrene[90] (see Sect. III.4. a.):

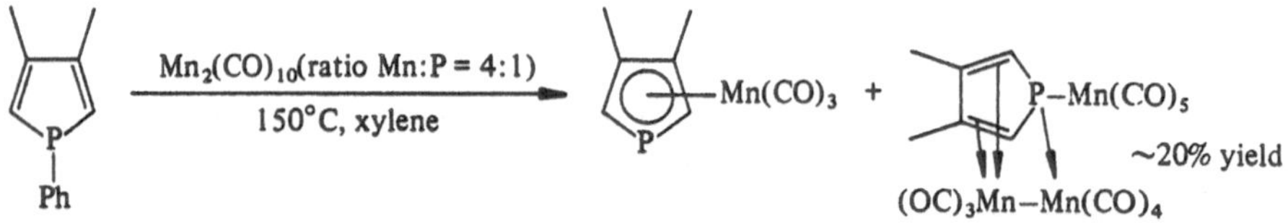

The formula was tentatively proposed on the basis of analytical and spectroscopic data, but no X-ray crystal structure analysis has been carried out. Alternate structures such as LII are possible.

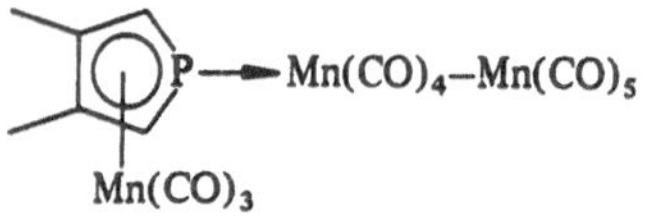

LII

IV. Complexes of Arsoles and Stiboles

The coordination chemistry of arsoles and stiboles is much less developed than that of phospholes. This is partly due to their reduced availability. Indeed, the simplest and most general method for preparing phospholes, starting from conjugated diene-dihalophosphine cycloadducts, cannot be transposed for synthesizing arsoles or stiboles. Scheme 3 summarizes the various routes to arsoles and stiboles which have been described in the literature.

It should be noted that the known arsoles and stiboles are at least 1,2,5-trisubstituted and that bismoles are presently unknown.

Arsoles as such have been reacted with iron and molybdenum carbonyls. Pentaphenylarsole gives only the π-complex LIII with $Fe(CO)_5$ at 150 °C, whereas under the same experimental conditions, pentaphenylphosphole gives only the σ-complex LIV[11]. This reflects the expected lower reactivity of the arsole lone pair. 1-Phenyl-2,5-dimethylarsole with $Fe_2(CO)_9$ in

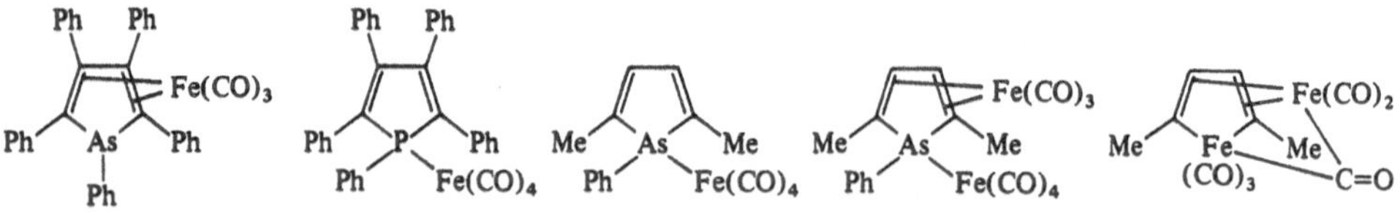

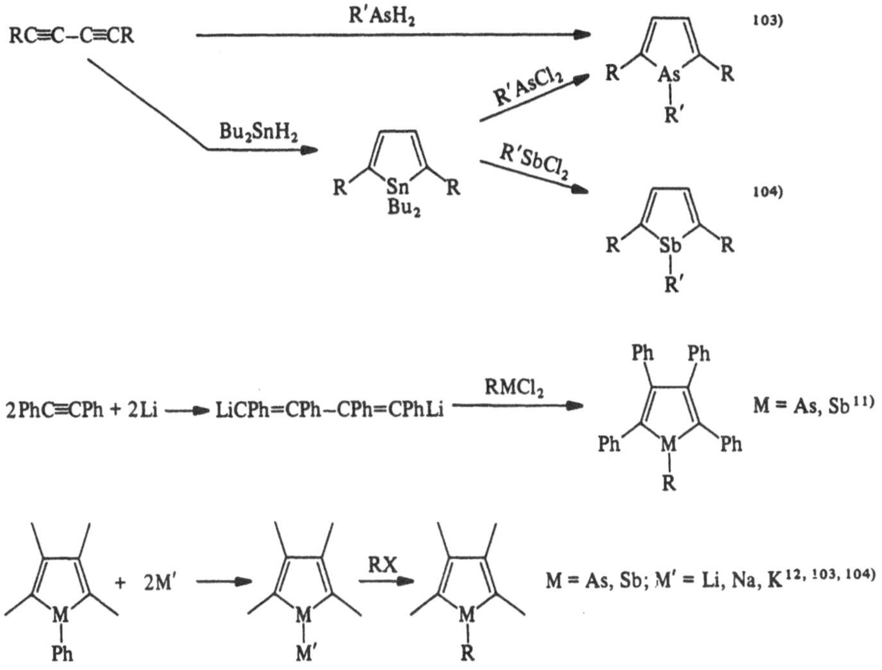

$RC\equiv C-C\equiv CR$ $\xrightarrow{R'AsH_2}$

$\xrightarrow{R'AsCl_2}$ 103)

$\xrightarrow{Bu_2SnH_2}$

$R'SbCl_2$

104)

$2\,PhC\equiv CPh + 2\,Li \longrightarrow LiCPh=CPh-CPh=CPhLi \xrightarrow{RMCl_2}$ M = As, Sb[11]

$+ 2M' \longrightarrow \xrightarrow{RX}$ M = As, Sb; M' = Li, Na, K[12, 103, 104]

Scheme 3. Syntheses of arsoles and stiboles

refluxing benzene gives the σ-complex LV in 80% yield[105]. The same arsole with $Fe_3(CO)_{12}$ in boiling toluene also gives a small amount (∼9%) of the σ, π-complex LVI[105]. At higher temperature in boiling xylene, a ferrole LVII is also obtained[105]. Such a replacement of the heteroatom by iron is unknown with phospholes and was previously only described with thiophenes[106]. It underlines the weakness of the arsenic-carbon bond when compared with the phosphorus-carbon bond. Complexes LVIII and LIX were also obtained by reacting cycloheptatriene-tricarbonyl-molybdenum with the corresponding arsole at room temperature in pentane[107].

LVIII cis $L_2Mo(CO)_4$

LIX fac $L_3Mo(CO)_3$

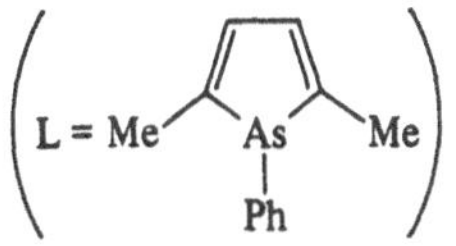

$\left(L = Me \quad \overset{\displaystyle As}{\underset{\displaystyle Ph}{}} \quad Me \right)$

This summarizes all that is known on arsole complexes. On the other hand, no stibole complex has yet been reported in the literature, but much more has been done on arsolyl and stibolyl complexes.

Two approaches allow the synthesis of 1-electron arsolyl and stibolyl complexes:

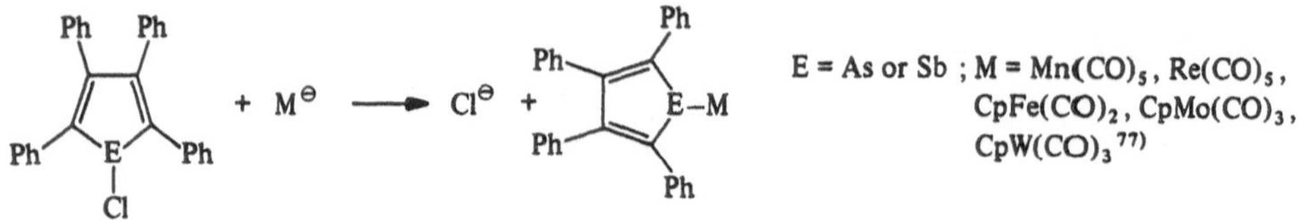

E = As or Sb ; M = Mn(CO)$_5$, Re(CO)$_5$, CpFe(CO)$_2$, CpMo(CO)$_3$, CpW(CO)$_3$ [77]

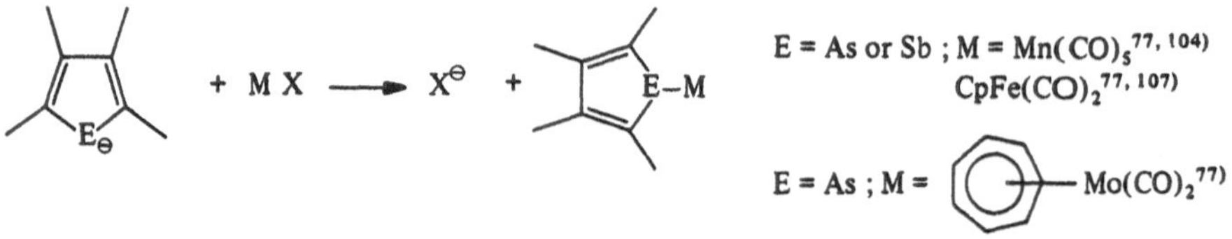

E = As or Sb ; M = Mn(CO)$_5$ [77, 104], CpFe(CO)$_2$ [77, 107]

E = As ; M = Mo(CO)$_2$ [77]

These complexes are stable at room temperature but they are readily cleaved by halogens[77]:

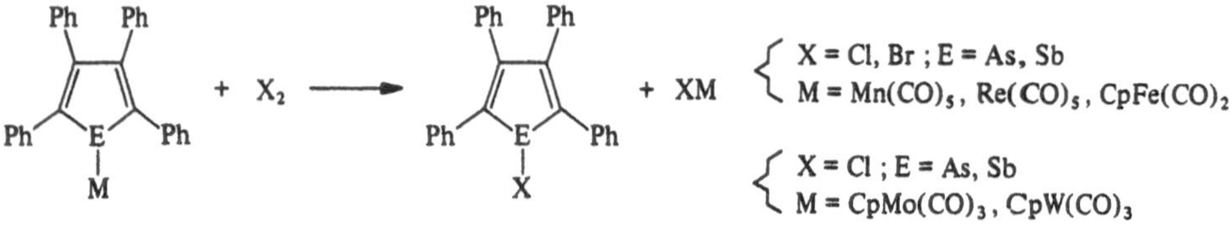

$\begin{cases} X = Cl, Br ; E = As, Sb \\ M = Mn(CO)_5, Re(CO)_5, CpFe(CO)_2 \end{cases}$

$\begin{cases} X = Cl ; E = As, Sb \\ M = CpMo(CO)_3, CpW(CO)_3 \end{cases}$

In one case, it has been possible to displace carbonyls by triphenyl phosphine without breaking the E-M bond[77]:

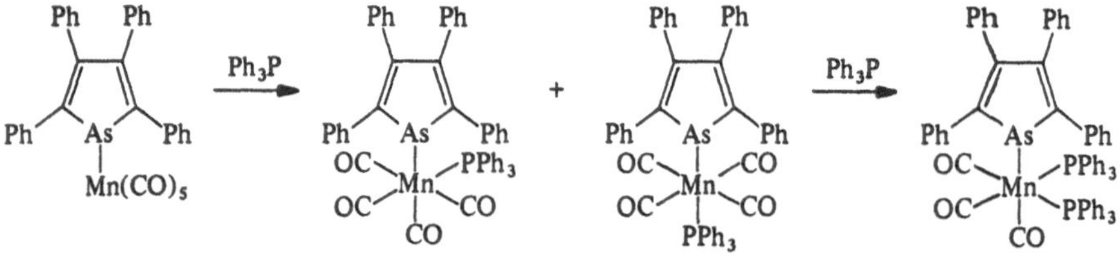

At high temperature, the arsolyl complexes lose carbon monoxide to give new species in which arsolyls act as 3-electron bridging ligands:

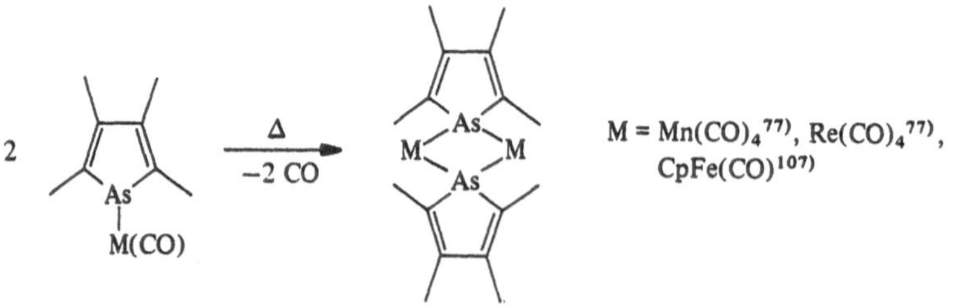

M = Mn(CO)$_4$ [77], Re(CO)$_4$ [77], CpFe(CO) [107]

The thermolysis of the corresponding stibolyl complexes leads either to complete decomposition[77] or directly to stibametallocenes[104] depending upon the substitution scheme on the ring, so that no complex is known in which a stibolyl acts as a 3-electron bridging ligand. At even higher temperature, 1-electron arsolyl complexes directly react to produce arsametallocenes:

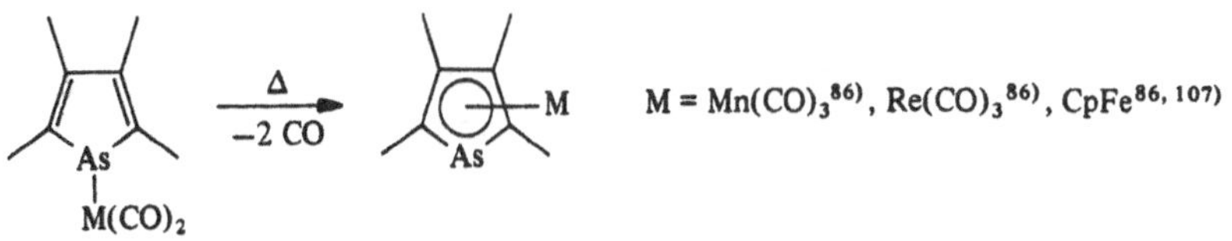

$$\text{(structure)} \xrightarrow[-2\,CO]{\Delta} \text{(structure)}-M \qquad M = Mn(CO)_3{}^{86)},\ Re(CO)_3{}^{86)},\ CpFe^{86,\ 107)}$$

A stibacymantrene has been made recently by using the same approach[104]:

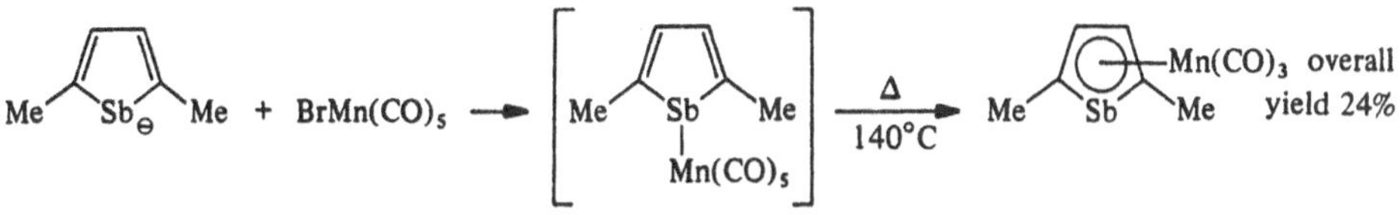

$$Me\!-\!\text{(ring)}_{Sb_\ominus}\!-\!Me + BrMn(CO)_5 \longrightarrow \left[Me\!-\!\text{(ring)}_{\underset{Mn(CO)_5}{Sb}}\!-\!Me \right] \xrightarrow[140°C]{\Delta} Me\!-\!\text{(ring)}_{Sb}\!-\!Me \ \text{—Mn(CO)}_3 \ \begin{array}{l}\text{overall}\\ \text{yield } 24\%\end{array}$$

Arsametallocenes have also been prepared by many other routes which are summarized below:

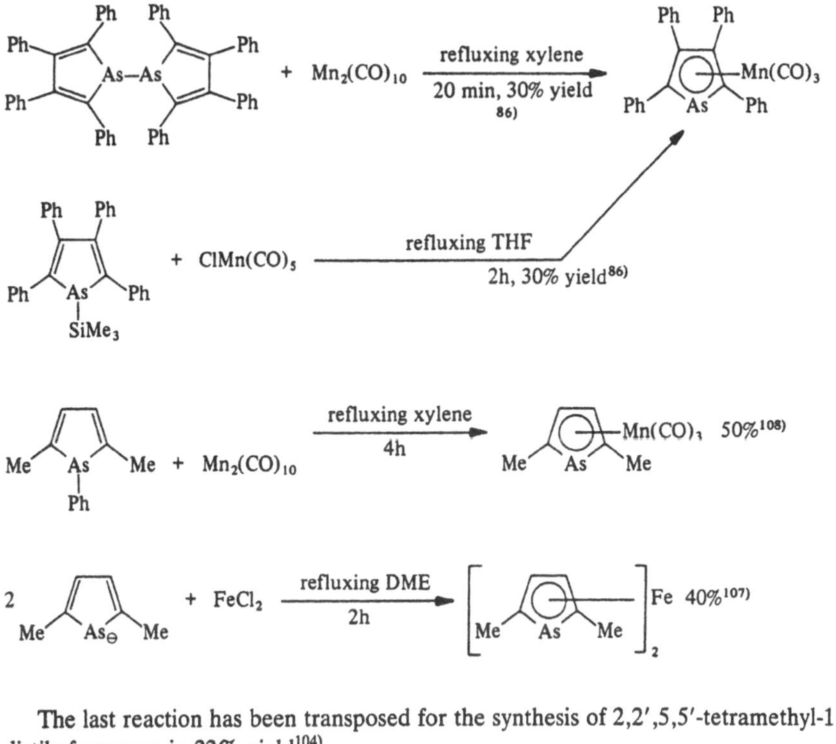

$$\begin{array}{c}Ph\quad Ph\\ Ph\!-\!\text{(bicyclic As—As structure)}\!-\!Ph\\ Ph\qquad\qquad Ph\\ Ph\quad Ph\end{array} + Mn_2(CO)_{10} \xrightarrow[\substack{\text{20 min, 30\% yield}\\ 86)}]{\text{refluxing xylene}} \begin{array}{c}Ph\quad Ph\\ \text{(ring)}\!-\!Mn(CO)_3\\ Ph\quad As\quad Ph\end{array}$$

$$\begin{array}{c}Ph\quad Ph\\ \text{(ring)}\\ Ph\quad As\quad Ph\\ \underset{SiMe_3}{|}\end{array} + ClMn(CO)_5 \xrightarrow[\text{2h, 30\% yield}^{86)}]{\text{refluxing THF}}$$

$$\begin{array}{c}\text{(ring)}\\ Me\quad As\quad Me\\ \underset{Ph}{|}\end{array} + Mn_2(CO)_{10} \xrightarrow[4h]{\text{refluxing xylene}} Me\!-\!\text{(ring)}_{As}\!-\!Me \ \text{—Mn(CO)}_3 \quad 50\%^{108)}$$

$$2\ \begin{array}{c}\text{(ring)}\\ Me\quad As_\ominus\quad Me\end{array} + FeCl_2 \xrightarrow[2h]{\text{refluxing DME}} \left[Me\!-\!\text{(ring)}_{As}\!-\!Me \right]_2 Fe \quad 40\%^{107)}$$

The last reaction has been transposed for the synthesis of 2,2',5,5'-tetramethyl-1,1'-distibaferrocene in 22% yield[104].

Two arsametallocenes have been characterized by X-ray crystal structure analysis *viz* tetraphenylarsacymantrene[109] and 2,2',5,5'-tetramethyl-1,1'-diarsaferrocene[110]. Their structures show close analogies with those of the corresponding phosphametallocenes. Since arsenic is bulkier than phosphorus, the dihedral angle between the two rings is larger in 1,1'-diarsaferrocenes than in their phosphorus analogs (6.2° vs 3.8° in the cases more precisely studied). The only surprise comes from the respective orientation of the two rings around their common axis. In 2,2',5,5'-tetramethyl-1,1'-diarsaferrocene they are eclipsed; the arsenic atom of one ring superposes with the arsenic atom of the other, whereas, in the phosphorus analog, the phosphorus atom of one ring superposes with a β carbon atom of the other[87]. This anomaly is probably due to packing effects in the crystals and suggests that the rotation barriers of the rings around their axes are low.

Only very limited chemistry has been performed with arsa – and stibametallocenes. First, it has been shown that arsacymantrenes[108] and diarsaferrocenes[104] can be acylated at ring carbons:

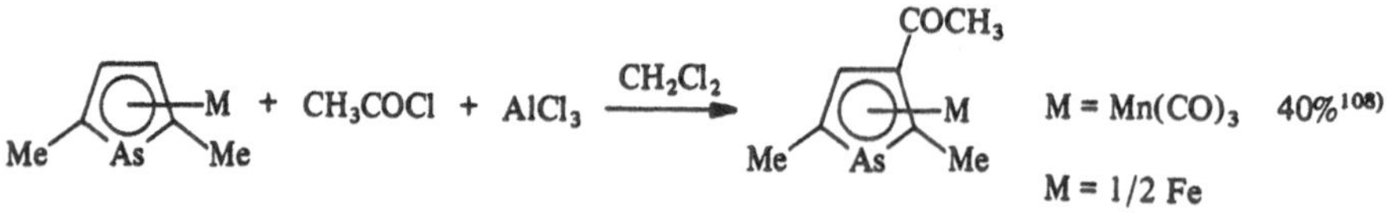

$M = Mn(CO)_3$ 40%[108]

$M = 1/2$ Fe

The diarsaferrocenes also undergo deuterium exchange in d_1-trifluoroacetic acid[104] On the contrary, 2,2',5,5'-tetramethyl-1,1'-distibaferrocene is destroyed by strong Lewis acids[104].

Finally, some carbonyl exchanges have been described with arsacymantrenes[86]:

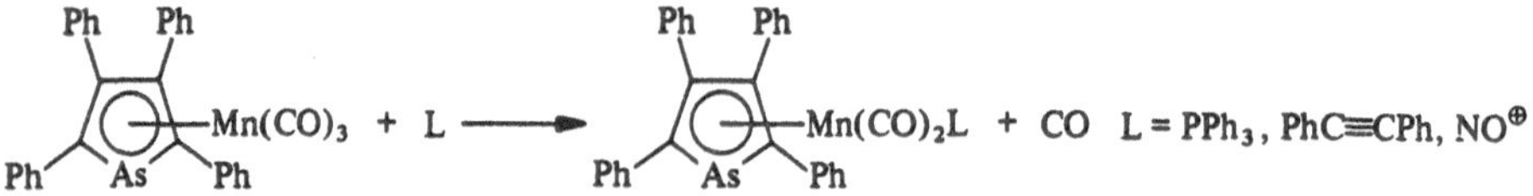

$L = PPh_3$, $PhC{\equiv}CPh$, $NO^{\oplus}$

V. A Brief Comparison Between Azametallocenes and Their Heavier Analogs

To close this review, it seems appropriate to mention azametallocenes. These η^5-pyrrolyl complexes have been known for a long time and their chemistry has been recently reviewed[111]. This chemistry appears to be radically different from that of phospha-, arsa- and stibametallocenes. We can briefly summarize the comparison by the following statements:

1) The π-complexing ability of the pyrrole ring appears to be lower than that of its homologs; for example, it has never been possible to prepare 1,1'-diazaferrocenes. This might be due to an excess of electron density on iron.
2) The nucleophilicity of nitrogen increases upon π-complexation of pyrroles. Azaferrocene gives stable salts with various acids (pK$_a$ 4–5 in aqueous methanol) and can be quaternized by methyl iodide[112]. A comparison has already been made between the chemical properties of phosphorus in phosphametallocenes and phosphorins. In the same vein, it is interesting to remember that pyridine is basic whereas phosphorin is not. The parallels azametallocene-pyridine and phosphametallocene-phosphorin suggest very different orderings of the orbitals in aza- and phospha-metallocenes as is the case in pyridines and phosphorins.
3) It seems impossible to functionalize the π-complexed pyrrole ring by electrophilic substitution without decomplexation. For example, the acetylation of azacymantrene causes the decomplexation of the dienic portion of the ring[113]:

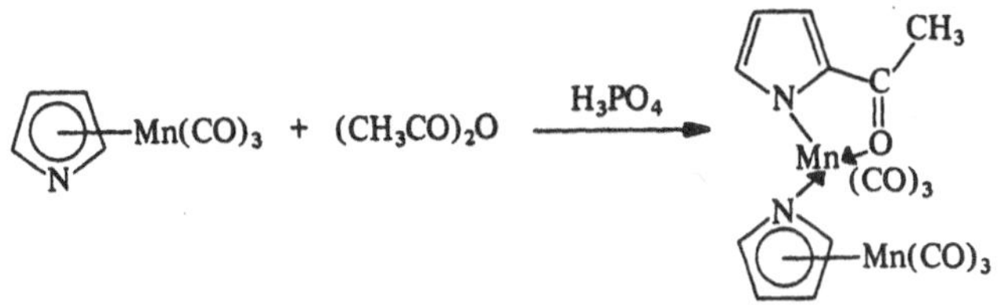

4) Nucleophilic attack at nitrogen seems more difficult. For example butyllithium reacts at the carbonyls of azacymantrenes[114]:

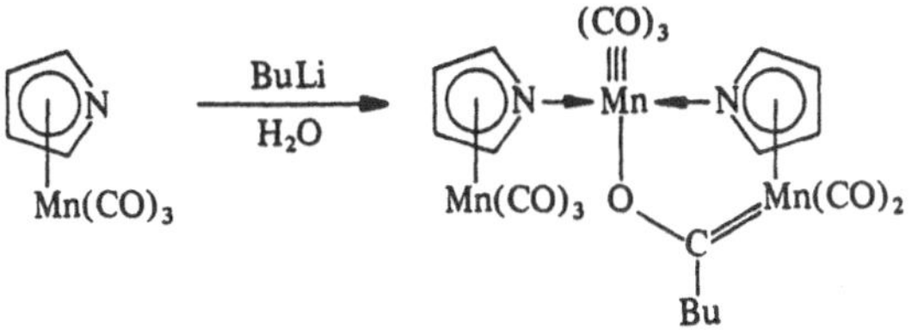

5) As with their phosphorus analogs, azametallocenes can play the role of 2-electron ligands by using the lone pair on nitrogen[115].

VI. Conclusion and Perspectives

From all the data summarized in this review it clearly appears that the potential of phospholes in coordination chemistry is enormous. As 2-electron ligands their behavior is already original. Contrary to what could be expected on the basis of their potentially aromatic structure, metal-phosphole bonds are strong due to three contributing factors: a) the phosphole does not undergo much structural reorganization upon coordination, b) phospholes are sterically undemanding and, c) phospholes have a moderate π-acceptor capacity. Besides, it is very easy to modulate their coordination ability over a wide range by just changing the ring substitution. For example, methyl groups in the 3 and particularly in both the 3 and 4 positions highly increase their nucleophilicity. The former is primarily the result of inductive effects and the latter the combined result of inductive and internal ring strain effects. On the contrary, phenyl substitution, particularly in the 3 and 4 positions significantly increases their electrophilicity. Practically, phospholes mainly coordinate to soft Lewis acids and the donor ability follows the general sequence

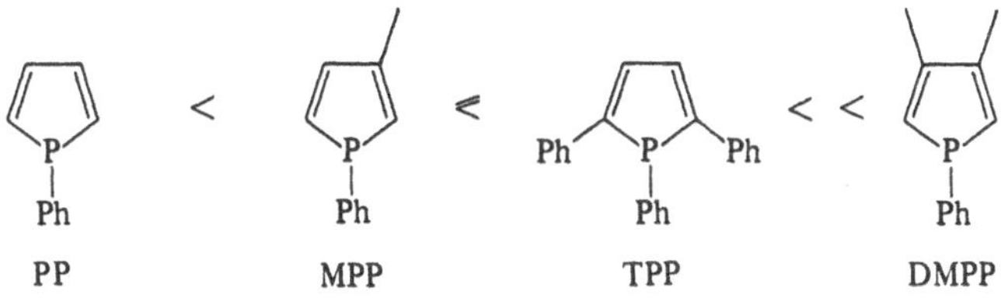

with the latter having similar donor properties to Me_2PPh. Due to their sterically small character and their polarizability, giving rise to an increased propensity for π-back donation, phospholes probably stabilize low oxidation state metals better than classical phosphines. As a final advantage, phosphole complexes are generally highly soluble and their structure is easily determined by a combination of ^{1}H, ^{13}C{^{1}H} and ^{31}P{^{1}H}NMR.

Certainly, many other original possibilities are offered by phospholes through their dienic system.

As 6-electron donors they are able to control the formation and the breaking of metal-metal bonds by just varying the steric bulk of their exocyclic P-substituent. As phospholium salts they stabilize unusual types of conjugated diene π-complexes, probably through steric protection of the metal by the phosphonium units.

A completely different type of chemistry can be built around the phospholyl anions. They can yield classical η^1- and μ^2-phosphido complexes. In the first case, an unusual stability is obtained through partial delocalization of the remaining lone pair on phosphorus and steric protection by 2,5-diphenyl-substitution. But, of course, the main interest in this area is centered on the η^5-phospholyl complexes, i.e., the phosphametallocenes. To date, they are the only known phosphorus heterocycles displaying an extensive aromatic chemistry. In the future this could mean an easy access to functional phospholes and to phosphorus analogs of complex molecules incorporating the phosphole nucleus such as phosphaporphyrins. Through their lone pair on phosphorus, phosphametallocenes can also act as a very special type of 2-electron phosphorus ligand. Their steric bulk, their strong π-acceptor capacity, and their high asymmetry when disymmetrically substituted all could favor the discovery of new structural types of complexes and new types of catalysts.

As a final statement, it must be stressed that much remains to be accomplished with phospholes in coordination chemistry. Practically nothing is known on the C-unsubstituted phosphole complexes; coordination with metals such as Ti, Zr, V, W, Ta, Os, Ir, Cu, Ag, Au, Zn, Cd, Hg ... has not been seriously investigated. Catalytic activity of phosphole complexes has not been systematically studied. Phosphametallocene complexes are unknown except with Fe(CO)$_4$. Undoubtedly, new exciting developments are at hand.

VII. References

1. Holah, D. G., Hughes, A. N., Wright, K.: Coord. Chem. Rev. *15*, 239 (1975)
2. Mathey, F.: Top. Phosphorus Chem. *10*, 1 (1980)
3. Hughes, A. N.: in: New Trends in Heterocyclic Chemistry, (eds.) Mitra, R. B., Ayyangar, N. R., Gogte, V. N., Acheson, R. M., Cromwell, N., Elsevier, Amsterdam 1979
4. Quin, L. D.: The Heterocyclic Chemistry of Phosphorus, Wiley Interscience, New York 1981
5. Mathey, F., Mankowski-Favelier, R.: Bull. Soc. Chim. Fr. 4433 (1970); Org. Magn. Resonance *4*, 171 (1972)
6. Quin, L. D., Borleske, S. G., Engel, J. F.: J. Org. Chem. *38*, 1858 (1973)
7. Breque, A., Mathey, F., Savignac, Ph.: Synthesis 983 (1981)
8. Campbell, I. G. M., Cookson, R. C., Hocking, M. B., Hughes, A. N.: J. Chem. Soc. 2184 (1965)
9. Märkl, G., Potthast, R.: Angew. Chem. Int. Ed. Engl. *6*, 86 (1967)
10. Egan, W., Tang, R., Zon, G., Mislow, K.: J. Amer. Chem. Soc. *93*, 6205 (1971)
11. Braye, E. H., Hubel, W., Caplier, I.: ibid. *83*, 4406 (1961)
12. Braye, E. H., Caplier, I., Saussez, R.: Tetrahedron *23*, 5523 (1971)

13. Coggon, Ph., McPhail, A. T.: J. Chem. Soc., Dalton Trans., 1888 (1973)
14. Schäfer, W., Schweig, A., Märkl, G., Hauptmann, H., Mathey, F.: Angew. Chem. Int. Ed. Engl. *12*, 145 (1973)
15. Schäfer, W., Schweig, A., Mathey, F.: J. Amer. Chem. Soc. *98*, 407 (1976)
16. Ozbirn, W. P., Jacobson, R. A., Clardy, J. C.: Chem. Commun. 1062 (1971)
17. Quin, L. D., Borleske, S. G., Engel, J. F.: J. Org. Chem. *38*, 1858 (1973)
18. Gray, G. A., Nelson, J. H.: Org. Mag. Res. *14*, 14 (1980)
19. Bundgaard, T., Jakobsen, H. J.: Tetrahedron Lett. 3353 (1972)
20. Santini, C. C., Fischer, J., Mathey, F., Mitschler, A.: J. Amer. Chem. Soc. *102*, 5809 (1980)
21. Cookson, R. C., Fowles, G. W. A., Jenkins, D. K.: J. Chem. Soc. 6406 (1965)
22. Jenkins, J. H., Moss, J. R., Shaw, B. L.: J. Chem. Soc. (A) 2796 (1969)
23. Mathey, F.: J. Organometal. Chem. *93*, 377 (1975)
24. Ziegler, M. L., Haas, H., Sheline, R. K.: Chem. Ber. *98*, 2454 (1965)
25. Holah, D. G., Hughes, A. N., Hui, B. C., Tse, P.: J. Heterocyclic Chem. *15*, 1239 (1978);
    Holah, D. G., Hughes, A. N., Wright, K.: Inorg. Nucl. Chem. Lett. *9*, 1265 (1973);
    Holah, D. G., Hughes, A. N., Hui, B. C., Krupa, N., Wright, K.: Phosphorus *5*, 145 (1975)
26. Cotton, F. A., Mague, J. T.: Inorg. Chem. *3*, 1094 (1964)
27. Cotton, F. A., Lippard, S. J., Mague, J. T.: ibid. *4*, 508 (1965)
28. Mathey, F.: J. Organometal. Chem. *139*, 77 (1977)
29. Le Bozec, H., Gorgues, A., Dixneuf, P.: ibid. *174*, C24 (1979)
30. Mathey, F., Muller, G.: ibid. *136*, 241 (1977)
31. Wilkes, L. M., Nelson, J. H., Mc Cusker, L. B., Seff, K., Mathey, F.: Inorg. Chem. *22*, 2476 (1983)
32. Lupin, M. S., Shaw, B. L.: J. Chem. Soc. (A) 741 (1968)
33. Barnard, C. F. J., Daniels, J. A., Jeffery, J., Mawby, R. J.: J. Chem. Soc. Dalton 953 (1976)
34. Wilkes, L. M., Nelson, J. H., Mitchener, J. P., Babich, M. W., Riley, W. C., Helland, B. J., Jacobson, R. A., Cheng, M. Y., Mc Cusker, L. B., Seff, K.: Inorg. Chem. *21*, 1376 (1982)
35. La Placa, S. J., Ibers, J. A.: ibid. *4*, 778 (1965)
36. Brown, L. D., Barnard, C. F. J., Daniels, J. A., Mawby, R. J., Ibers, J. A.: ibid. *17*, 2932 (1978)
37. Raston, C. L., White, A. H.: J. Chem. Soc. Dalton 2422 (1975)
38. Cotton, F. A., Faut, O. D., Goodgame, D. M., Holm, R.: J. Amer. Chem. Soc. *83*, 1780 (1961);
    Sestili, L., Furlani, C., Festuccia, G.: Inorg. Chim. Acta *4*, 542 (1970)
39. Mathey, F., Muller, G., Demay, C., Lemke, H.: Informations Chimie 191 (1978)
40. Shell pat., Netherlands 6,604,094 (1966), Chem. Abstr. *66*, 65 101 r (1967);
    van Winkle, J. L., Morris, R. C., Mason, R. F., Shell pat., Germany, 1,909,620 (1969), Chem. Abstr. *72*, 3033 k (1970)
41. Holah, D. G., Hughes, A. N., Hui, B. C.: Can. J. Chem. *50*, 3714 (1972)
42. McAuliffe, C. A. (ed.), Transition Metal Complexes of Phosphorus, Arsenic and Antimony Ligands, Halsted Press, New York, N.Y. 1973
43. Bright, A., Mann, B. E., Masters, C., Shaw, B. L., Slade, R. M., Stainbank, R. E.: J. Chem. Soc. (A) 1826 (1971)
44. Mann, B. E., Masters, C., Shaw, B. L.: ibid. 1104 (1971)
45. Budd, D., Chuchman, R., Holah, D. G., Hughes, A. N., Hui, B. C.: Can. J. Chem. *50*, 1008 (1972)
46. Quin, L. D., Bryson, J. G., Engel, J. F.: Phosphorus *2*, 205 (1973)
47. Holah, D. G., Hughes, A. N., Hui, B. C., Tse, P.: J. Heterocyclic Chem. *15*, 89 (1978)
48. LaMar, G. N., Sherman, E. O.: J. Amer. Chem. Soc. *92*, 2691 (1970)
49. Walton, R. A.: J. Chem. Soc. (A) 365 (1966)
50. MacDougall, J. J., Nelson, J. H., Mathey, F., Mayerle, J. J.: Inorg. Chem. *19*, 709 (1980)
51. MacDougall, J. J., Holt, E. M., De Meester, P., Alcock, N. W., Mathey, F., Nelson, J. H.: ibid. *19*, 1439 (1980)
52. MacDougall, J. J., Mathey, F., Nelson, J. H.: ibid. *19*, 1400 (1980)
53. Redfield, D. A., Nelson, J. H.: ibid. *12*, 15 (1973)
54. MacDougall, J. J., Nelson, J. H., Babich, M. W., Fuller, C. C., Jacobson, R. A.: Inorg. Chim. Acta *27*, 201 (1978)

55. Nelson, J. H., MacDougall, J. J., Alcock, N. W., Mathey, F.: Inorg. Chem. 21, 1200 (1982)
56. MacDougall, J. J., Nelson, J. H., Mathey, F.: ibid. 21, 2145 (1982)
57. Favez, R., Roulet, R., Pinkerton, A., Schwarzenbach, D.: ibid. 19, 1356 (1980) and references therein
58. Bell, N. A., Goldstein, M., Jones, T., Nowell, I. W.: Inorg. Chim. Acta 43, 87 (1980)
59. Gallagher, M. J., Graddon, D. P., Sheikh, A. R.: Aust. J. Chem. 29, 759 (1976)
60. Rosalky, J. M., Metz, B., Mathey, F., Weiss, R.: Inorg. Chem. 16, 3307 (1977)
61. Santini, C. C., Fischer, J., Mathey, F., Mitschler, A.: ibid. 20, 2848 (1981)
62. Mathey, F., Thavard, D.: Can. J. Chem. 56, 1952 (1978)
63. Mathey, F.: Bull. Chem. Soc. Fr. 2783 (1973)
64. Rose, D., Wilkinson, G.: J. Chem. Soc. (A) 1791 (1970)
65. Gilbert, J. D., Rose, D., Wilkinson, G.: ibid. 2765 (1970)
66. Barrow, M. J., Davidson, J. L., Harrison, W., Sharp, D. W. A., Sim, G. A., Wilson, F. B.: Chem. Comm. 583 (1973)
67. Barrow, M. J., Freer, A. A., Harrison, W., Sim, G. A., Taylor, D. W., Wilson, F. B.: J. Chem. Soc., Dalton 197 (1975)
68. Yasufuku, K., Hamada, A., Aoki, K., Yamazaki, H.: J. Am. Chem. Soc. 102, 4363 (1981)
69. Lindner, E., Rau, A., Hoehne, S.: J. Organomet. Chem. 218, 41 (1981)
70. Deschamps, B., Fischer, J., Mathey, F., Mitschler, A.: Inorg. Chem. 20, 3252 (1981)
71. Deschamps, B., Fischer, J., Mathey, F., Mitschler, A., Ricard, L.: Organometallics, 312 (1982)
72. Kaufmann, G., Mathey, F.: Phosphorus 4, 231 (1974)
73. Guimon, C., Pfister-Guillouzo, G., Mathey, F.: Nouveau J. Chim. 3, 725 (1979)
74. Quin, L. D., Orton, W. L.: J. Chem. Soc., Chem. Commun. 401 (1979)
75. Issleib, K., Vollmer, R., Oehme, H., Meyer, H.: Tetrahedron Lett. 441 (1978)
76. Braye, E. H., Joshi, K. K.: Bull. Soc. Chim. Belges 80, 651 (1971)
77. Abel, E. W., Towers, C.: J. Chem. Soc., Dalton Trans. 814 (1979)
78. Ashe III, A. J., Chan, Woon-Tung, Smith, T. W.: Tetrahedron Lett. 2537 (1978)
79. Mathey, F.: ibid. 4155 (1976)
80. Breque, A., Mathey, F., Santini, C.: J. Organometal. Chem. 165, 129 (1979)
81. Mathey, F., Mitschler, A., Weiss, R.: J. Am. Chem. Soc. 99, 3537 (1977)
82. Kilcast, D., Thomson, C.: Tetrahedron 27, 5705 (1971)
83. Beckwith, A. J. L., Leydon, R. J.: Aust. J. Chem. 19, 1853 (1966) and references therein
84. Mathey, F., Mercier, F., Charrier, C., Fischer, J., Mitschler, A.: J. Am. Chem. Soc. 103, 4595 (1981)
85. Mathey, F.: J. Organometal. Chem. 154, C13 (1978)
86. Abel, E. W., Clark, N., Towers, C.: J. Chem. Soc., Dalton Trans. 1552 (1979)
87. De Lauzon, G., Deschamps, B., Fischer, J., Mathey, F., Mitschler, A.: J. Amer. Chem. Soc. 102, 994 (1980);
see also: De Lauzon, G., Mathey, F., Simalty, M.: J. Organometal. Chem. 156, C33 (1978)
88. Meunier, P., Gautheron, B.: ibid. 193, C13 (1980)
89. Charrier, C., Bonnard, H., Mathey, F., Neibecker, D.: ibid. 231, 361 (1982)
90. Mathey, F., Mitschler, A., Weiss, R.: J. Amer. Chem. Soc. 100, 5748 (1978)
91. Wiest, R., Rees, B., Mitschler, A., Mathey, F.: Inorg. Chem. 20, 2966 (1981)
92. Poizat, O., Sourisseau, C.: J. Organometal. Chem. 213, 461 (1981)
93. Chan, S., Goldwhite, H., Keyzer, H., Rowsell, D. G., Tang, R.: Tetrahedron 25, 1097 (1969);
Bowers, M. T., Beaudet, R. A., Goldwhite, H., Tang, R.: J. Amer. Chem. Soc. 91, 17 (1969)
94. Deberitz, J., Nöth, H.: J. Organometal. Chem. 49, 453 (1973);
Vahrenkamp, H., Nöth, H.: Chem. Ber. 106, 2227 (1973)
95. Deberitz, J., Nöth, H.: ibid. 103, 2541 (1970);
Vahrenkamp, H., Nöth, H.: ibid. 105, 1148 (1972)
96. Ashe III, A. J., Sharp, R. R., Tolan, J. W.: J. Amer. Chem. Soc. 98, 5451 (1976)
97. Klebach, Th. C., Lourens, R., Bickelhaupt, F.: ibid. 100, 4886 (1978)
98. Khetrapal, C. L., Kunwar, A. C., Mathey, F.: J. Organometal. Chem. 181, 349 (1979)
99. Poizat, O., Sourisseau, C., Calvarin, G., Chhor, K., Pommier, C.: Mol. Cryst., Liq. Cryst. 73, 159 (1981)
100. De Lauzon, G., Deschamps, B., Mathey, F.: Nouveau J. Chim. 4, 683 (1980)

101. Breque, A., Mathey, F.: J. Organometal. Chem. *144*, C9 (1978)
102. Fischer, J., Mitschler, A., Ricard, L., Mathey, F.: J. Chem. Soc., Dalton Trans. 2522 (1980)
103. Märkl, G., Hauptmann, H.: Tetrahedron Lett. 3257 (1968)
104. Ashe III, A. J., Diephouse, T. R.: J. Organometal. Chem. *202*, C95 (1980)
105. Thiollet, G., Mathey, F.: Inorg. Chim. Acta *35*, L331 (1979)
106. Kaesz, H. D., King, R. B., Manuel, T. A., Nichols, L. D., Stone, F. G. A.: J. Amer. Chem. Soc. *82*, 4749 (1960)
107. Thiollet, G., Mathey, F., Poilblanc, R.: Inorg. Chim. Acta *32*, L67 (1979); Thiollet, G.: Thesis, Toulouse (1979)
108. Thiollet, G., Poilblanc, R., Voigt, F., Mathey, F.: Inorg. Chim. Acta *30*, L294 (1978)
109. Abel, E. W., Nowell, I. W., Modinos, A. G. J., Towers, C.: J. Chem. Soc., Chem. Commun. 258 (1973)
110. Chiche, L., Galy, J., Thiollet, G., Mathey, F.: Acta Cryst. *B36*, 1344 (1980)
111. Pannell, K. H., Kalsotra, B. L., Párkányi, C.: J. Heterocyclic Chem. *15*, 1057 (1978)
112. Joshi, K. K., Pauson, P. L., Qazi, A. R., Stubbs, W. H.: J. Organometal. Chem. *1*, 471 (1964)
113. Pyshnograeva, N. I., Setkina, V. N., Andrianov, V. G., Struchkov, Yu. T., Kursanov, D. N.: J. Organometal. Chem. *128*, 381 (1977)
114. Pyshnograeva, N. I., Setkina, V. N., Andrianov, V. G., Struchkov, Yu. T., Kursanov, D. N.: ibid. *206*, 169 (1981)
115. Pyshnograeva, N. I., Setkina, V. N., Andrianov, V. G., Struchkov, Yu. T., Kursanov, D. N.: ibid. *157*, 431 (1978)
116. Nief, F., Charrier, C., Mathey, F., Simalty, M.: Phosphorus and Sulfur *13*, 259 (1982)

Author-Index Volumes 1–55

Structure and Bonding

Editors: *M. J. Clarke,*
J. B. Goodenough, J. A. Ibers,
C. K. Jørgensen, J. B. Neilands,
D. Reinen, R. Weiss, R. J. P. Williams

Springer-Verlag
Berlin
Heidelberg
New York
Tokyo